VALENCY AND MOLECULAR STRUCTURE

Valency and Molecular Structure

Fourth Edition

E. Cartmell, BSc, CChem, FRIC
formerly Senior Lecturer in Inorganic Chemistry,
University of Southampton

and

G.W.A. Fowles, BSc, PhD, DSc, CChem, FRIC
Professor of Inorganic Chemistry,
University of Reading

BUTTERWORTHS
LONDON - BOSTON
Sydney - Wellington - Durban - Toronto

The Butterworth Group

United Kingdom	**Butterworth & Co (Publishers) Ltd**
London	88 Kingsway, WC2B 6AB
Australia	**Butterworths Pty Ltd**
Sydney	586 Pacific Highway, Chatswood, NSW 2067
	Also at Melbourne, Brisbane, Adelaide and Perth
South Africa	**Butterworth & Co (South Africa) (Pty) Ltd**
Durban	152–154 Gale Street
New Zealand	**Butterworths of New Zealand Ltd**
Wellington	26–28 Waring Taylor Street, 1
Canada	**Butterworth & Co (Canada) Ltd**
Toronto	2265 Midland Avenue Scarborough, Ontario M1P 4S1
USA	**Butterworth (Publishers) Inc**
Boston	19 Cummings Park, Woburn, Mass. 01801

First edition	1956	Second edition	1961
Reprinted	1957	Reprinted	1962
Reprinted	1958	Reprinted	1963
Reprinted	1959	Reprinted	1964
Reprinted	1959	Third edition	1966
Reprinted	1960	Reprinted	1970
		Fourth edition	1977

ISBN 0 408 70809 3

British Library Cataloguing in Publication Data

Cartmell, Edward
Valency and molecular structure. – 4th ed.
1. Valence (Theoretical chemistry)
I. Title. II. Fowles, Gerald Wilfred Albert
541′.224 QD469 77–30013

ISBN 0–408–70809–3

[illegible]eset [illegible] produced by Scribe Design, Chatham, Kent
[illegible]ted [illegible]gland by Chapel River Press, Andover, Hants.

Preface to the Fourth Edition

Successive editions of undergraduate textbooks tend to get bigger and bigger. So much has happened in chemistry since the first edition of this book appeared twenty years ago that there is every temptation for expansion. We have tried to resist this temptation. This book has been completely rewritten. Some material in earlier editions (e.g. the chapter on Bohr theory) has been removed, while some sections (e.g. structures of crystalline solids) have been expanded. In this connection we are grateful to the Royal Institute of Chemistry for permission to include material written by one of us for an R.I.C. Monograph on Crystal Chemistry.

Details of structures have been revised and extended using the best available values for bond lengths and bond angles. Recent developments are perhaps most noticeably seen in the chapter on complex compounds where the discussion of, for example, compounds of co-ordination number five and seven takes up considerably more space than it did in earlier editions. While these editions included comprehensive references for all the structures discussed, this is no longer possible in a book of this size and we have therefore frequently quoted important review articles from which references to earlier work can be obtained. References to recent work have, however, been included as before.

Undergraduate courses now lay considerable stress on spectroscopic methods and we have included a new chapter to serve as an introduction to the spectroscopy of complex compounds.

E.C.
G.W.A.F.

Contents

1

HISTORICAL INTRODUCTION

1.1 Electricity and Chemical Bonding

The idea that chemical forces are electrical in nature can be traced back to the beginning of the nineteenth century. In 1800, Nicholson and Carlisle decomposed water into hydrogen and oxygen by passing an electric current through it, and in the next few years many other electrolytic decompositions were reported. Perhaps the outstanding examples were Sir Humphry Davy's isolation of sodium and potassium from caustic soda and caustic potash, respectively, in 1807. Davy, indeed, suggested that the forces governing chemical combination were electrical in nature, the electrification of the combining particles being produced by contact. Electrochemical ideas were also developed by Berzelius in his Dualistic Theory (1812). He assumed that every atom possessed two electric poles of opposite sign, electropositive atoms having the positive pole in excess, electronegative atoms the negative pole in excess. The combination of an element with oxygen might produce a basic oxide (e.g. CuO) with a residual positive polarity, or an acidic oxide (e.g. SO_3) with a residual negative polarity. These oxides could combine because of the attraction of the residual opposite charges: thus

$$\overset{+}{CuO} + \overset{-}{SO_3} = CuSO_4$$

The theory, when subsequently applied to organic chemistry, could not, however, explain the fact that the substitution of negative chlorine for positive hydrogen in many organic molecules produced comparatively little change in chemical properties (cf. CH_3COOH and CCl_3COOH). Moreover, it could not account for Faraday's Laws of Electrolysis.

Electrochemical theories of chemical combination were then neglected for many years; the advances in theoretical organic chemistry associated with the names of Laurent, Gerhardt, Frankland, Williamson and Kekulé did not refer specifically to the electrical nature of combining forces in molecules. The electrical theory was, however, restated by Helmholtz in his Faraday Lecture of 1881, when he said '. . . the very mightiest among the chemical forces are of electric origin. The atoms cling to their electric charges and opposite charges cling to each other'.

Arrhenius, in 1884 and 1887, published a theory of 'electrolytic dissociation' in which he proposed the idea that salts (e.g. sodium chloride) in dilute aqueous solution were dissociated into positive (e.g. sodium) ions and negative (e.g. chloride) ions. This theory of easily dissociated molecules or of 'ionizable' atoms was used by Werner (1891), who discussed the constitution of compounds of the type $CoCl_3,6NH_3$ in terms of 'principal' and 'auxiliary' valencies. The auxiliary valencies were exerted in the co-ordination of a number of atoms, molecules or radicals to the central metal atom, in an 'inner sphere of combination', and the principal valencies represented the attachment of ionizable atoms of groups in an 'outer sphere of combination'. Thus Werner represented $CoCl_3,6NH_3$ as formula (1), where the dotted lines represent the 'auxiliary' bonds and the

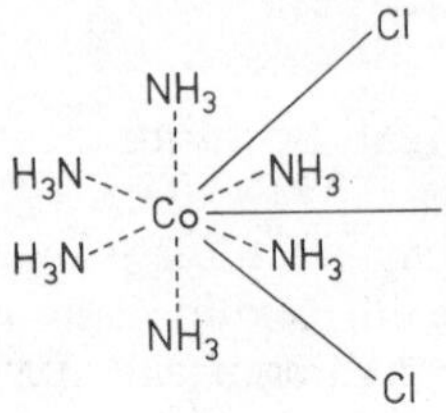

(1)

full lines the attachment of the ionizable atoms. Werner's theory was subsequently of great importance in the study of isomerism in inorganic compounds, and in other ways.

J.J. Thomson's (1897) recognition of the negatively charged electron as a constituent of all atoms, and his measurements of the ratio of its charge to its mass, provided a new stimulus to the electrochemical theories. He realized that the chemical properties of elements depended in some way on the arrangement of their electrons, and he suggested that electropositive atoms were those which could achieve a stable electronic state by losing one or two electrons, whereas electronegative atoms achieved stable states by acquiring one or more electrons.

Sir William Ramsay's Presidential Address to the Chemical Society of London in 1908 stressed the role of the electron in valency. Thus he said "... they (the electrons) serve as the 'bonds of union' between atom and atom". He also suggested that in molecules such as NaCl and Cl_2 the electron might form a 'cushion' between the two atoms. Ramsay, and other chemists of the period, believed that the maximum number of electrons involved in compound formation was eight; thus nitrogen, with five available electrons, could acquire three more by combining with three hydrogen atoms to form ammonia, NH_3, whereas NH_4 could only be obtained by removing one electron and forming a positively charged NH_4^+ ion.

Abegg (1904) had postulated that elements had two valencies, a 'normal' and a 'contravalency'. These were of opposite polarity, and the total sum of the two valencies was always eight. J. Newton Friend (1908) distinguished three types of valency – negative, positive and residual – where the negative valency of an atom was defined as being equal to the number of electrons with which it could combine, and the positive valency was related to the loss of electrons from the atom. His use of the term 'residual valency' is of great interest in the light of later developments, for he suggested that when an element exerted its residual valency, it simultaneously parted with, and received, an electron. The hydrogen molecule was represented as H⇄H, the arrows indicating the directions in which the electrons were transferred.

J.J. Thomson made a further important contribution to valency theory in 1914. He emphasized the difference between polar molecules (e.g. NaCl) and non-polar ones (e.g. most organic substances), and observed that the electropositive valency of an element was equal to the number of electrons that could easily be separated from it, while the negative valency was equal to the difference between eight and the positive valency.

1.2 The Lewis–Langmuir Electron-pair Bond

The development of the Rutherford theory of the nuclear atom, and Bohr's work on the hydrogen atom structure, paved the way for a more comprehensive 'electronic theory of valency'. The foundations of this modern theory were laid in two independent publications by W. Kossel and by G.N. Lewis in 1916. Kossel, who was mainly concerned with polar (ionizable) molecules, pointed out that in the periodic system, a noble gas element always separates an alkali metal and a halogen, and that the formation of a negative ion by the halogen atom, and a positive ion by the alkali metal atom, would give each of these atoms the structure of a noble gas. The noble gases were assumed to possess a particularly stable configuration of eight electrons in an outer shell (two electrons in the case of helium). No compounds of these elements were known at that time – indeed until very recently they were usually called the 'inert' gases. Many compounds of xenon and krypton have now been made, but it seems unlikely that stable compounds of helium, neon and argon can be formed and these elements might well still be called inert gases. The stable ions in a compound such as NaCl are, on this view, held together by electrostatic attraction and form what is now called an 'electrovalent' bond or link. The electrovalency of the ion is defined as the number of unit charges on the ion; thus magnesium has a positive electrovalency of two, while chlorine has a negative electrovalency of one. We shall discuss the formation and properties of electrovalent (now usually called ionic) bonds in some detail in Chapter 9.

Lewis discussed atomic structure in terms of a positively charged 'kernel' (i.e. the nucleus plus the 'inner' electrons) and an outer shell that could contain up to eight electrons. He assumed that these outer electrons were arranged at the corners of a cube surrounding the kernel; thus, the single electron in the outer shell of sodium would occupy one corner of the cube, whereas all eight corners would be occupied by the electrons in the outer shell of an inert gas. This octet of electrons represented a particularly stable electronic arrangement, and Lewis suggested that, when atoms were linked by chemical bonds, they achieved this stable octet of electrons in their outer shells. Atoms such as sodium and chlorine could achieve an outer octet by the transfer of an electron from sodium to chlorine, forming Na^+ and Cl^- ions, respectively. This was essentially the mechanism proposed by Kossel, but Lewis proposed a second mechanism to account for the formation of non-polar molecules. Here, there was no transfer of electrons from one atom to another (and thus no ion formation), but the bond resulted from the sharing of a pair of electrons, each atom contributing one electron to the pair.

This theory was considerably extended by Langmuir (1919), although he abandoned the idea of the stationary cubical arrangement of the outer electrons; he introduced the term 'covalent bond' to describe the Lewis 'electron-pair' bond or link. We can illustrate the Lewis–Langmuir theory by considering the chlorine molecule, Cl_2. Chlorine, with the electronic configuration (using the modern notation introduced in Chapter 5) $(Ne)3s^2 3p^5$, is one electron short of the inert gas configuration of argon, $(Ne)3s^2 3p^6$, and the formation of the stable diatomic chlorine molecule, Cl_2, results from the sharing of electrons by the two chlorine atoms. If we represent the outer electrons of one chlorine atom by dots, and those of the other by crosses, we can write

```
  ●●                  XX                  ●●       XX

●  Cl  ●    +    X    Cl   X     =     ●  Cl  ●   Cl  X
●                          X           ●      X       X

  ●●                  XX                  ●●       XX
```

where Cl represents the chlorine nucleus and the inner electrons. The atoms linked in this way need not be identical; thus, in carbon tetrachloride, each of the four outer electrons of the carbon pairs with an electron from a chlorine atom to form four covalent bonds. The Lewis–Langmuir theory would represent the structure as

```
                 XX
             X        X
             X   Cl   X

                 X●

      XX                    XX
   X       X             X       X
   X  Cl   ●      C      ●  Cl   X
      XX                    XX

                 X●
             X        X
             X   Cl   X
                 XX
```

Double and triple bonds are considered to involve the sharing of four and six electrons, respectively, so that the structures of ethene (ethylene) and ethyne (acetylene) would be written

```
H               H
 X            X●
●                                ●       X X X     X
    C   X X   C       and      H    C           C     H
        X X                      X       X X X     ●
 X             ●
  ●           X
H               H
```

In these molecules (Cl_2, CCl_4, C_2H_4 and C_2H_2) the electron pair for each single bond is provided by the two combining atoms, each contributing one electron. Perkins (1921) postulated a related type of link in which both electrons for the electron-pair bond come from only one of the two combining atoms. An example is provided in the combination of ammonia with boron trimethyl, $B(CH_3)_3$. Ammonia may be written

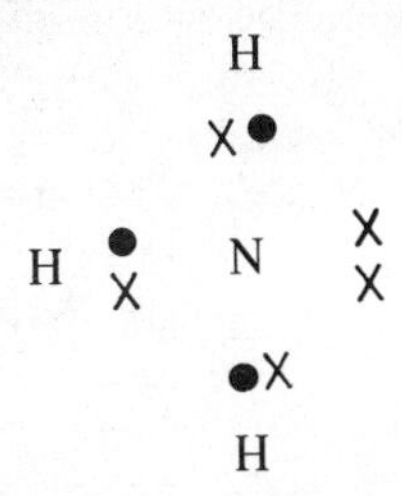

each hydrogen atom pairing its electron with a nitrogen electron to form three electron-pair bonds; the two unused outer electrons on the nitrogen atom form a so-called 'lone pair'. In boron trimethyl, however, there are

Me
Me B
Me

(where Me represents the methyl radical

H
H C H)

only six electrons around the boron atom, since the boron atom itself only has three outer electrons, which it uses to form three electron-pair bonds with the carbon atoms of the three methyl groups, as shown above. A compound $H_3N \rightarrow BMe_3$ can now be formed. The nitrogen atom is the 'donor' and the boron atom the 'acceptor' in this operation; the arrow indicates the relationship donor → acceptor. It should be emphasized that 'donation' is a special case of electron 'sharing', and no complete transfer of electrons takes place; nevertheless, the nitrogen atom has, in effect, 'lost' one electron (it now only shares two electrons both of which it had entirely to itself), and the boron atom has, so to say, 'gained' an electron. The formation of this type of bond thus involves a charge displacement, producing what is called an electric 'dipole' in the molecule. (We refer again to this term in Chapter 11.) Nitrogen acquires a 'formal' positive charge, and boron a 'formal' negative charge in $H_3N \rightarrow BMe_3$. Apart from this charge displacement the bond, once formed, does not differ in any way from the normal covalent bond of the Lewis–Langmuir theory.

The Lewis–Langmuir electron-pair or covalent bond is now often referred to as the homopolar bond, whereas the complete transfer of electrons, resulting in ion formation (e.g. Na^+, Mg^{2+}, Cl^-) gives rise to the electrovalent, or ionic, or heteropolar, bond by the attraction of opposite charges. Bonds formed by the Perkins mechanism were originally called 'co-ordinate' links, but the terms 'donor' or 'dative' bond have also been used. The production of formal charges on the atoms linked in this way is emphasized by the term 'semi-polar' bond, which implies that the bond has something of the character of the polar (i.e. ionic) link; another term introduced by Palmer (1944) – the co-ionic bond – also indicates that the bond has characteristics of both the covalent and the ionic bond. In bond diagrams, the covalent bond may be represented either by the colon, as in Cl:Cl, or, more usually, by the dash, Cl–Cl. Co-ordinate links may be represented, as we saw in the $H_3N{\rightarrow}BMe_3$ example, by a single arrow showing the direction of the electronic charge displacement. An alternative method is to write $H_3\overset{+}{N}{-}\overset{-}{B}Me_3$, where + and - are used to indicate the formal charges. The different names given above are often a source of confusion to beginners, so they are collected together for clarity as follows:

Polar = Heteropolar = Ionic = Electrovalent
Non-polar = Homopolar = Covalent
Semi-polar = Co-ordinate = Dative = Co-ionic = Donor–Acceptor

1.3 Quantum Theory of Chemical Bonding

The modern theory of the covalent bond with which this book is mainly concerned started with the recognition at the beginning of the century that the classical laws of physics did not apply to radiation. The quantum theory of radiation was developed by M. Planck (1900) to account for the way in which radiation from a 'black body' varied with wavelength (see p.9). His concept of corpuscles or 'quanta' of radiant energy was applied by A. Einstein (1905) to explain the photoelectric effect (p.11), and by N. Bohr (1913) to explain the spectrum of atomic hydrogen (p.13).

The link between radiation, described in classical terms by a wave theory, and energy quanta was formalized by L. de Broglie (1924); he related wavelength (λ), characteristic of a wave theory, with momentum p ($= m \times v$), characteristic of a particle theory. The de Broglie relationship

$$\lambda = h/p = h/mv$$

received experimental verification with the discovery of electron diffraction in 1927. We shall see that the fundamental constant of nature h, the Planck constant, is involved in the the theory of all these very different phenomena – black-body radiation, the photoelectric effect, the spectrum of atomic hydrogen and the de Broglie relationship.

The implications of a wave theory of electrons were worked out independently by W. Heisenberg (1925), E. Schrödinger (1926) and P. Dirac (1928). They developed quantum mechanical theories of electron behaviour in atoms and their methods were then applied to electrons in molecules by H. Heitler and F. London who, for the first time, calculated, approximately, the strength of the electron-pair bond in the hydrogen molecule. Their method – the so-called valence-bond method – was developed by L. Pauling and by J.C. Slater, while at the same time (1927) Ø. Burrau, R.S. Mulliken and J.E. Lennard-Jones were developing the complementary 'molecular-orbital' method of treating valency problems.

The discovery of the diffraction of X-rays by crystals (M. von Laue, W. Friedrich and P. Knipping, 1912) and the subsequent development of crystal structure determination by W.H. and W.L. Bragg led to a very satisfactory theory of bonding in electrovalent (ionic) compounds in terms of electrostatic interactions between charged ions.

In fact the 'chemical bond' is rarely purely covalent or purely ionic. The detailed quantum-mechanical theory of the covalent bond has to take account of partial 'ionic character' of the bond, just as the more simple electrostatic treatment of the ionic bond breaks down when the bond has 'covalent character'. The quantum theory of electrons in molecules provides a method of calculating the charge distribution in any system, the energies of the electrons and the bond lengths, angles and dissociation energies. The calculations involved are of a formidable complexity even for a simple diatomic molecule such as H_2, and it is only in recent years, with the development of very large computers, that detailed calculations on larger molecules have become possible.

1.4 Bibliography

1. NEWTON FRIEND, J., *The Theory of Valency*, Longmans, London (1909)
2. PARTINGTON, J.R., *A Short History of Chemistry*, Macmillan, London (1939)
3. BERRY, A.J., *Modern Chemistry – Some Sketches of its Historical Development*, Cambridge University Press, London (1946)

2

THE EXPERIMENTAL FOUNDATION OF THE QUANTUM THEORY

2.1 Energy Units

The international system of units (S.I.) expresses fundamental physical quantities such as mass, time, thermodynamic temperature and amount of substance in terms of the units metre (m), kilogram (kg), second (s), kelvin (K) and mole (mol), respectively.

Energy is expressed in joules (J). Many chemists have been brought up in the C.G.S. system and have been accustomed to expressing energy in calories or kilocalories, which are not S.I. units. Many existing reference books also list energies in calories. The appropriate conversion factor is

$$1 \text{ 'thermochemical' calorie} = 4.184 \text{ J}$$

The energies of electrons in atoms can very conveniently be expressed in electron-volts (eV), where 1 eV is the energy acquired by an electron when it is accelerated by a potential difference of one volt:

$$1 \text{ eV} = 1.6021 \times 10^{-19} \text{ J}$$

In radiation theory, wavelength λ and frequency ν are related by

$$\nu = c/\lambda$$

where c is the velocity of electromagnetic radiation in a vacuum ($c = 2.9979 \times 10^8$ m s^{-1}). The frequency unit, ν, is called the hertz (Hz). The wave number, $\bar{\nu}$, is often used in spectroscopy.

$\bar{\nu} = 1/\lambda$ and the wavenumber unit is thus the reciprocal metre, m^{-1}; in practice, chemists normally find it more convenient to use the reciprocal centimetre, cm^{-1}. A wave number $\bar{\nu}$ is related to a frequency ν by

$$c \text{ (m s}^{-1}) \times \bar{\nu} \text{ (m}^{-1}) = \nu \text{ (Hz)}$$

We can now use the Planck expression for the quantum of energy (see p.11), $E = h\nu = hc\bar{\nu}$, to associate energy and wavenumber units. Substitution of $h = 6.6256 \times 10^{-34}$ J s and $c = 2.9979 \times 10^8$ m s^{-1} in this expression gives

$$E \text{ (J)} = 19.863 \times 10^{-26} \ \bar{\nu} \text{ (m}^{-1})$$

These values relate to single atoms. Chemists usually refer to energy changes per mole of substance, where 1 mole is the amount of substance that contains as many elementary particles (electrons, atoms, molecules, etc.) as there are atoms in 0.012 kg of carbon-12. This number is the Avogadro number N where

$$N = 6.0225 \times 10^{23} \text{ atoms mol}^{-1}$$

The relationship between energy and wavenumber now becomes

$$E \text{ (J mol}^{-1}) = 19.863 \times 10^{-26} \times 6.0225 \times 10^{23} \; \bar{\nu} \text{ (m}^{-1}) = 0.1196 \; \bar{\nu}$$

and

$$E \text{ (eV)} = 1.6021 \times 10^{-19} \times 6.0225 \times 10^{23} = 9.649 \times 10^{4} \; E \text{ (J mol}^{-1})$$

As examples of these interconversions we conclude this section with numerical values for the energies corresponding to radiation in the infrared, the visible and the ultraviolet regions of the electromagnetic spectrum. (Wavelengths in these regions are conveniently quoted in ångstrom units (Å), where 1 Å = 10^{-10} m, or in nanometres (nm) where 1 nm = 10^{-9}m.)

Wavelength/ Å	*Wavenumber*/ cm^{-1}	*Energy*/ kJ mol^{-1}	*Energy*/ eV atom^{-1}
10 000 (infrared)	10^4	119.63	1.2398
5 000 (visible)	2×10^4	239.26	2.4796
2 000 (ultraviolet)	5×10^4	598.14	6.1990

Bond lengths in molecules are now quoted in picometres (pm), where 1 pm = 10^{-12} m. A bond length of, for example, 1.54 Å in the earlier literature would now be quoted as 154 pm.

2.2 Black Body Radiation

The rate of production of energy by a hot surface depends upon the temperature, the nature and the area; a dull black surface radiates more energy per second than a polished one of the same area and temperature. The blacker the surface the greater the radiation, so that the maximum radiation at a given temperature will be that from a 'perfectly black' surface. The dependence of radiant energy upon temperature alone can therefore be examined if such a radiating surface is available. The radiation is called 'black-body radiation' and although it is impossible to make a 'perfectly black' surface, radiation with characteristics very close to those expected from such a surface can be obtained experimentally from a small opening in the wall of a furnace kept at constant temperature. If radiation from such a source is dispersed by a prism system and allowed to fall on a sensitive energy detector such as a thermocouple, the distribution of the energy among different wavelengths can be studied. The classical experiments in this field were those of Lummer and Pringsheim at the end of the nineteenth century. A typical result is shown in *Figure 2.1*, where E corresponds to the radiant energy emitted per unit wavelength interval per unit area per second; $E_\lambda d\lambda$ will measure the energy radiated between the wavelengths λ and $\lambda + d\lambda$. It will be seen that only a small amount of the total energy is radiated at very short or at very long wavelengths, and that the curve passes through a maximum.

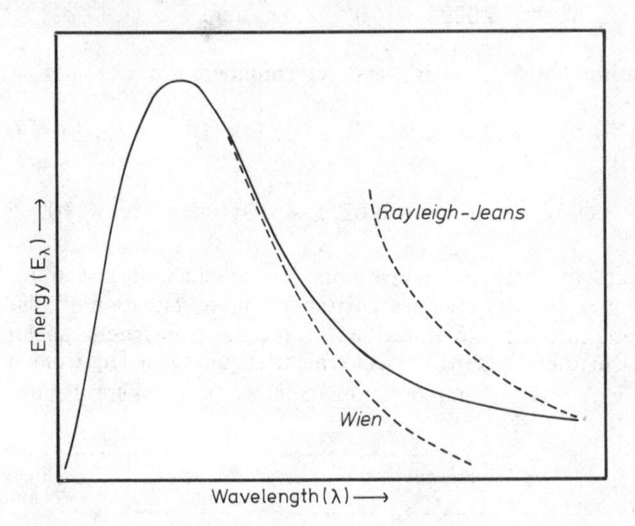

Figure 2.1 ***Distribution curves for black-body radiations***

Two important attempts were made to explain these results. Wien, using the methods of classical thermodynamics, derived the following expression

$$E_\lambda d\lambda = c_1 \lambda^{-5} \exp\{-(c_2/\lambda T)\} d\lambda \tag{2.1}$$

where λ is the wavelength, T the absolute temperature, and c_1 and c_2 are constants. E passes through a maximum value as λ increases, and with a suitable selection of values for the constants c_1 and c_2 good agreement with experiment is obtained for small values of λT. At long wavelengths, however, in the infrared region of the spectrum, the agreement is not satisfactory. The dotted curve in *Figure 2.1* shows the results predicted by the Wien theory compared with the experimental results given by the full line.

The problem was tackled from a different point of view by Rayleigh and by Jeans. They worked out the number of 'modes of vibration' of electromagnetic waves in an enclosure, and assumed that the total energy was equally distributed among these modes. This gave the relationship

$$E_\lambda d\lambda = \frac{c_1}{c_2} \cdot \frac{T}{\lambda^4} d\lambda \tag{2.2}$$

There is good agreement with experiment at long wavelengths, where the Wien equation breaks down, but the equation is unsatisfactory at short wavelengths. It can be seen from equation 2.2 that as λ decreases E_λ increases, so that most of the radiant energy should be emitted at short wavelengths. Now both these methods correctly use the long-accepted techniques of classical thermodynamics and mechanics; if the results contradict experimental observations, then the fundamental assumptions must be at fault. Max Planck's great achievement was the formulation of new assumptions.

He started work on this problem by deriving a purely empirical equation which would fit the experimental results. This equation was

$$E_\lambda \mathrm{d}\lambda = \frac{c_1 \lambda^{-5}}{\exp(c_2/\lambda T) - 1} \mathrm{d}\lambda \tag{2.3}$$

Now it can be shown that equation 2.3 reduces to the Wien expression (equation 2.1) for small values of the product λT and to the Rayleigh–Jeans expression (equation 2.2) for large values of λT. For small values of λT the exponential function $\exp(c_2/\lambda T)$ is very much larger than unity; the denominator in equation 2.3 thus reduces to $\exp(c_2/\lambda T)$ and we get equation 2.1. Again, on expansion of the exponential function

$$\exp(c_2/\lambda T) = 1 + \frac{c_2}{\lambda T} + \tfrac{1}{2}\left(\frac{c_2}{\lambda T}\right)^2 + \ldots$$

only the first two terms are significant for large values of λT and equation 2.3 reduces to equation 2.2. Planck now showed that any theory which was to produce equation 2.3 would have to relate the energy E with the frequency ν of the radiation by the equation $E = h\nu$, where h is a constant. (We shall in general refer to frequencies rather than wavelengths; they are related by $\nu = c/\lambda$ where c is the velocity of propagation.)

Such a relationship marks a complete departure from classical theory. The Rayleigh–Jeans method assumes that the electric oscillators associated with the electromagnetic radiation can have any energy values between zero and infinity. The Planck hypothesis states that the energy of these oscillators cannot vary continuously; they can only have definite amounts of energy, the so-called 'quanta', $h\nu$, $2h\nu$, . . . $nh\nu$, where ν is the frequency, n an integer and h a universal constant, now known as Planck's constant. Any change in the energy of the oscillating system can only be in discrete amounts – one or more quanta. If this assumption is made, the constants c_1 and c_2 in the empirical equation 2.3 can be expressed in terms of fundamental constants, and the equation becomes

$$E_\lambda \mathrm{d}\lambda = \frac{2\pi c^2 h}{\lambda^5} \cdot \frac{1}{\exp(ch/\lambda kT) - 1} \mathrm{d}\lambda \tag{2.4}$$

where c is the velocity of light and k the Boltzmann constant ($k = R/N$, where R is the gas constant and N the Avogadro number). By using the results of measurements of radiant energy and equation 2.4, Planck found $h = 6.61 \times 10^{-34}$ J s. The modern value is $h = 6.625\,54 \times 10^{-34}$ J s.

2.3 The Photoelectric Effect

Electrons may be emitted when light falls on a metal surface. In some cases (e.g. the alkali metals), light in the visible region of the spectrum can eject these so-called 'photoelectrons', but for most metals ultraviolet radiation must be used to produce the effect; there is a critical frequency, ν_0, for each metal, below which no photoelectrons are emitted. Experiments show that

(a) the energy of the photoelectrons is independent of the intensity, but proportional to the frequency of the incident radiation, and
(b) the number of photoelectrons emitted per second is proportional to the intensity of the incident radiation.

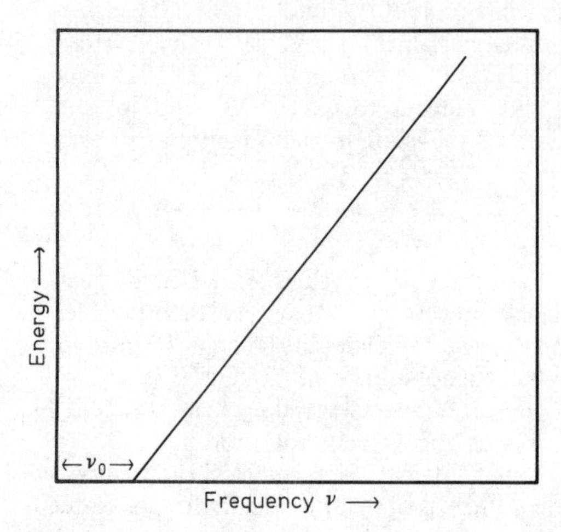

Figure 2.2 Variation of the maximum energy of photoelectrons with frequency of the incident radiations

Figure 2.2 shows how the maximum energy of the photoelectrons varies with the frequency of the incident radiation. The results may be expressed by the equation

$$\text{maximum energy} = \tfrac{1}{2}mv^2 = \text{constant}\ (\nu - \nu_0) \tag{2.5}$$

where m is the mass and v the velocity of the electron. Such a relationship cannot be derived from classical electromagnetic theory, however, for this predicts that the energy of the photoelectrons should vary with the intensity but should be independent of the frequency. Here, therefore, we have another example of the breakdown of classical radiation theory.

In 1905 Einstein showed that the difficulties could be resolved if the quantum postulates of Planck were applied to the photoelectric effect. Instead of regarding the incident light as radiation of frequency ν, he considered it to be a stream of corpuscles, now called photons, each possessing energy $h\nu$, where h is the Planck constant. Each of these photons gives up its energy to an electron in the metal, and part of the energy is used in just removing the electron from the metal surface. The remainder appears as the kinetic energy of the photoelectron. Thus

$$h\nu = W + \tfrac{1}{2}mv^2 \tag{2.6}$$

W is called the 'work function'; it represents the energy which is needed just to remove the electron from the metal surface. Comparing equation 2.6 with equation 2.5 representing the experimental results we can write

$$\tfrac{1}{2}mv^2 = h\nu - W = h(\nu - \nu_0) \qquad (2.7)$$

The value of h obtained from Millikan's (1917) experiments on the photoelectric effect, 6.56×10^{-34} J s, agrees well with that obtained from the radiation measurements.

2.4 The Bohr Theory of the Hydrogen Atom

The general acceptance of the quantum theory came with the development of a theory by Niels Bohr which accounted for some well-known empirical relationships between the frequencies of the lines observed in the spectrum of hydrogen. When an electric discharge is passed through a tube containing hydrogen at low pressure, light is emitted, and its spectrum is found to consist of sharp lines. Four lines can be seen by the naked eye, and many others can be recorded photographically. *Figure 2.3* represents the

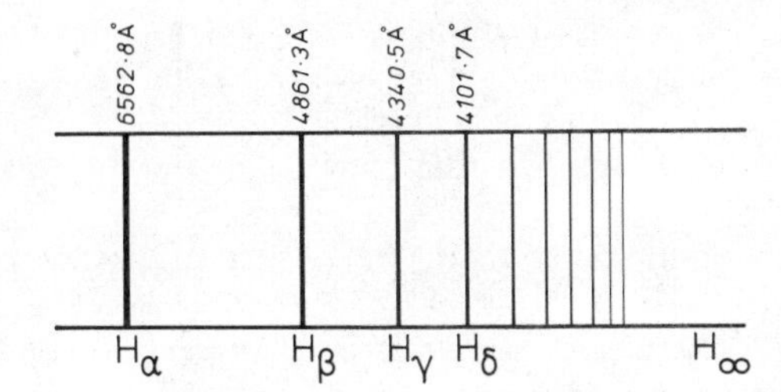

Figure 2.3 Atomic spectrum of hydrogen in the visible regions

lines as they appear on looking into the spectroscope eyepiece. (Note that the horizontal frequency scale is not linear; the dispersion of the instrument increases as the frequency increases.) It was shown by Balmer in 1885 that the wavelengths of many of these lines were given by a simple empirical formula

$$\lambda = \frac{bn^2}{n^2 - 4} \qquad (2.8)$$

where b is a numerical constant and n an integer, e.g. 3, 4, 5, ... etc. This expression can be rearranged to give frequencies, ν, by using the relationship $\nu = c/\lambda$, where c is the velocity of light, giving

$$\nu = \frac{c(n^2 - 4)}{bn^2} = Rc\left\{\frac{1}{2^2} - \frac{1}{n^2}\right\}$$

where R, the so-called Rydberg constant, is equal to $4/b$. Later work showed that the frequencies of all the lines in the hydrogen spectrum could be expressed by a general equation

$$\nu = Rc\left\{\frac{1}{n_1^2} - \frac{1}{n_2^2}\right\} \qquad (2.9)$$

When $n_1 = 2$ and $n_2 = 3, 4, 5, \ldots$ the Balmer lines are obtained; other series are obtained by giving n_1 other fixed values such as 1, 3 or 4, and giving n_2 the values $(n_1 + 1)$, $(n_1 + 2)$, $(n_1 + 3)$ ... etc.

Now if we apply classical electromagnetic theory to the problem of the structure of the hydrogen atom, we come across a fundamental difficulty. The Rutherford theory of atomic structure had postulated a small positively charged nucleus containing most of the mass of the atom, surrounded by electrons. A simple mechanical model of this system is that of electrons moving in orbits around the nucleus, like planets round the sun in the solar system. The negatively charged electrons are held in their orbits by electrostatic attraction to the positively charged nucleus. This attraction varies inversely as the square of the distance, just as the gravitational force varies in the case of the solar system; the mathematical theory developed by Kepler and Newton for the motion of heavenly bodies should therefore apply to the motion of electrons in atoms.

There is this difficulty, however. Classical electromagnetic theory states that an accelerating electric charge must radiate energy, so that an electron moving in an orbit around a nucleus must therefore radiate energy, and move in a decreasing spiral path until it finally disappears into the nucleus. The quantum theory of radiation avoids this difficulty.

Planck's quantum theory introduced two important ideas:

(a) that electric oscillators can only possess discrete quanta of energy, and

(b) that radiation is emitted only when the oscillator changes from one quantized state to another of lower energy.

Bohr now (1913) adapted these ideas to the quite different system of the hydrogen atom, and postulated that the motion of the electron was restricted to a number of discrete circular orbits, with the nucleus at the centre. An electron moving in one of these orbits has a constant energy, and radiation is emitted only when the electron moves into an orbit of lower energy. Bohr also adapted Planck's expression for the relationship between frequency and energy, for he assumed that the frequency of the emitted radiation is given by

$$\nu = (E_2 - E_1)/h$$

where E_2 and E_1 are the energies of the electron in two different orbits, and h is Planck's constant. The size of the orbits was determined by an arbitrary assumption that the angular momentum of the electron about the nucleus is an integral multiple of $h/2\pi$. A quantizing condition of this form had previously been put forward by J.W. Nicholson (1912), and Bohr was able to use it, together with the postulates already mentioned, to derive the empirical Balmer Law in the form of equation 2.9. The form of this equation results from the nature of the assumptions made in the theory, and the Rydberg constant R is given by

$$R = me^4/8\epsilon ch^3$$

The success of the Bohr theory lay in the fact that substitution of known values for the physical constants in this expression gave a value for the Rydberg constant in agreement with the experimental value.

We shall not describe the Bohr theory in detail since it was superseded in 1926 by the new quantum theory of Heisenberg and Schrödinger. The Bohr theory expression for the possible energies E_n of the electron in the hydrogen atom

$$E_n = -\frac{me^4}{8\epsilon^2 h^2 n^2} \text{ J} \tag{2.10}$$

(m = electron mass, e = electron charge, h = Planck constant, ϵ = permittivity of a vacuum and n the principal quantum number ($n = 1, 2, \ldots \infty$)) is, however, exactly the same as that derived from the new quantum theories.

The energies of electrons in atoms are conveniently expressed in electron-volt units (eV). Substitutions of appropriate values for the constants in equation 2.10 gives, for hydrogen

$$E_n = -\frac{13.6}{n^2} \text{ eV} \tag{2.11}$$

The negative sign in the expression arises from the way in which an arbitrary energy zero is defined. Potential energy can only be defined in relation to an arbitrary zero, and for atoms this is taken to be when the electron is at infinite distance from the nucleus. The negatively charged electron is attracted by the positively charged nucleus, so that energy has to be supplied to move the electron from its equilibrium position in the atom to infinity. Thus the electron with energy E_n has a negative energy; this becomes less negative as n increases until at $n = \infty$, $E_\infty = 0$.

2.5 Energy Level Diagrams

These ideas are conveniently illustrated by energy level diagrams, where energy is plotted on a vertical scale, and the possible energies are shown as horizontal lines (levels). *Figure 2.4* shows such a diagram for the electron in a hydrogen atom. The state for which $n = 1$ ($E_1 = -13.6$ eV) is called the *ground state*, and states for which $n > 1$ are called *excited states*. In an assembly of hydrogen atoms the electrons will usually be found in the ground state, but if energy is put into the assembly (e.g. by heating, or passing an electric discharge) the number of atoms having an electron in one or other of the excited states increases. Sooner or later, however, these 'excited' electrons will revert to the more stable ground state. In doing so they will emit radiation of frequency ν given by the Bohr relationship

$$\Delta E = h\nu$$

where ΔE is the energy difference between the levels involved in the electron transition (e.g. $\Delta E = E_2 - E_1$).

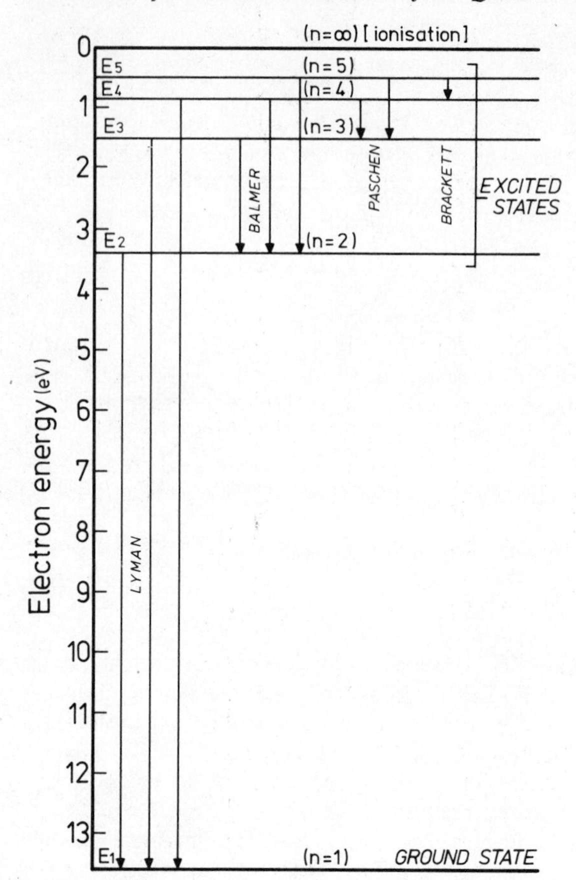

Figure 2.4 Energy level diagram for the hydrogen atom

Some possible transitions are shown in *Figure 2.4*, where they are represented by vertical lines. Thus all the lines of the Balmer series are derived from transitions that place the electron in the E_2 level.

Transitions such as $E_2 \rightarrow E_1$ correspond to very large values of ΔE and, consequently, high values for the frequency ν of the emitted radiation. This transition gives rise to the first line of the Lyman series which appears in the far-ultraviolet (high-frequency) region of the electromagnetic spectrum. *Figure 2.4* also shows series associated with the names of Paschen and Brackett which give rise to lines in the infrared region of the spectrum.

Equation 2.11 can be modified to cover all *one-electron* atoms (e.g. He^+, Li^{2+}, Be^{3+}, B^{4+}, etc.) by writing $E_n = -13.6Z^2/n^2$, where Z is the atomic number (the number of protons in the atomic nucleus). Thus the energy of the first *excited* state of the He^+ system ($Z = 2$)

$$E_2 = \frac{-13.6}{2^2} \times 2^2 = -13.6 \text{ eV}$$

is the same as that of the *ground* state of the hydrogen atom ($Z = 1$). This dependence on Z^2 illustrates the important influence that nuclear charge has on electron energies in atoms.

2.6 Extensions of the Bohr Theory

The Bohr theory was extremely successful in accounting for the general features of the hydrogen atom spectrum. More detailed experimental study of atomic spectra revealed additional lines which could only be explained by modification and extension of the Bohr theory. One refinement arose from the recognition that *nuclear* mass is involved in the expression for the Rydberg constant, and we write

$$R = \mu e^4/8\epsilon ch^3$$

where μ, the *reduced mass*, replaces m; $\mu = mM/(m + M)$, where m = the electron mass and M the nuclear mass. It is as a result of this that the existence of some isotopes can be detected from atomic spectra observations. In 1931, Urey, Brickwedde and Murphy discovered deuterium by observing a satellite of the Balmer H_α line from gas obtained from the residue left after the evaporation of a large volume of liquid hydrogen. This line, separated by 179 pm from the H_α line, corresponds to an atom with a nucleus of mass double that of the hydrogen nucleus. The change in R is not very large: R_H = 109 677.58 cm^{-1}, R_D = 109 707.42 cm^{-1}.

The Bohr theory was, however, unable to account for the complexities of the atomic spectra of many-electron atoms; the large numbers of observed lines require additional energy levels, and these require additional assumptions in extending the Bohr theory. The simple model of an electron moving in a closed circular or elliptical orbit around the nucleus is now inadequate. Complete explanation of the observed spectra of many-electron systems had to await the development of the new quantum theory.

3

ELEMENTARY QUANTUM THEORY

3.1 Particles and Waves

We saw in earlier chapters that although the phenomena of the interference and diffraction of light required the use of a wave theory of radiation, the photoelectric effect could only be explained by assuming that light travelled as particles (photons). Now many of the phenomena of optics – reflection and refraction, for example – can be quite adequately explained by a particle theory; it is only when experiments are done with apertures or obstacles of very small dimensions that diffraction effects are obtained, and a wave theory is needed. We thus get a duality of behaviour; light radiation can be considered either as a stream of photons or as a wave motion. It was pointed out in 1924 by de Broglie that a similar duality might exist in the case of the electron. All the experiments performed on electrons from their discovery up to this time could be explained by assuming they were small particles of a certain mass and charge, but de Broglie showed, by arguments based on the theory of relativity, that if an electron of momentum p could be described by a wave theory in terms of a wavelength λ, then

$$\lambda = h/p \tag{3.1}$$

This wavelength can conveniently be expressed in terms of electron energy rather than momentum. The kinetic energy, $\frac{1}{2}mv^2$, of an electron of mass m and velocity v is given by

$$\tfrac{1}{2}mv^2 = E - V$$

where E is the total energy and V the potential energy. Thus

$$v^2 = 2(E - V)/m$$

and

$$\lambda = \frac{h}{mv} = \frac{h}{m\sqrt{\{2(E - V)/m\}}} = \frac{h}{\sqrt{\{2m(E - V)\}}} \tag{3.2}$$

An electron with kinetic energy of 100 eV and no potential energy will therefore have an associated wavelength given by

$$(6.626 \times 10^{-34})/\sqrt{(2 \times 9.109 \times 10^{-31} \times 100 \times 1.6021 \times 10^{-19})}$$

$$= 1.23 \times 10^{-10} \text{ m}$$

which is of the same order of magnitude as the spacings of atoms in crystals. It was realized that it might be possible to use crystals as diffraction gratings for electrons, just as they can be used to diffract X-rays, and such electron diffraction effects were in fact observed experimentally in 1927. Davisson and Germer examined electrons scattered from metal surfaces at small angles, and Thomson and Reid photographed the diffraction rings produced when electrons were fired through a very thin metal foil. In each case the wavelengths calculated from the diffraction patterns fitted the equation.

3.2 The New Quantum Theory and the 'Uncertainty Principle'

A new quantum theory of the electron is therefore required. The Bohr theory succeeded because it provided both a simple physical picture of the motion of an electron in an atom and a method of calculating its possible energies. Any new theory must do all this, but it must also account for diffraction effects and reduce the number of arbitrary assumptions required by the older theory.

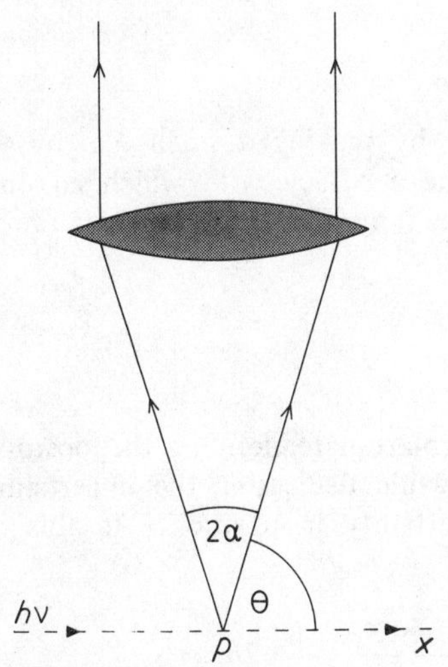

Figure 3.1 The 'γ-ray' microscope

We shall first of all discuss the problem of using a wave theory to describe the position of particles. This at once highlights a fundamental difficulty which lies at the root of all electron theory.

This difficulty, usually referred to as the 'Heisenberg uncertainty principle', is that it is impossible to state, simultaneously and precisely, both the position and the momentum of an electron. Suppose we devise a hypothetical experiment to measure the position and velocity of an electron. We could set up two 'γ-ray' microscopes that could 'see' electrons and measure the time taken for the electron to pass from one to the other. P, in *Figure 3.1*, represents the electron. It can only be observed if a photon incident upon it is scattered into the aperture of the microscope, i.e. within the cone of angle $2a$. Now a photon of frequency ν will have an associated wavelength $\lambda = c/\nu$, and by equation 3.1, $c/\nu = h/p$, whence $p = h\nu/c$. If the photon is scattered in a direction making an angle θ with the x axis, the electron will receive a component of momentum along the x axis of

$$\frac{h\nu}{c}(1 - \cos\theta)$$

and since the electron will be detected for any value of the angle between $90^\circ \pm a$, the momentum may have any value between

$$\frac{h\nu}{c}[1 - \cos(90 - a)] \quad \text{and} \quad \frac{h\nu}{c}[1 - \cos(90 + a)]$$

i.e. between

$$\frac{h\nu}{c}(1 - \sin a) \quad \text{and} \quad \frac{h\nu}{c}(1 + \sin a)$$

If we define this spread of values as the 'uncertainty' in the value of p, and denote it by Δp, then

$$\Delta p = \frac{2h\nu}{c}\sin a$$

We could try to reduce this uncertainty by making a small, i.e. by using a microscope of smaller aperture, but the accuracy with which an object can be located by a microscope is defined by the Rayleigh equation for the resolving power

$$\Delta x = \frac{c}{\nu \sin a}$$

where Δx is the uncertainty in x, the co-ordinate defining the position of the electron. Thus a smaller aperture, while decreasing the uncertainty in the momentum, would increase the uncertainty in position. In this experiment

$$\Delta x \,.\, \Delta p = \frac{2h\nu}{c}\sin a \,.\, \frac{c}{\nu \sin a} = 2h$$

In general, the product $\Delta x \,.\, \Delta p$ is of the order of magnitude of the Planck constant, h. This is one way of expressing the Heisenberg uncertainty prinple (1927).

We can illustrate its importance by a rough calculation. Suppose that we could locate the position of an electron with an 'uncertainty' of 0.1 pm, i.e. $\Delta x = 10^{-13}$ m. Substituting values for Δx and h in the above equation then gives

$$\Delta p \approx 2 \times 6.6 \times 10^{-34}/10^{-13} = 13.2 \times 10^{-21} \text{ kg m s}^{-1}$$

This uncertainty in momentum would be quite negligible in macroscopic systems, but it is far from negligible in systems containing electrons, since we are then dealing with masses of the order of 10^{-30} kg. Precise statements of the position and momentum of electrons have to be replaced by statements of the probability that the electron has a given position and momentum. The introduction of probabilities into the description of electronic behaviour is a direct consequence of the operation of the uncertainty principle; a small uncertainty in position implies a high probability that the electron is at a given point. This probability concept can be further illustrated if we reconsider the electron diffraction experiments of Thomson. If a single electron is sent through the diffraction apparatus it obviously cannot interfere with itself to give a diffraction pattern, and the Heisenberg principle tells us that we cannot follow its course precisely. We can say, however, that there is a certain probability that it will take a particular path, and that the observed diffraction rings are regions in which there is a high probability of locating the electron. A very useful, although less rigorous, description is to say that the diffraction rings are regions of high electron density. We recall that the photon theory of light interrelates photon density (number of photons per unit volume) with light intensity; high intensity implies a large number of photons per unit volume. In wave theory, intensity is measured as the square of a quantity called the 'amplitude', and this quantity can be obtained from the solution of a 'wave equation' which describes the particular system. If, therefore, we use a wave theory to calculate the variations in intensity in a diffraction experiment, we can express the results in terms of a particle theory by equating the square of the amplitude with the particle density. The new quantum theory describes the behaviour of a beam of electrons in the same way. The dark rings on the photographic plate in an electron diffraction experiment reveal positions of high electron density; these densities are related to the square of the amplitude factor obtained by solving a suitable wave equation.

3.3 Waves and Wave Equations

It is easy to visualize a wave motion in terms of the disturbance that spreads over the surface of a pond when a stone is dropped into it, or in terms of the vibrations of a violin string; it is not so easy to get a physical picture of the sound waves in an organ pipe, or the electromagnetic waves of light radiation. All these wave motions, however, are characterized by a transmission of energy from one point to another without permanent displacement of the intervening medium; thus waves can be made to

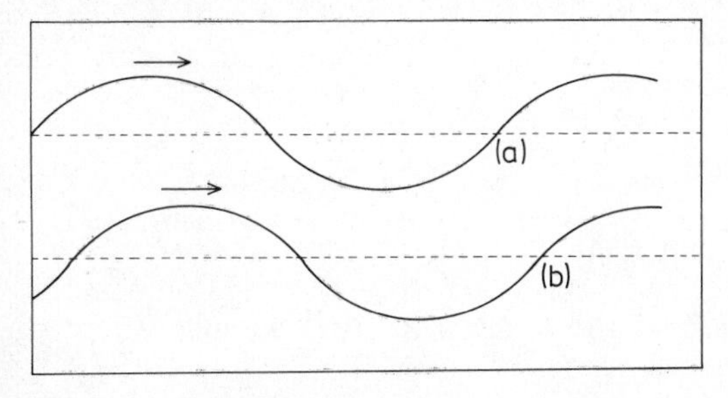

Figure 3.2 Progressive waves

travel down a stretched string, but the string itself does not undergo permanent translation. These wave motions can, moreover, all be described by wave equations of similar form.

Let us consider a simple case, the wave motion produced by moving up and down one end of a long horizontal string. The appearance (or 'profile') of the string at some instant will be as shown in *Figure 3.2*(*a*), and at a later instant by *Figure 3.2*(*b*). The disturbance is thus transmitted along the string in the form of a progressive wave. The equation for this wave motion, which can be shown to be

$$\frac{\partial^2\phi}{\partial x^2} = \frac{1}{c^2}\cdot\frac{\partial^2\phi}{\partial t^2} \tag{3.3}$$

has a solution of the form

$$\phi = a \sin 2\pi(x/\lambda - \nu t) \tag{3.4}$$

where ϕ is the amplitude of the disturbance at a distance x along the string, λ is the wavelength, ν the frequency, a a constant (the maximum value of the amplitude) and c the velocity of the progression. (The reader can check that equation 3.4 is a solution of equation 3.3 by appropriate differentiation of equation 3.4 and substitution in equation 3.3.) Equation 3.3 is 'linear', i.e. the function ϕ and its differential coefficients are always of the first order. An important characteristic of such linear equations is that if ϕ_1 and ϕ_2 are any two solutions, the linear combination $a_1\phi_1 + a_2\phi_2$, where a_1 and a_2 are arbitrary constants, is also a solution. This can be seen by substitution in equation 3.3; thus

$$\frac{\partial^2\phi_1}{\partial x^2} = \frac{1}{c^2}\cdot\frac{\partial^2\phi_1}{\partial t^2}, \qquad \frac{\partial^2\phi_2}{\partial x^2} = \frac{1}{c^2}\cdot\frac{\partial^2\phi_2}{\partial t^2}$$

and

$$\frac{\partial^2(a_1\phi_1 + a_2\phi_2)}{\partial x^2} = a_1\frac{\partial^2\phi_1}{\partial x^2} + a_2\frac{\partial^2\phi_2}{\partial x^2}$$

$$= \frac{1}{c^2}\left\{a_1\frac{\partial^2\phi_1}{\partial t^2} + a_2\frac{\partial^2\phi_2}{\partial t^2}\right\}$$

$$= \frac{1}{c^2}\cdot\frac{\partial^2(a_1\phi_1 + a_2\phi_2)}{\partial t^2}$$

This is an example of the important 'Principle of Superposition', which is used extensively in the quantum theory of valency (see p.53). We can apply the principle at this stage to the problem of the vibration of a string stretched between two fixed points. A progressive wave can be made to travel down the string from left to right, the amplitude being given by the equation

$$\phi_1 = a \sin 2\pi(x/\lambda - \nu t) \tag{3.5}$$

When the wave reaches the fixed end of the string it is reflected, and travels back from right to left, with amplitude ϕ_2 given by

$$\phi_2 = a \sin 2\pi(x/\lambda + \nu t) \tag{3.6}$$

This wave will interfere with the wave travelling from left to right, and the resulting wave motion has an amplitude ϕ given by

$$\phi = \phi_1 + \phi_2 = a \sin 2\pi(x/\lambda - \nu t) + a \sin 2\pi(x/\lambda + \nu t)$$

which by simple trigonometry gives

$$\phi = 2a \sin 2\pi x/\lambda \, . \cos 2\pi\nu t \tag{3.7}$$

Inspection of equation 3.7 shows that ϕ will be zero when $\sin(2\pi x/\lambda)$ is zero, i.e. when

$$\frac{2\pi x}{\lambda} = n\pi \quad \text{and} \quad x = \frac{n\lambda}{2} \tag{3.8}$$

where n is an integer. These waves are called 'standing' or 'stationary' waves; the amplitude is always zero at the particular values of x given by equation 3.8, whereas with 'progressive' waves the amplitude at any value of x is continually changing. We shall find that wave equations appropriate to the description of the behaviour of electrons in atoms are analogous to those describing 'stationary' waves in ordinary mechanics, and it will therefore be useful to consider this example in rather more detail.

We can for convenience write equation 3.7 in the form

$$\phi = f(x) \, . \cos 2\pi\nu t$$

Simple differentiation then gives

$$\frac{\partial\phi}{\partial x} = \cos 2\pi\nu t \, . \frac{\partial f(x)}{\partial x}$$

$$\frac{\partial^2\phi}{\partial x^2} = \cos 2\pi\nu t \, . \frac{\partial^2 f(x)}{\partial x^2}$$

$$\frac{\partial\phi}{\partial t} = -f(x) \, . 2\pi\nu \sin 2\pi\nu t$$

$$\frac{\partial^2\phi}{\partial t^2} = -f(x) \, . 4\pi^2\nu^2 \cos 2\pi\nu t$$

Substitution of these values in equation 3.3 gives

$$\cos 2\pi\nu t \,.\, \frac{\partial^2 f(x)}{\partial x^2} = \frac{1}{c^2} \,.\, (-f(x) \,.\, 4\pi^2\nu^2 \cos 2\pi\nu t)$$

i.e.

$$\frac{\mathrm{d}^2 f(x)}{\mathrm{d}x^2} = -\frac{4\pi^2\nu^2}{c^2} f(x) \tag{3.9}$$

if c is constant. We have thus eliminated the variable t from our original wave equation 3.3, and equation 3.9 does not, therefore, contain partial differentials.

Now $c = \lambda\nu$, where λ is the wavelength and ν the frequency, and we can therefore write equation 3.9 in the form

$$\frac{\mathrm{d}^2 f(x)}{\mathrm{d}x^2} = -\frac{4\pi^2}{\lambda^2} f(x) \tag{3.10}$$

It is important to realize that not every possible solution of equation 3.10 is a physically acceptable one. We can perhaps best make this clear by considering the solution of the simplest of all differential equations

$$\mathrm{d}y/\mathrm{d}x = 1$$

The solutions of this are $y = x + a$, where a, the arbitrary integration constant, can have any value we choose to give it. A graphical representation of these solutions would be an infinite number of parallel straight lines; a is given by the intercept on the y axis in each case. In the same way the solution of equation 3.10 involves an arbitrary constant, and we get a large number of solutions $f(x_1)$, $f(x_2)$, $f(x_3)$, etc. The only acceptable ones will be those that satisfy what are known as 'continuity' and 'boundary' conditions which state that $f(x)$ must be continuous, finite and single-values, for every value of x between given boundaries.

Thus the curve of *Figure 3.3*(*a*) represents a function which would not be an acceptable solution of the wave equation, since there is a discontinuity at $x = x'$, and since, in addition, the function has an infinite value at this point. The 'single-valued' limitation is self-explanatory, for the amplitude can have only one value at a given position x'. The boundary conditions are imposed by the physical constraints acting on the vibrating system; thus, in the case of the stretched string, $f(x)$ must be zero at each end. *Figure 3.3*(*b*) represents a solution that is not acceptable since $f(x)$ is not zero at the right-hand fixed end, but some acceptable solutions are shown in *Figure 3.4.* Functions which are acceptable solutions in this sense are called 'Eigenfunctions'; the corresponding values of λ are called 'Eigenvalues'. (These words are German–English hybrids; the German expression 'eigen-funktion', or the English translation 'characteristic function' would be preferable, but the hybrids are now too firmly established.) We shall often, however, refer simply to 'wave-functions', implying that these are, in fact, eigen-functions.

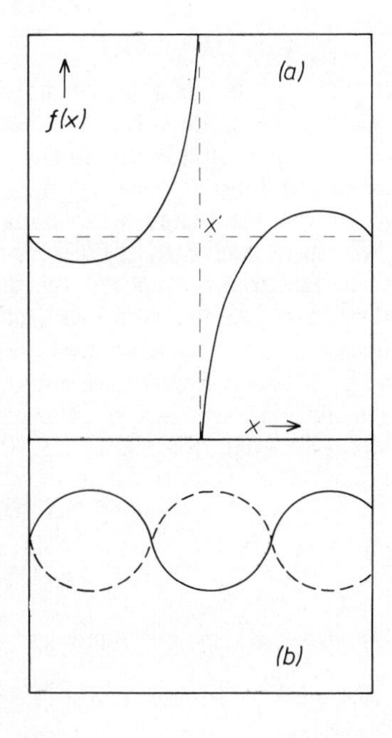

Figure 3.3 Unacceptable solutions of the wave equation for vibrations of a stretched string

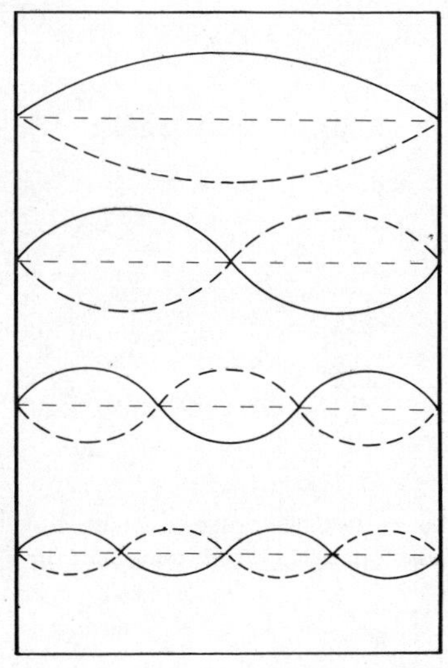

Figure 3.4 Acceptable solutions of the wave equation for vibrations of a stretched string

3.4 The Wave Equation for Electrons

We have now reached the stage where we can formulate a wave equation for electrons. Historically, the wave mechanical theory of the hydrogen atom began in 1927, when Schrödinger wrote down a wave equation to describe electron behaviour, the form of the equation being determined by mathematical intuition. This equation was then solved to give numerical values for quantities that could be determined experimentally, and agreement between calculation and experiment was then taken to confirm the correctness of the postulated wave equation. Experience shows, however, that chemists usually respond without enthusiasm to the bald initial statement '. . . the behaviour of electrons in atoms can be represented by the differential equation. . .'. We shall therefore develop a simple treatment which is certainly not rigorous in the mathematical sense of the word, but which may help to show that the form of the postulated wave equation is not entirely arbitrary. First of all we must realize that the electron in, say, a hydrogen atom is under certain constraints imposed by the attraction of the nucleus. The wave equation that describes the electron's motion must therefore be an equation analogous to that used to describe a standing wave system, since this also represents a constrained system, e.g. the vibrating string fixed at its ends. If the electron moved in one dimension only, the appropriate equation would be that of equation 3.10, which we now write

$$\frac{d^2\psi}{dx^2} = -\frac{4\pi^2}{\lambda^2}\psi \qquad (3.11)$$

where ψ has been written for $f(x)$. This equation can be extended to describe motion in three dimensions, when it becomes

$$\frac{\partial^2\psi}{\partial x^2} + \frac{\partial^2\psi}{\partial y^2} + \frac{\partial^2\psi}{\partial z^2} + \frac{4\pi^2}{\lambda^2}\psi = 0 \qquad (3.12)$$

in which ψ is a function of the Cartesian co-ordinates x, y and z. This may be written more concisely as

$$\nabla^2\psi + \frac{4\pi^2}{\lambda^2}\psi = 0 \qquad (3.13)$$

where $\nabla^2\psi$ is written for

$$\frac{\partial^2\psi}{\partial x^2} + \frac{\partial^2\psi}{\partial y^2} + \frac{\partial^2\psi}{\partial z^2}$$

The inverted delta symbol ∇ is read as 'del'. We now use the de Broglie relationship, equation 3.1; if the electron has an associated wavelength λ, given by

$$\lambda = \frac{h}{p} = \frac{h}{mv}$$

then substitution in equation 3.13 gives

$$\nabla^2\psi + \frac{4\pi^2 m^2 v^2}{h^2}\psi = 0 \qquad (3.14)$$

We wish to use our wave equation to calculate values for the energy states of the hydrogen atom, so we make use of the fact that kinetic energy, $\frac{1}{2}mv^2$, is given by $E - V$, where E is the total energy and V the potential energy (V for the hydrogen atom electron is $-e^2/4\pi\epsilon r$). Substituting

$$v^2 = \frac{2}{m}(E - V)$$

in equation 3.14, we get

$$\nabla^2\psi + \frac{8\pi^2 m}{h^2}(E - V)\psi = 0 \qquad (3.15)$$

which is the celebrated Schrödinger equation describing the behaviour of the electron in the hydrogen atom. We emphasize again that the above treatment is in no way a 'proof' of the Schrödinger equation; it merely shows that if the de Broglie relationship is assumed, and if the motion of the electron is analogous to a system of standing waves, equation 3.15 is the type of wave equation to be expected. It was the genius of Schrödinger to arrive at equation 3.15 by intuition, and to justify this by solving the equation to give values for E in agreement with experiment. We shall discuss the way in which this was done in the next chapter.

3.5 Operator Form of the Schrödinger Equation

The Schrödinger equation can be rearranged to the form

$$\left\{-\frac{h^2}{8\pi^2 m}\nabla^2 + V\right\}\psi = E\psi \qquad (3.16)$$

and shortened to

$$\mathcal{H}\psi = E\psi \qquad (3.17)$$

$\mathcal{H}$ is called a 'Hamiltonian Operator' and it defines the operation or sequence of operations to be performed on the function ψ. The result of carrying out these operations on function ψ is the same as multiplying ψ by the electronic energy E.

The function $\psi(x,y,z)$ is called an *atomic orbital.* It is analogous to the function $f(x)$ which represented the amplitude in the case of the vibration of a stretched string. We can therefore refer to ψ as an amplitude function. Now we saw in the first two sections of this chapter that if the behaviour of electrons is represented by a wave equation, we can equate the square of the function representing the amplitude in the wave equation with either (a) the electron density, or (b) the probability that the electron will be found in a given volume element. We thus get a physical significance for the function $\psi^2(x,y,z)$, in that

$$\psi^2 \,.\, \mathrm{d}x \,.\, \mathrm{d}y \,.\, \mathrm{d}z \quad (= \psi^2 \mathrm{d}v)$$

measures the probability that the electron will be found in the volume element $\mathrm{d}v$ surrounding the point whose co-ordinates are (x,y,z). The other interpretation, that $\psi^2\mathrm{d}v$ represents the electron density in the volume element $\mathrm{d}v$, cannot be justified so rigorously as the probability interpretation, but it has proved to be very useful in practice. We have seen that difficulty arises when the concept of electron density is applied to systems containing only a single electron. What, for example, is the meaning of a statement that the electron density at a particular point is 0.2? We are tempted to recall John Dalton's celebrated remark 'Thou knowest no man can split an atom', and apply it to the electron. The concept may perhaps be made more plausible if we consider a hypothetical experiment. Suppose we could photograph the position of an electron at any particular instant; three-dimensional photography would enable us to assign coordinates x, y, z to the electron and to plot it as a point in a three-dimensional diagram. Now the electron is in rapid motion, and a photograph taken a fraction of a second later would reveal the electron in a new position. Several million photographs would produce an array of dots in our diagram which would resemble a cloud, dense in regions where a large number of points are crowded together, diffuse in regions where there are very few

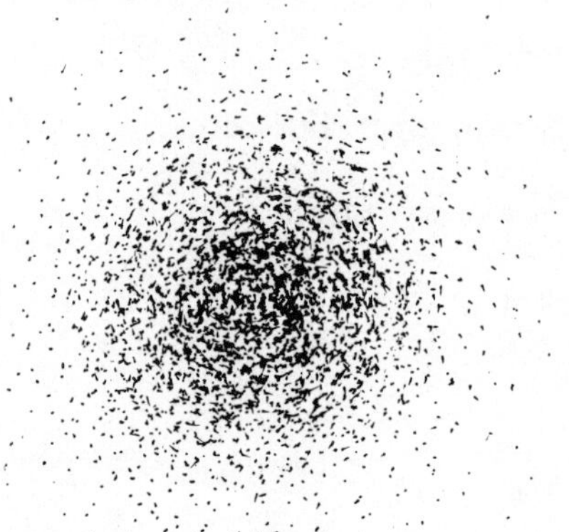

Figure 3.5 Charge cloud for the hydrogen atom electron in its state of lowest energy

points. *Figure 3.5* represents a section through such a charge cloud for the electron in its state of lowest energy in the hydrogen atom. Although a few points will be at a considerable distance from the origin, it is possible to construct a surface which will enclose a large proportion of the points, say 95%. We shall see in the next chapter that analogous electron charge-cloud distributions of different shapes can be drawn for other energy states of the hydrogen atom electron.

Equation 3.17 is of universal application to an electron in any environment – atomic or molecular. The Hamiltonian $\mathcal{H}$ involves a potential energy term V which arises from interactions between all the charged particles in the system according to the Coulomb law. Thus, for the single electron in a hydrogen atom, at a distance r from the nucleus

$$V = -\frac{e^2}{4\pi\epsilon r}$$

4

THE HYDROGEN ATOM

4.1 Introduction

In this chapter we are going to consider acceptable solutions of the Schrödinger equation for the single electron in a hydrogen atom. In this particular case the equation can be solved completely, but because the mathematical treatment is rather complicated and would take up a disproportionate amount of space, we are going to limit our discussion to a qualitative description.

4.2 Polar Co-ordinates

The problem we are concerned with is that of calculating the 'amplitude' of electron waves at various points in a hydrogen atom. These points can be defined by drawing a set of Cartesian (x, y and z) axes through an origin at the nucleus of the atom and locating points by x, y and z co-ordinates. However, it turns out that the mathematics involved are much simpler if we use an alternative way of specifying position, namely

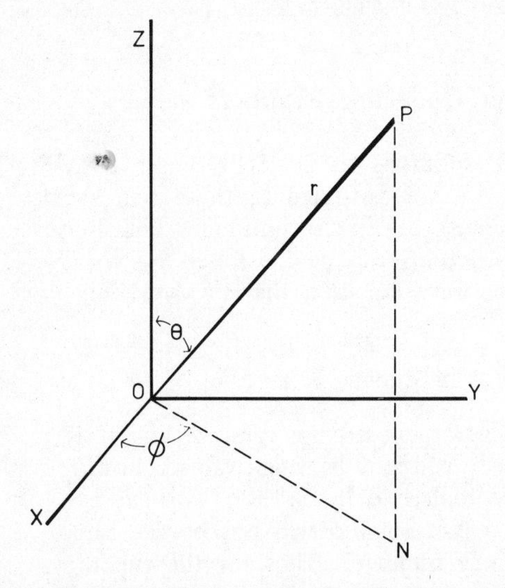

Figure 4.1 The spherical polar co-ordinate system

the polar co-ordinate system. These are shown in *Figure 4.1*. The co-ordinates of a point P are defined by the three polar co-ordinates r, θ and ϕ. Thus for an atom we locate the origin (0) at the nucleus, and r is the distance of the point in question (P) from this origin. The other two polar co-ordinates are angles. θ is the angle between OP and the z axis; ϕ is the angle between the x axis and the line ON, where N is the foot of the perpendicular from P on the xy plane.

Some readers may be helped if we provide an everyday example of the use of polar co-ordinates, namely specifying the location of a particular point on the Earth's surface. The distance from the Earth's centre is the radial co-ordinate, r, and there are two angular co-ordinates, latitude and longitude. Latitude is the angle between the equatorial plane (x,y) and the line drawn to the point from the Earth's centre; it can be seen that latitude is $90 - \theta$. Longitude (ϕ) is the angle between the reference plane passing through Greenwich (i.e. the xz plane) and the meridional plane containing the point.

The illustration of the use of polar co-ordinates can be taken further if we imagine that we have a 'perfect' planet in which the oceans cover the Earth to a uniform depth, to give a 'flooded planet'. Disturbances in the ocean then result in a wave pattern being set up, the allowed patterns following the one-dimensional concepts outlined for a stretched string (p.25); boundary conditions apply so that the waves must 'join up'. The location of a particular disturbance is then described in terms of r, θ and ϕ in the usual way.

Returning now to electrons in atoms, we find that acceptable solutions of the Schrödinger equation are best written as the product of two functions: one of these is the so-called radial function, $f(r)$ or $R(r)$, which involves only the r co-ordinate; the other, the angular function $Y(\theta,\phi)$, only involves the two angles θ and ϕ. Hence we can write the total amplitude function, $\psi(r,\theta,\phi)$, as a product of the radial and angular functions, i.e.

$$\psi(r,\theta,\phi) = R(r) \times Y(\theta,\phi)$$

4.3 Acceptable Solutions – the Quantum Numbers

There will be an infinite number of solutions to the Schrödinger equation, but, as we saw in Chapter 3, we are restricted to those that satisfy the requirement that the functions must be finite, continuous and single-valued. The operation of these restrictions can be illustrated by considering one possible solution to the wave equation that involves only the angle ϕ. Such an example is

$$f(\phi) = \Phi = a \sin m\phi$$

This function is finite since it can never exceed the value a, and it is continuous, being a sine function. It will not be single-valued, however, unless m is an integer. Thus if we replace ϕ by $2\pi + \phi$ we have returned to the same point in space, but $\sin m\phi$ will not be the same as $\sin m(2\pi + \phi)$ if m is not a whole number. Thus for the function Φ to be an acceptable solution to the Schrödinger equation, it is necessary for m to be 0, ±1, ±2, ±3, etc.

m is called the *magnetic quantum number*, and we see that it arises as a direct consequence of the requirement that a wave motion appropriate for the description of the motion of an electron in an atom should be single-valued. The requirements (finite, continuous and single-valued), when applied to solutions involving the r and θ co-ordinates, introduce, respectively, the additional quantum numbers n and ℓ.

These three quantum numbers, n, ℓ and m, are inter-related in respect of the values they can take:

n, the *principal quantum number*, 1, 2, 3, . . .

ℓ, the *azimuthal* or *orbital angular momentum quantum number*, 0, 1, 2, . . . $(n - 1)$

m, the *magnetic quantum number*, 0, ±1, ±2, . . . $\pm\ell$

Hence each wave function, ψ (referred to as an orbital), is defined by the allotted values of n, ℓ and m, i.e. $\psi_{n,\ell,m}$. It is convenient to use a shorthand to describe such orbitals, such that orbitals for ℓ = 0, 1, 2, 3,... etc. are referred to as *s*, *p*, *d* and *f*, respectively. Moreover, the numerical value of n is placed before the letter specifying the ℓ value, so that an orbital for $n = 2$ and $\ell = 0$ is described as a 2*s* orbital. The shorthand is summarized as follows:

$n = 1$	$\ell = 0$	$m = 0$	1*s*
$n = 2$	$\ell = 0$	$m = 0$	2*s*
	$\ell = 1$	$m = 0, \pm 1$	2*p* (three 2*p* a.o.'s)
$n = 3$	$\ell = 0$	$m = 0$	3*s*
	$\ell = 1$	$m = 0, \pm 1$	3*p* (three 3*p* a.o.'s)
	$\ell = 2$	$m = 0, \pm 1, \pm 2$	3*d* (five 3*d* a.o.'s)

4.4 The Radial Functions $R(r)$

These functions are independent of θ and ϕ; they are 'spherically symmetrical', so that we get the same value of $R(r)$ at a given distance r from the nucleus no matter what values are given to θ and ϕ.

Figure 4.2 shows plots of the $R(r)$ function against r drawn to the same r scale for 1*s* (n = 1, ℓ = 0), 2*s* (n = 2, ℓ = 0), 3*s* (n = 3, ℓ = 0), 2*p* (n = 2, ℓ = 1), 3*p* (n = 3, ℓ = 1) and 3*d* (n = 3, ℓ = 2) orbitals of the hydrogen atom electron. It will be seen that:

(a) All the *s* functions have their maximum values at the nucleus ($r = 0$) and that the value of $R(r)$ initially drops very steeply as r increases. This is because the mathematical expression for each solution includes a negative exponential expression of the type e^{-r}; functions of this type decrease rapidly from a maximum value at $r = 0$ to zero at $r = \infty$.

(b) $R(r)$ for the 2*s* orbital becomes zero at a particular value of r between $r = 0$ and $r = \infty$; at this point, the so-called nodal point, $R(r)$ changes sign from positive to negative. The 3*s* orbital has two nodal points between 0 and ∞ and, in general, the number of such nodes in a *ns* orbital is given by $n - 1$.

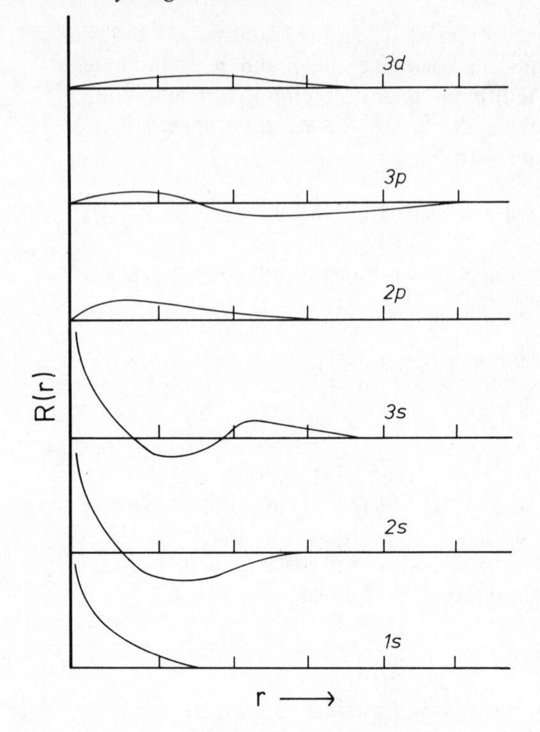

Figure 4.2 Plot of radial function R(r) against r

(c) The p and d radial functions are all zero at $r = 0$ and $r = \infty$ and the number of nodes between these limits is given by $n - \ell - 1$. Thus there are no nodes in this region in the $2p$ and $3d$ radial functions, but there is one node in the $3p$ function.

(d) At distances close to the nucleus the $R(r)$ function for s orbitals is greater than that for p and d orbitals of the same principal quantum number.

4.5 Radial Distribution Functions

The radial functions $R(r)$ have no physical significance in themselves, but the square of the function, multiplied by a volume element $\mathrm{d}v$, $R^2(r)\mathrm{d}v$, measures the probability that the electron will be in this volume element $\mathrm{d}v$ at a point that is a distance r from the nucleus. This value will, however, be the probability for particular angles θ and ϕ that are needed to specify the position of the electron. A more useful value is the probability of finding the electron at a distance r from the nucleus irrespective of the values of θ and ϕ. Instead of the volume element $\mathrm{d}v$ we now consider the volume of a spherical shell of thickness $\mathrm{d}r$ and radius r: this will be $4\pi r^2\,\mathrm{d}r$ since the surface area of a sphere of radius r is $4\pi r^2$. The *radial distribution function*, $4\pi r^2 R^2(r)\mathrm{d}r$, thus measures the probability of finding the electron in a spherical shell of thickness $\mathrm{d}r$ at various distances r from the nucleus.

Figure 4.3 shows plots of this function for the hydrogen $1s$, $2s$ and $3s$ orbitals. The functions differ from the simple radial functions $R(r)$ in that

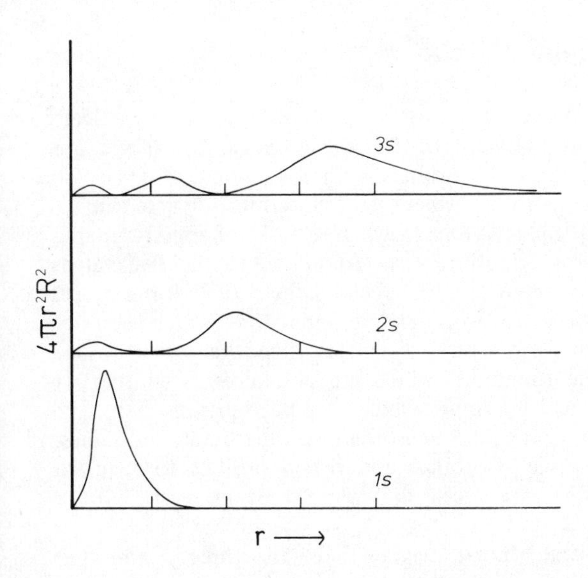

Figure 4.3 Plot of radial distribution function $4\pi r^2 R^2 (r)\mathrm{d}r$ against r for s electrons

they are always zero at $r = 0$. The number of 'peaks' in the distribution function for the s orbitals is equal to the n value, i.e. 1 for 1s, 2 for 2s, etc.

Figure 4.4 shows radial distribution functions for 2p, 3p and 3d orbitals. Here the number of peaks is $(n - 1)$ for p orbitals and $(n - 2)$ for d orbitals.

These functions are particularly useful when discussing the screening effect of electrons in many-electron atoms (see Chapter 5, p.39) and the peaks of maximum probability correspond to concentric 'shells' strongly resembling the 'orbits' of the Bohr theory.

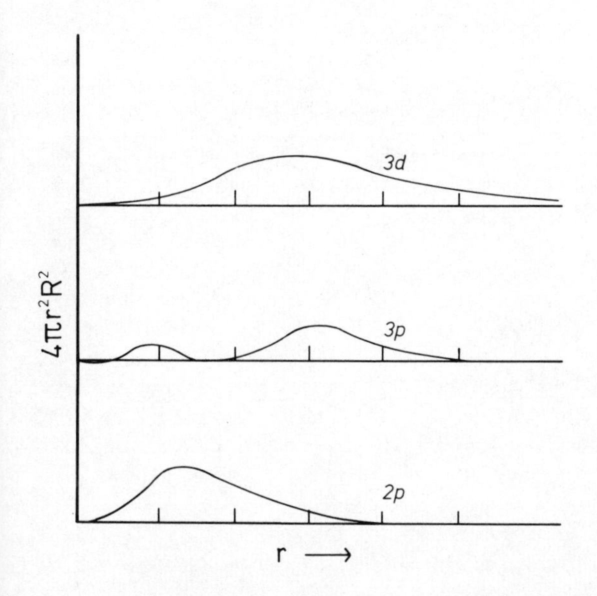

Figure 4.4 Plot of radial distribution function against r for p electrons

4.6 The Angular Functions $Y(\theta,\phi)$

We have already seen that wave functions for all *s* orbitals are spherically symmetrical, i.e. they are independent of the angles θ and ϕ. There are, however, *three* angular functions for orbitals with $n = 2$ and $\ell = 1$. These are the $2p_0$, $2p_{+1}$ and $2p_{-1}$ orbitals where the subscripts refer to the possible values of the *m* quantum number when $\ell = 1$. Similarly there are five angular functions for 3*d* orbitals corresponding to the five values for the *m* quantum number when $\ell = 2$. The mathematical form of these solutions of the wave equation involves complex functions which cannot readily be represented in diagram form. Chemists therefore prefer to take linear combinations of these functions, which are also allowed solutions of the wave equation, to get real solutions which can be represented by 'polar diagrams'. These functions cannot be identified with particular *m* values, but there must always be three *p* orbitals and five *d* orbitals for a given principal quantum number *n*. We recall that *n* must be at least 2 for *p* orbitals and at least 3 for *d* orbitals.

Figures 4.5 and *4.6* illustrate 'polar diagrams' for the three *p* and the five *d* orbitals. In these diagrams the length of the line OP gives the value

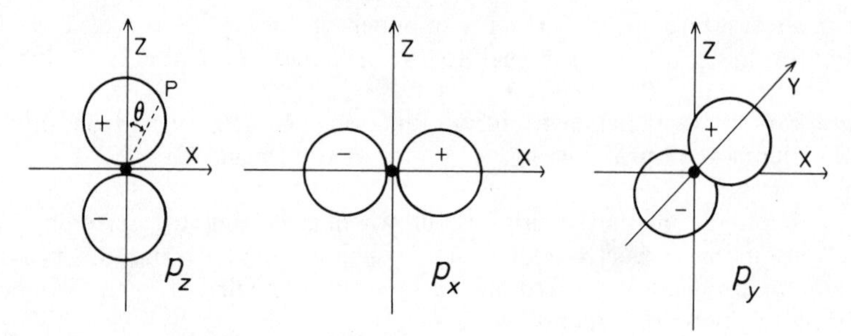

Figure 4.5 Polar diagrams for p electrons

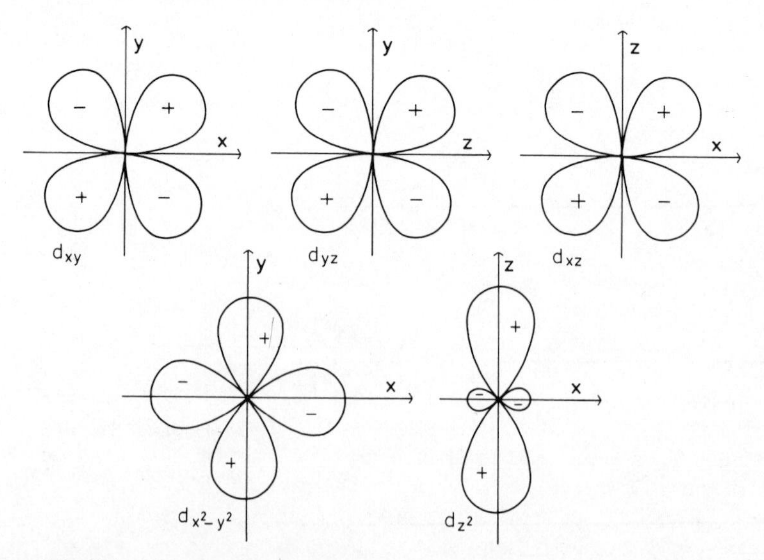

Figure 4.6 Polar diagrams for d electrons

of the $Y(\theta,\phi)$ function for particular values of θ and ϕ. The $+z$ axis corresponds to $\theta = 0$, $\phi = 0$, the x axis to $\theta = \pi/2$, $\phi = 0$, and the y axis to $\theta = \pi/2$, $\phi = \pi/2$. Each diagram shows how $Y(\theta,\phi)$ = OP varies with one of the two angles, the other angle remaining fixed. Thus the diagram labelled p_z shows how OP varies with θ, ϕ being fixed at 0°. The p_z function has its maximum value along the z axis, and is zero when $\theta = \pi/2$; this is because the mathematical expression for this particular function includes a cosine θ term which is 1 for $\theta = 0$ and zero for $\theta = \pi/2$. If this two-dimensional diagram is rotated about the z axis we get a solid surface, and the distance from O to a point P on this surface will now give the value of $Y(\theta,\phi)$ for all values of θ and ϕ. In this case the plane xy will be a nodal plane. In these diagrams + and – denote the algebraic signs of the angular function. Since cos θ is +1 for $\theta = 0$ and –1 for $\theta = \pi$ the part above the xy plane is positive and the part below negative.

The p_x orbital is aligned along the x axis with yz as the nodal plane, while p_y is aligned along the y axis with xz as the nodal plane.

Three of the five d orbitals have similar polar diagrams, differing only in the plane which contains the axes along which the function has its maximum values. The subscripts xy, yz, xz indicate these axes, which are oriented to make angles of 45° with the Cartesian axes x, y and z. The fourth d orbital, labelled with the subscript $x^2 - y^2$, has a similar shape to that of the first three, but it is aligned to have maximum values *along* the x and the y axes. In each case there are *two* nodal planes (e.g. xz and yz for d_{xy}) and the algebraic signs of the function alternate + and – as shown in the diagram.

The d_{z^2} orbital has a different shape with lobes aligned along the z axis together with a 'tyre' or 'annulus' surrounding the z axis with maximum extension in the xy plane.

4.7 Orbitals, Probability Distributions and Charge Clouds

We have now reached the stage where we can combine the radial and the angular functions to get the complete wave function or atomic orbital. We saw in Chapter 3 that

$$\psi^2 dv \quad [= R^2(r) \times Y^2(\theta,\phi)]$$

gives the probability that the electron will be found in the volume dv surrounding a point of co-ordinates r, θ, ϕ or, though less precisely, the 'electron density' of this volume element. It might at first sight seem impossible to draw diagrams to represent these probability functions since the radial function only becomes zero when r is infinite. However, we have already seen that the radial function also always contains a negative exponential term which decreases in value very steeply as r increases. This makes it possible to represent the functions by finite diagrams. We can construct boundary surfaces such that they contain a volume within which there is a 99% probability of finding the electron. Alternatively we could construct a surface which would enclose 99% of the total electronic charge.

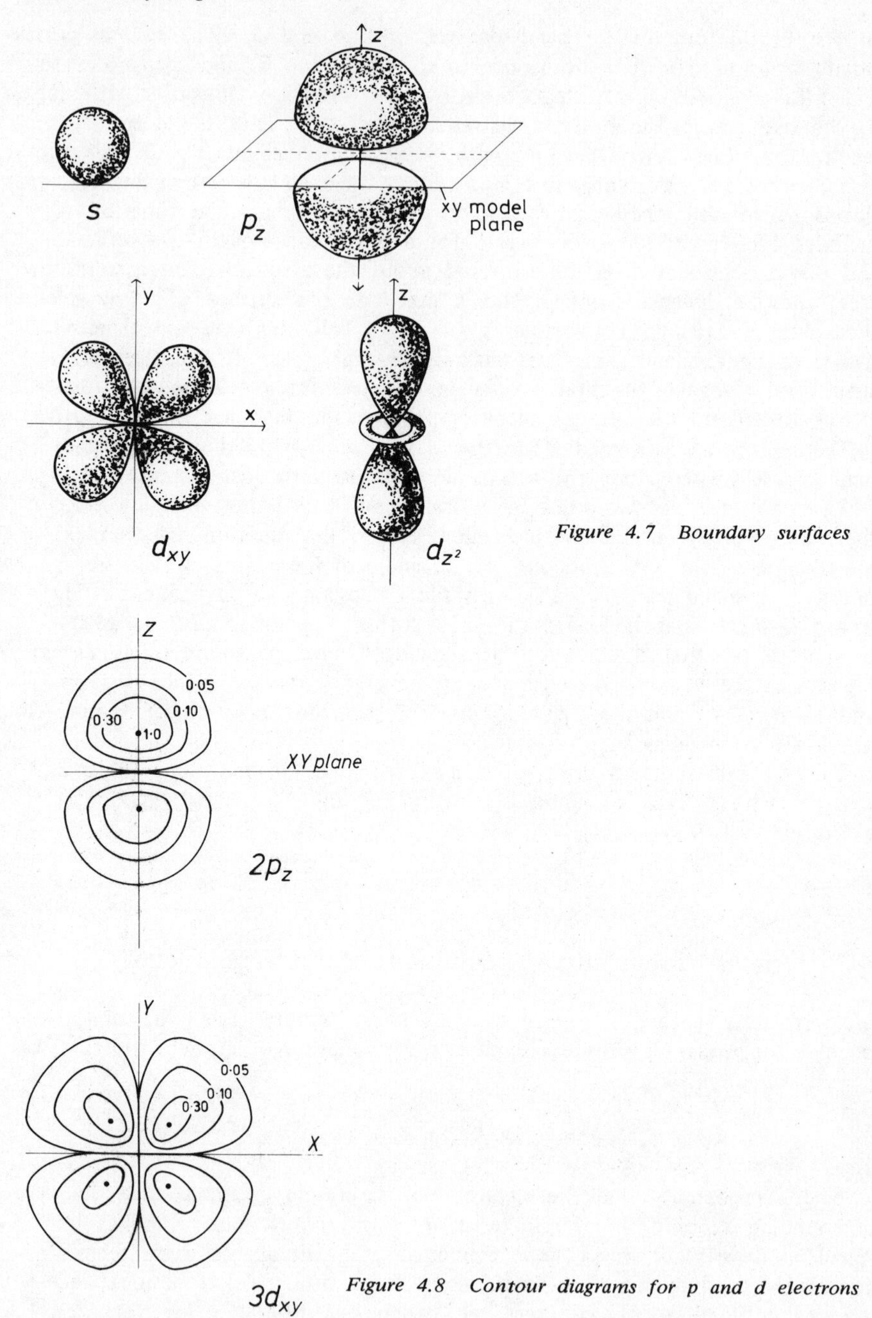

Figure 4.7 Boundary surfaces

Figure 4.8 Contour diagrams for p and d electrons

The geometry of these diagrams is controlled essentially by the angular function, which establishes the nodal planes, but the variation in probability or in charge density as r changes will very much be controlled by the radial function. *Figure 4.7* represents some of these boundary surfaces.

The most informative way of representing the ψ^2 function is to make use of electron density contour diagrams. These are obtained by computing values of 'electron density' at a large number of points of various co-ordinates x, y and z. Points with the same value of electron density are then joined by 'contour' lines which give a visual indication of regions of high density and therefore of high probability. *Figure 4.8* shows contour diagrams for p and d orbitals.

There is often confusion between diagrams representing atomic orbitals and diagrams representing electron (charge-cloud) densities. Thus the correct *orbital* diagram for a p orbital is that shown in *Figure 4.5*: two spheres in contact with opposite algebraic signs above and below the nodal plane. The diagrams for the charge densities, however, are everywhere positive since they represent the square of the orbital function. The geometry of the p orbitals also is no longer that of spheres in contact. However, when we come to consider, in later chapters, the way in which atomic orbitals may be combined to form molecular orbitals, the diagrams we shall use will be those for the orbitals illustrated in *Figures 4.5* and *4.6* and the algebraic signs of the functions in the various lobes will be important.

4.8 Energy Levels

We must now discuss the energy of the hydrogen atom electron in the various possible orbitals. An expression for the energy E is obtained in the course of solving the Schrödinger equation. This solution is

$$E_n = -me^4/8\epsilon^2 n^2 h^2$$

which is exactly the same as the expression obtained on the old Bohr theory.* The normal or ground state for the electron in the hydrogen atom is that for which $n = 1$, the $1s$ orbital. It will be noticed that the expression for the energy does not involve the ℓ or the m quantum numbers. Thus all orbitals of the hydrogen atom electron having the same principal quantum number (e.g. $3s$, $3p$ and $3d$) have the same energy. They are said to be 'degenerate'. We shall see in the next chapter that this degeneracy is removed as soon as we go from hydrogen to atoms with more than one electron.

5

QUANTUM THEORY AND THE PERIODIC CLASSIFICATION

5.1 The Wave Equation for Many-electron Atoms

Difficulties at once arise when we try to apply the methods of the preceding chapter to atoms containing more than one electron. In these atoms, any particular electron is attracted to the oppositely charged nucleus, but it is also repelled by the other electrons present, and the expression for the potential energy of such an electron involves terms containing interelectronic distances. It is the presence of these terms in the wave equations that makes an exact solution impossible, and approximation methods have to be used. One widely used method is that developed by D.R. and W.H. Hartree. In effect, they reduce the problem to that of a single electron moving in a spherically symmetrical field of force provided by the nucleus and the averaged effect of all the other electrons. The complete wave function for the atom is then written as the product of a number of one-electron wave functions. Some typical results of these lengthy calculations are shown in *Figure 5.1*, where the radial distribution function $4\pi r^2 \psi^2$ is plotted against r for the sodium ion, Na^+. The shaded portion shows the electron distribution in Na^+, and the broken line marked $3s$ shows the distribution curve for the outermost electron in the neutral sodium atom. The diagram shows that in the Na^+ ion the electron charge cloud effectively forms two concentric shells quite close to the nucleus, and there is evidently a strong resemblance to the orbits of the Bohr theory. Solution of the wave equation again introduces the quantum numbers n and ℓ which retain the significance they had in the case of hydrogen; thus the $3s$ distribution curve is that of an electron with principal quantum number $n = 3$ and aximuthal quantum number $\ell = 0$. Similar distribution curves have been worked out by Hartree and his pupils for a large number of ions and neutral atoms. They all reveal a characteristic arrangement of electrons in concentric shells, the inner electrons in general being close to the nucleus.

5.2 Energy Levels

The energies associated with different orbitals in many-electron atoms differ from the corresponding energies in the hydrogen atom. Orbitals with the same principal quantum number n but with different ℓ values (e.g. $2s$, $2p$) no longer have the same energy, since the energy needed to remove an electron from the atom now depends on the nuclear charge, the orbital from which the electron is removed and on the screening effect of the other electrons. The screening effect can best be discussed in terms of the

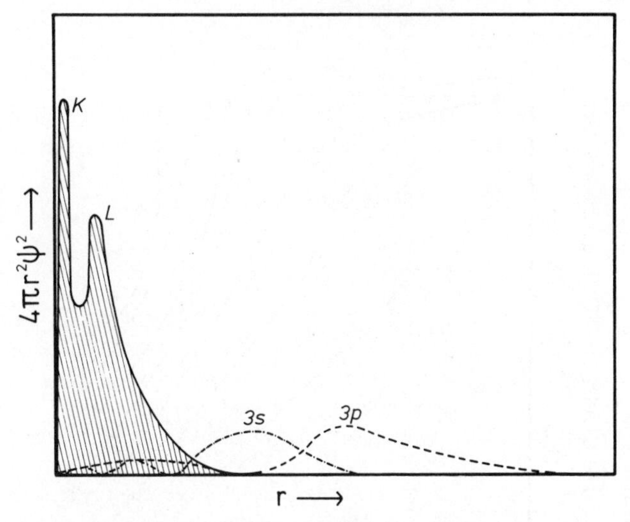

Figure 5.1 Distribution function for the sodium ion, Na^+

radial distribution functions. *Figure 5.1* shows that the 3*s* distribution function for sodium overlaps the inner electron distribution shown by the shaded portion of the diagram to a considerable extent. This means, in effect, that the 3*s* electron moves under the influence of the full nuclear charge. The diagram also shows that the 3*p* distribution has little overlap with the inner electrons, which therefore screen the 3*p* electron from the full nuclear charge. The 3*d* orbital is almost completely shielded. The effectiveness of the screening by inner electrons will depend upon the number of electrons present, and the way in which they are distributed among the various available orbitals. General trends in changes of orbital energy with increasing atomic number can be predicted, but the actual sequence of energy levels has to be determined for each atom in turn. *Figure 5.2* shows how the energies change for a sequence of atomic numbers. The *s* orbital energies are most strongly affected by the increase in atomic number, whereas the 3*d* orbital energies are almost the same as in hydrogen ($13.6/3^2$ eV) since the 11 protons in the nucleus are completely screened by the 10 electrons in the Na^+ ion core.

As more and more orbitals are occupied, so the effective screening changes. *Figure 5.2* shows that at about $Z = 20$ the 3*d* orbital begins to overlap inner electron orbitals. Screening is then less effective and changes in nuclear charge have a big effect on orbital energy. For $Z = 21$ the 3*d* level lies below the 4*s* level in the energy level diagram.

5.3 Electron Spin

So far we have defined atomic orbitals using three quantum numbers *n*, *ℓ* and *m*. In 1928 P.A.M. Dirac produced a wave equation which linked quantum theory with relativity theory, and this treatment introduces a fourth quantum number: the spin quantum number m_s which is allowed

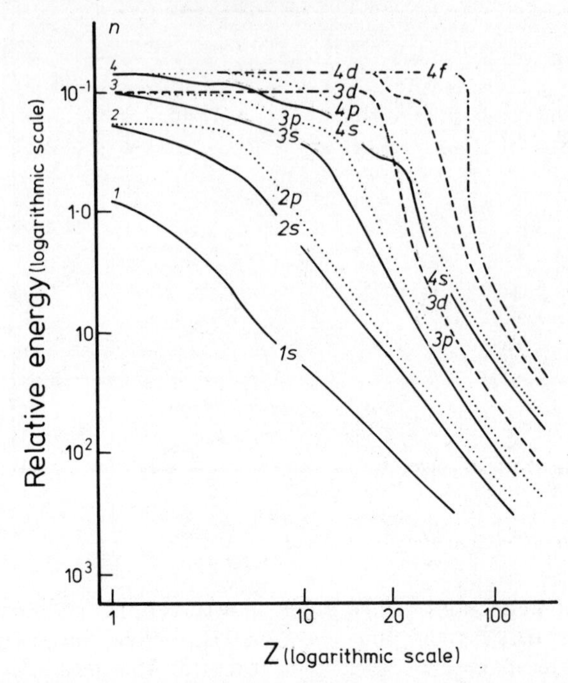

Figure 5.2 Relative energies of atomic orbitals as a function of nuclear charge z

only two values, +½ and −½. The quantum numbers n, ℓ and m retain the same significance that they have in the Schrödinger theory: n largely determines the energy associated with the orbital; ℓ is linked with the geometrical shape of the orbital and with 'orbital' angular momentum; m is associated with different components of the orbital angular momentum along a defined axis. It is impossible to give a simple physical picture of electron spin – it is linked with a complicated oscillatory spiral motion of the electron – but m_s defines the two allowed values of spin angular momentum. An electron possessing angular momentum has an associated magnetic moment, so that the moment arising from 'orbital' angular momentum can be reinforced or diminished by interaction with the moment arising from one or othpr of the two values of the 'spin' angular momentum' Electrons with the same m_s value have 'parallel' spins.

5.4 The Exclusion Principle

We can now consider the way in which the electrons in an atom distribute themselves among the various orbitals. The first guiding principle is that electrons will tend to go into the orbitals of lowest energy. Thus the single electron in a hydrogen atom will go into a $1s$ orbital; the electronic configuration in hydrogen is therefore described briefly as $1s^1$, where the superscript shows that there is a single electron in an orbital of principal quantum number $n = 1$ and azimuthal quantum number $\ell = 0$. It might at first sight be thought that in other atoms all the electrons will also crowd into the $1s$ orbital, since this has the lowest energy. This is indeed the case in helium, where the configuration is summarized by $1s^2$. However, the interpretation of the atomic spectrum of helium requires these two electrons to have opposite spins; if they had parallel spins the appearance of the spectrum would be quite different. The lithium spectrum can be interpreted if we treat it as a one-electron system. The observed frequencies can be related to a number of transitions of the outer electron between orbitals of different energy, and principal quantum number $n = 2$ and aximuthal quantum number $\ell = 0$ can be assigned without difficulty to this outer electron. We therefore assign the configuration $1s^2 2s^1$ to the lithium atom in its stable state, noting that apparently the $1s$ orbital can accommodate only two electrons. Here we have an example of the working of what is probably the most important principle in theoretical chemistry. It was first explicitly stated by the Swiss physicist Pauli in 1925, and it is now referred to as the Pauli Exclusion Principle. He pointed out that the correlation of experimental observations on line spectra with the quantum theory required the introduction of a postulate, viz.

> that in any system, whether a single atom or a molecule, no two electrons could be assigned the same set of four quantum numbers.

Thus, the two electrons in helium, possessing the same three quantum numbers $n = 1$, $\ell = 0$, $m = 0$ must have different values of m_s, and no other electrons can enter the $1s$ orbital. No proof of the exclusion principle exists, but so far, nothing in nature has appeared to contradict it, and its acceptance brings a new order into the whole of science.

It follows from the exclusion principle that, if there are a number of electrons with parallel spins in an atom, these electrons must be in different orbitals. They tend to be as far away from each other as possible, and the direction of the bonds formed by such electrons is thus largely determined by the operation of the exclusion principle. *Table 5.1* shows the maximum number of electrons that can be accommodated in the different orbitals when the exclusion principle is taken into account.

When $n = 1$ there is only a single s orbital into which two electrons of opposite spin can go, but when $n = 2$ we can have both s and p type orbitals. The three p orbitals, $2p_x$, $2p_y$ and $2p_z$ (related to the three possible m values, ±1, 0), can each hold two electrons with opposite spins,

Table 5.1 The distribution of electrons in orbitals

n	ℓ	m	m_s	*Total number of electrons*	
1	0	0	±½	2	
2	0	0	±½	2	8
	1	±1	±½, ±½		
		0	±½	6	
3	0	0	±½	2	18
	1	±1	±½, ±½		
		0	±½	6	
	2	±2	±½, ±½		
		±1	±½, ±½	10	
		0	±½		

making a total of eight electrons for the full set of orbitals with this principal quantum number. It can be shown, in the same way, that for $n = 3$ and $n = 4$ the corresponding maximum numbers of electrons are 18 and 32, respectively. The orbitals associated with the various energy levels are shown

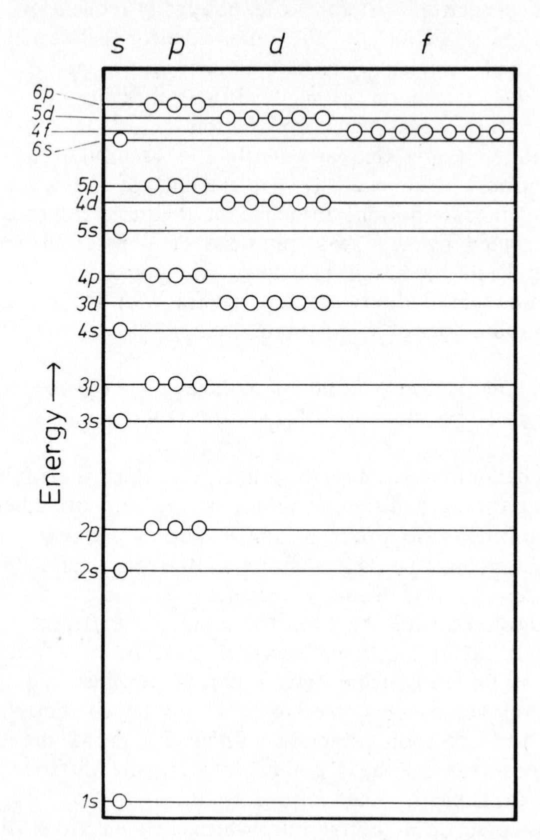

Figure 5.3 Energy levels of atomic orbitals

in diagrammatic form in *Figure 5.3* (no geometrical significance is to be attached to the use of circles; they simply represent orbitals into which a maximum of two electrons can go).

5.5 Ionization Energies

The distribution of electrons in the various orbitals, determined in an empirical way by the operation of the exclusion principle, is supported experimentally by the analysis of atomic spectra and also by the measurement of the so-called ionization potentials or ionization energies. The ionization potential is defined as the energy needed to displace the most easily removable electron from its equilibrium position in the atom to infinity. This energy can be determined by electrical methods, which usually give a value quoted in electron volts – hence the term ionization potential. Chemists, however, preferring to quote energies in kilojoules per mole, tend nowadays to speak of ionization energies. Many-electron atoms have a series of ionization energies. The first ionization energy is the energy needed to move

Table 5.2 Ionization energies

Atom	*First ionization energy*/ kJ mol^{-1}
H	1312
He	2372
Li	520
Be	900
B	800
C	1086
N	1403
O	1314
F	1681
Ne	2081
Na	495
Mg	738
Al	577
Si	787
P	1060
S	1000
Cl	1255
Ar	1520

one electron from a neutral atom to infinity, e.g. for He → He^+, I_1 = 2372 kJ mol^{-1}. The second ionization energy is defined as the energy needed to remove an electron from a singly charged ion to infinity, e.g. $He^+ \rightarrow He^{2+}$, I_2 = 5250 kJ mol^{-1}. I_2 is greater than I_1 since the negatively charged electron has to be dragged away from an ion with a net positive charge.

Table 5.2 gives numerical values for the first ionization energies of elements in the first two rows of the Periodic Table, and *Figure 5.4* plots these values against atomic number.

We see that the energy needed to remove an electron from a lithium atom is very much less than that needed to remove an electron from helium. The configuration Li, $1s^2 2s^1$, explains this. Ionization of a lithium atom requires the removal of the electron in a 2*s* orbital, which is effectively screened from the nuclear charge by the 'inner' electrons in the 1*s* orbital. The electrons in a helium atom, configuration $1s^2$, are not effectively screened, and so very much more energy is needed to remove one of them.

The next break in the first ionization energy plot comes at beryllium. We see that it takes more energy to remove an electron from this atom than it does to remove one from its neighbours lithium and boron. The configuration $1s^2 2s^2$ is evidently more stable than that for lithium. The exclusion principle would require the electron configuration for boron to be $1s^2 2s^2 2p^1$ (see *Figure 5.3*), and the experimental value for the first ionization energy shows that an electron in a 2*p* orbital is more effectively screened from the nuclear charge than one in the 2*s* orbital, which is paired with another 2*s* electron of opposite spin.

5.6 The 'Building-up' ('Aufbau') Principle and the Periodic Classification

We are now in a position to make use of the Exclusion Principle, and the experimental information available from atomic spectra and ionization energy

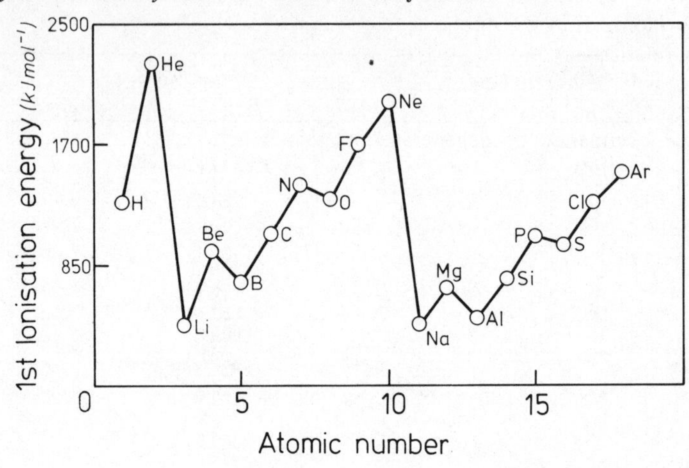

Figure 5.4 First ionization energies of the elements in the first two short periods

measurements, to assign electron configurations to atoms, taking them in order of increasing atomic number. In doing so, we shall also use another important generalization, called the 'Building-up' or 'Aufbau' principle. This states that in determining the configuration of an atom of atomic number Z, we first of all write down the configuration of the atom of atomic number $(Z - 1)$; we then have to consider only the allocation of quantum numbers to the one additional electron, assuming this electron to go into the available orbital of lowest energy. So far we have assigned H, $1s^1$; He, $1s^2$; Li, $1s^2 2s^1$; Be, $1s^2 2s^2$; and B, $1s^2 2s^2 2p^1$. The filling up of the three $2p$ orbitals continues in this way from carbon to neon. In each case the inner electron distribution remains unchanged as $1s^2$ – the helium configuration – and it will be convenient in future to write (He) to represent this arrangement. We therefore arrive at the following electron configurations: carbon, (He)$2s^2 2p^2$; nitrogen, (He)$2s^2 2p^3$; oxygen, (He)$2s^2 2p^4$; fluorine, (He)$2s^2 2p^5$; and neon (He)$2s^2 2p^6$. These configurations are consistent with the observed atomic spectra and ionization energies of the elements. *Figure 5.4* clearly shows that the neon configuration represents a particularly stable arrangement from which it is difficult to remove an electron, and the very much smaller value for the ionization energy of sodium indicates that the electron removed is in an orbital well screened from the nuclear charge, i.e. $3s$.

Low ionization energy values such as those shown for the alkali metals can be correlated with particularly stable electron configurations for the ion which remains when the electron has been removed, e.g. Li^+ ($1s^2$), Na^+ ($1s^2 2s^2 2p^6$), etc. The low value for boron reflects the stability of the configuration B^+, $1s^2 2s^2$, while the low value for oxygen compared with its adjacent elements is perhaps related to the configuration of the O^+ ion, $1s^2 2s^2 2p_x{}^1 2p_y{}^1 2p_z{}^1$, in which there is one electron in each of the three p orbitals. We shall see later that analogous configurations, where the available d or f orbitals are just half-filled, are more stable than arrangements containing one electron more or less than this number.

Study of atomic spectra also leads to another very useful empirical principle, usually known as Hund's rule: that when a set of p (or d or f)

orbitals is being filled with electrons, the distribution is such that the electrons retain parallel spins as far as possible. Thus the two $2p$ electrons in carbon will be arranged $2p_x{}^1 2p_y{}^1$, since the exclusion principle will prevent them from retaining parallel spins if both are in a single p orbital. This single occupancy of degenerate orbitals which have different spatial orientations reduces the electrostatic repulsion between the electrons. The electronic configuration $(He)2s^2 2p_z{}^2$ is known, but it corresponds to an 'excited' state of the carbon atom, with an energy 121 kJ mol^{-1} greater than the stable state $(He)2s^2 2p_x{}^1 2p_y{}^1$. The stable nitrogen configuration, when written out in full, will therefore be $(He)2s^2 2p_x{}^1 2p_y{}^1 2p_z{}^1$, the configuration $(He)2s^2 2p_x{}^2 2p_y{}^1$ referring to an excited state 230 kJ mol^{-1} greater in energy. Pairing of electrons with opposed spins will have to take place in oxygen, and the structure is written $(He)2s^2 2p_x{}^2 2p_y{}^1 2p_z{}^1$. For neon we get a completely filled set of p orbitals in which there are no 'unpaired' electrons.

Reference to *Figure 5.3* shows that the next orbitals to be occupied will be the $3s$, $3p$ and $4s$ ones, although it must be noted that for elements of atomic number greater than 20 the relative positions of the $4s$ and $3d$ energy levels is changed. This change will be discussed in the next section. We can now write down the configurations of the atoms from sodium to calcium inclusive, noting that the inner electron arrangement in each case is that of neon which will be written for conciseness as (Ne). Thus we get sodium, $(Ne)3s^1$, magnesium, $(Ne)3s^2$, aluminium, $(Ne)3s^2 3p^1, \ldots,$ and so on up to argon; $(Ne)3s^2 3p^6$. Potassium and calcium have the inner electron arrangement of argon, and their configurations therefore become $(Ar)4s^1$ and $(Ar)4s^2$, respectively. We can now see how the chemical resemblances between different elements, so brilliantly expressed by Mendeléeff in the periodic system, are paralleled by resemblances in electronic structure. In particular, we observe that the alkali metals have one unpaired electron in the outermost s orbital, that the alkaline earth metals have two electrons in the outer s orbital and, especially, that the noble gases are characterized by completely filled s and p type orbitals.

It has sometimes been stated that the periodic system could have been derived from observations on atomic spectra and application of the quantum theory. It seems at least doubtful that this could have been achieved. The men who developed the quantum theory and the methods of interpreting atomic spectra were brought up in a tradition in which the Mendeléeff classification was part of the scientific climate of the age; the sorting out of the complexities of atomic spectra would scarcely have been possible without the conscious, or unconscious, utilization of the periodic system. The quantum theory, with many triumphs to its credit, has not yet surpassed the achievement of Mendeléeff in predicting accurately the properties of the then undiscovered elements scandium, gallium and germanium. However, the quantum theory does resolve some of the defects of the old periodic system. (The anomalous position of argon and potassium, tellurium and iodine, and nickel and cobalt had already been rectified by the use of an arrangement based on atomic number rather than atomic weight, before the development of the new quantum theory.) The outstanding defects were the apparently arbitrary position of the noble gas elements, the difficulty of accommodating the rare earth (lanthanide) elements, and the existence in the same group of not very similar elements.

The last defect was recognized by separation into A and B sub-group elements, but this hardly provided an explanation of the difference between them. We have already seen that the quantum theory provides a logical position for the noble gases in the periodic system – they are atoms in which the *s* and *p* orbitals are completely filled. The resolution of the remaining defects will be discussed in the next section.

5.7 Transition Series: the '*d*-block' Elements

The next element to consider is scandium, atomic number 21. Analysis of its atomic spectrum shows quite clearly that the electronic structure cannot be similar to that of boron or aluminium, i.e. it cannot be, what might possibly have been expected, (Ar)$4s^2 4p^1$. Further study of the spectrum reveals that for scandium and for atoms of greater atomic number, the 3*d* orbital, so far unoccupied, is now more stable than the 4*p* one, and thus the electronic configuration of scandium is (Ar)$3d^1 4s^2$. *Figure 5.3* shows the position of the five 3*d* orbitals with respect to the neighbouring 4*s* and 4*p* orbitals.

Beginning with scandium, we get a series of ten elements (ending at zinc) which is characterized, in general, by two outer electrons in a 4*s* orbital, and from one to ten electrons in an inner 3*d* orbital. The filling up of the *d* orbitals is not quite regular; there is not much difference in energy between the 3*d* and 4*s* orbitals, and chromium and copper find their respective configurations of $3d^5 4s^1$ and $3d^{10} 4s^1$ to be more stable than the expected $3d^4 4s^2$ and $3d^9 4s^2$; a particularly stable state seems, in fact, to be formed either when the five *d* orbitals are completely filled with ten electrons, or when there is one electron in each of the five. The same general principles which operated in the first two periods still apply here: in a series of equivalent orbitals (e.g. the five 3*d*), electrons arrange themselves as far as possible with parallel spins. Thus, the stable (or 'ground') state of manganese ($3d^5 4s^2$) contains five unpaired electrons in the 3*d* orbitals. Manganese readily forms a stable divalent Mn^{2+} ion ($3d^5$), and iron ($3d^6 4s^2$) a stable tervalent Fe^{3+} ion ($3d^5$); the manganic ion Mn^{3+} ($3d^4$) and the ferrous ion Fe^{2+} ($3d^6$) are less stable. It is nowadays the custom to refer to this group of ten elements (Sc → Zn) as the 'First Transition Series', a term formerly restricted to the group VIII triad (iron, cobalt and nickel). There are considerable chemical similarities between neighbouring members of the series, a fact related to the similar configurations in the outermost orbital. The presence of an incompletely filled set of *d* orbitals also confers characteristic properties, such as variable valency and the existence of coloured paramagnetic ions; these will be discussed elsewhere in this book.

Zinc has an electronic structure with completely filled 3*d* and 4*s* orbitals, but there are still the three unoccupied 4*p* orbitals to consider. These become filled in the elements from gallium to krypton; gallium has the configuration (Ar)$3d^{10} 4s^2 4p^1$, and krypton (Ar)$3d^{10} 4s^2 4p^6$. Elements in the second long period from rubidium to xenon have configurations analogous to those just discussed. A second series of transition elements starts with yttrium which has the arrangement (Kr)$4d^1 5s^2$, but the filling up of the

$4d$ orbitals does not occur in an entirely regular manner. Details of the individual configurations are given in *Table 5.3.* The $4d$ orbitals are completely occupied in cadmium, and the $5p$ orbitals then fill up, giving, when fully occupied, the noble gas structure of xenon.

In the next long period, caesium, barium and lanthanum have the expected structures $(Xe)6s^1$, $(Xe)6s^2$ and $(Xe)5d^1 6s^2$. We should then expect to find that lanthanum was the first member of a third transition series of ten elements; in fact, this is not the case. The atomic spectrum of cerium does not closely resemble that of titanium or zirconium, elements in corresponding positions in the first two transition series. For an atom of high atomic number (Ce = 58) the energy of the $4f$ orbitals is lower than that of the $5d$ ones; in fact, cerium takes the configuration* $4f^2 5s^2 5p^6 5d^0 6s^2$, and we get a series of 15 elements from lanthanum to lutecium where the main electronic differences are in the number of electrons occupying an inner set of seven $4f$ orbitals. Chemical resemblances between these elements are very marked, since the two outer orbitals remain almost unchanged. These elements (formerly called rare earths, but now generally referred to as the lanthanides), thus fall naturally into place in an arrangement of the elements based on electronic configuration.

The filling up of the $4f$ orbitals as we go from lanthanum to lutecium is not entirely regular. Configurations in which the seven f orbitals are completely occupied ($4f^{14} 5s^2 5p^6$), or completely empty ($4f^0 5s^2 5p^6$), or with a single electron in each of the seven orbitals ($4f^7 5s^2 5p^6$), seem particularly stable, and the variable valency of some of the lanthanide elements can be explained in terms of the tendency to form these stable structures. After lutecium we reach hafnium which, with the configuration $4f^{14} 5s^2 5p^6 5d^2 6s^2$, has a normal transition series structure, resembling titanium and zirconium. This transition series is complete at mercury and in the later elements the $6p$ orbitals are occupied until we get the inert gas radon, with the configuration $4f^{14} 5s^2 5p^6 5d^{10} 6s^2 6p^6$. We come finally to the incomplete period, the first two members of which are the radioactive alkali metal francium, with configuration $(Rn)7s^1$, and radium, $(Rn)7s^2$. The configuration of the remaining elements cannot as yet be given with complete certainty. Some of these elements have been obtained in only microscopic amounts, and in any case the interpretation of their very complex spectra is far from complete. It seems, however, to be generally accepted that, possibly in thorium, and almost certainly in protactinium, $5f$ orbitals are being occupied. Configurations of individual actinide elements are given in *Table 5.3.* Actinium, with the configuration $5f^0 6s^2 6p^6 6d^1 7s^2$, forms the first member of the 'actinide' series. It strongly resembles lanthanum, the first member of the 'lanthanide' series, in its chemical and physical properties. Both elements form insoluble fluorides, hydroxides, oxalates and phosphates, and the crystal structures of corresponding compounds are very similar.

Before leaving this section we must emphasize that the electron configurations here described refer to isolated atoms or ions in the gaseous state. The configuration may well be different if the ion is in a solid structure or in solution.

*Here we quote only the configuration in the outer orbitals.

Table 5.3 Electron configurations for the ground states of the elements*

Element	*Atomic number*	*K*	*L*		*M*			*N*				*O*				*P*				*Q*
		1*s*	2*s*	2*p*	3*s*	3*p*	3*d*	4*s*	4*p*	4*d*	4*f*	5*s*	5*p*	5*d*	5*f*	6*s*	6*p*	6*d*	6*f*	7*s*
H	1	1																		
He	2	2																		
Li	3	2	1																	
Be	4	2	2																	
B	5	2	2	1																
C	6	2	2	2																
N	7	2	2	3																
O	8	2	2	4																
F	9	2	2	5																
Ne	10	2	2	6																
Na	11	2	2	6	1															
Mg	12	2	2	6	2															
Al	13	2	2	6	2	1														
Si	14	2	2	6	2	2														
P	15	2	2	6	2	3														
S	16	2	2	6	2	4														
Cl	17	2	2	6	2	5														
Ar	18	2	2	6	2	6														
K	19	2	2	6	2	6		1												
Ca	20	2	2	6	2	6		2												
Sc	21	2	2	6	2	6	1	2												
Ti	22	2	2	6	2	6	2	2												
V	23	2	2	6	2	6	3	2												
Cr	24	2	2	6	2	6	5	1												
Mn	25	2	2	6	2	6	5	2												
Fe	26	2	2	6	2	6	6	2												
Co	27	2	2	6	2	6	7	2												
Ni	28	2	2	6	2	6	8	2												
Cu	29	2	2	6	2	6	10	1												
Zn	30	2	2	6	2	6	10	2												
Ga	31	2	2	6	2	6	10	2	1											
Ge	32	2	2	6	2	6	10	2	2											
As	33	2	2	6	2	6	10	2	3											
Se	34	2	2	6	2	6	10	2	4											
Br	35	2	2	6	2	6	10	2	5											
Kr	36	2	2	6	2	6	10	2	6											
Rb	37	2	2	6	2	6	10	2	6			1								
Sr	38	2	2	6	2	6	10	2	6			2								
Y	39	2	2	6	2	6	10	2	6	1		2								
Zr	40	2	2	6	2	6	10	2	6	2		2								
Nb	41	2	2	6	2	6	10	2	6	4		1								
Mo	42	2	2	6	2	6	10	2	6	5		1								
Tc	43	2	2	6	2	6	10	2	6	(5)		(2)								
Ru	44	2	2	6	2	6	10	2	6	7		1								
Rh	45	2	2	6	2	6	10	2	6	8		1								
Pd	46	2	2	6	2	6	10	2	6	10										
Ag	47	2	2	6	2	6	10	2	6	10		1								
Cd	48	2	2	6	2	6	10	2	6	10		2								
In	49	2	2	6	2	6	10	2	6	10		2	1							
Sn	50	2	2	6	2	6	10	2	6	10		2	2							

*Numbers in parentheses are uncertain

Ele-ment	Atomic number	K	L		M			N				O				P				Q
		1s	2s	2p	3s	3p	3d	4s	4p	4d	4f	5s	5p	5d	5f	6s	6p	6d	6f	7s
Sb	51	2	2	6	2	6	10	2	6	10		2	3							
Te	52	2	2	6	2	6	10	2	6	10		2	4							
I	53	2	2	6	2	6	10	2	6	10		2	5							
Xe	54	2	2	6	2	6	10	2	6	10		2	6							
Cs	55	2	2	6	2	6	10	2	6	10		2	6			1				
Ba	56	2	2	6	2	6	10	2	6	10		2	6			2				
La	57	2	2	6	2	6	10	2	6	10		2	6	1		2				
Ce	58	2	2	6	2	6	10	2	6	10	(2)	2	6			(2)				
Pr	59	2	2	6	2	6	10	2	6	10	(3)	2	6			(2)				
Nd	60	2	2	2	2	6	10	2	6	10	(4)	2	6			(2)				
Pm	61	2	2	6	2	6	10	2	6	10	(5)	2	6			(2)				
Sm	62	2	2	6	2	6	10	2	6	10	6	2	6			2				
Eu	63	2	2	6	2	6	10	2	6	10	7	2	6			2				
Gd	64	2	2	6	2	6	10	2	6	10	(7)	2	6	(1)		2				
Tb	65	2	2	6	2	6	10	2	6	10	(8)	2	6	(1)		2				
Dy	66	2	2	6	2	6	10	2	6	10	(10)	2	6			(2)				
Ho	67	2	2	6	2	6	10	2	6	10	(11)	2	6			(2)				
Er	68	2	2	6	2	6	10	2	6	10	(12)	2	6			(2)				
Tm	69	2	2	6	2	6	10	2	6	10	13	2	6			2				
Yb	70	2	2	6	2	6	10	2	6	10	14	2	6			2				
Lu	71	2	2	6	2	6	10	2	6	10	14	2	6	1		2				
Hf	72	2	2	6	2	6	10	2	6	10	14	2	6	2		2				
Ta	73	2	2	6	2	6	10	2	6	10	14	2	6	3		2				
W	74	2	2	6	2	6	10	2	6	10	14	2	6	4		2				
Re	75	2	2	6	2	6	10	2	6	10	14	2	6	5		2				
Os	76	2	2	6	2	6	10	2	6	10	14	2	6	6		2				
Ir	77	2	2	6	2	6	10	2	6	10	14	2	6	7		2				
Pt	78	2	2	6	2	6	10	2	6	10	14	2	6	9		1				
Au	79	2	2	6	2	6	10	2	6	10	14	2	6	10		1				
Hg	80	2	2	6	2	6	10	2	6	10	14	2	6	10		2				
Tl	81	2	2	6	2	6	10	2	6	10	14	2	6	10		2	1			
Pb	82	2	2	6	2	6	10	2	6	10	14	2	6	10		2	2			
Bi	83	2	2	6	2	6	10	2	6	10	14	2	6	10		2	3			
Po	84	2	2	6	2	6	10	2	6	10	14	2	6	10		2	4			
At	85	2	2	6	2	6	10	2	6	10	14	2	6	10		2	5			
Rn	86	2	2	6	2	6	10	2	6	10	14	2	6	10		2	6			
Fr	87	2	2	6	2	6	10	2	6	10	14	2	6	10		2	6			1
Ra	88	2	2	6	2	6	10	2	6	10	14	2	6	10		2	6			2
Ac	89	2	2	6	2	6	10	2	6	10	14	2	6	10		2	6	(1)		(2)
Th	90	2	2	6	2	6	10	2	6	10	14	2	6	10		2	6	(2)		(2)
Pa	91	2	2	6	2	6	10	2	6	10	14	2	6	10	(2)	2	6	(1)		(2)
U	92	2	2	6	2	6	10	2	6	10	14	2	6	10	(3)	2	6	(1)		(2)
Np	93	2	2	6	2	6	10	2	6	10	14	2	6	10	(5)	2	6			(2)
Pu	94	2	2	6	2	6	10	2	6	10	14	2	6	10	(6)	2	6			(2)
Am	95	2	2	6	2	6	10	2	6	10	14	2	6	10	(7)	2	6			(2)
Cm	96	2	2	6	2	6	10	2	6	10	14	2	6	10	(7)	2	6	(1)		(2)
Bk	97	2	2	6	2	6	10	2	6	10	14	2	6	10	(8)	2	6	(1)		2
Cf	98	2	2	6	2	6	10	2	6	10	14	2	6	10	(10)	2	6			2
Es	99	2	2	6	2	6	10	2	6	10	14	2	6	10	11	2	6			2
Fm	100	2	2	6	2	6	10	2	6	10	14	2	6	10	12	2	6			2
Md	101	2	2	6	2	6	10	2	6	10	14	2	6	10	13	2	6			2
No	102	2	2	6	2	6	10	2	6	10	14	2	6	10	14	2	6			2
Lr	103	2	2	6	2	6	10	2	6	10	14	2	6	10	14	2	6	1		2

6

THE MOLECULAR-ORBITAL METHOD

6.1 Introduction

The molecular-orbital approach starts by considering the stable molecular system in which the nuclei are in their equilibrium positions, and produces molecular wave functions that describe the molecular energy states (molecular orbitals) into which electrons can be placed.

In the earlier chapters we have seen how this was done for atoms. Solution of the Schrödinger equation for the hydrogen atom gave a series of acceptable solutions which described allowed atomic energy states or atomic orbitals ($1s$, $2s$, $2p$, etc.). The ideas were extended to other atoms in which there was the same number of atomic orbitals as for hydrogen, although the energies of the atomic orbitals were modified to allow for the increased nuclear charge and the presence of the other electrons. Electrons were 'fed' into the available atomic orbitals one at a time, according to the 'Aufbau' or 'building-up' principle, by which each electron entered the orbital of lowest energy. The Pauli Exclusion Principle was taken into account, so that each orbital was restricted to a maximum of two electrons (and then their spin quantum numbers had to have different values), and Hund's Rule of Maximum Multiplicity was applied when several available orbitals had the same energy (i.e. were degenerate); thus the ground-state configuration of the nitrogen atom was $1s^2 2s^2 2p_x 2p_y 2p_z$ rather than $1s^2 2s^2 (2p_x)^2 (2p_y)$, giving the maximum number of 'unpaired spins'.

The molecular-orbital method uses the same approach, and if we restrict ourselves for the moment to diatomic molecules we shall see that the procedure is directly analogous to that used for atomic systems. Thus:

1. The molecular Schrödinger equation is 'solved' to provide acceptable molecular wave functions that describe molecular energy states (molecular orbitals).
2. These molecular orbitals will embrace both nuclei. (In the case of polynuclear molecules the molecular orbitals will in principle involve all the nuclei.)
3. The molecular orbitals will be associated with molecular quantum numbers.
4. The molecular wave functions will have the same significance as do atomic wave functions, i.e. $\psi^2 dv$ is proportional to the probability of finding the electron in a given volume of space dv.
5. Each molecular wave function corresponds to a definite energy value, and the sum of the energies of individual electrons in the molecular orbitals, after correction for interaction, represents the total energy of the molecule.

6. The 'Aufbau' principle may be applied. The electrons are fed into the available molecular orbitals one at a time, the molecular orbital or lowest energy being filled first.
7. Each electron has a spin, and by the Pauli Exclusion Principle (p.40) each molecular orbital can accommodate a maximum of two electrons, provided their spins are opposed.

These molecular orbitals may be obtained by several methods. To help the less-mathematically minded reader we start with a qualitative discussion to relate atomic orbitals and molecular orbitals, using the concept of a 'United Atom', and then give a more detailed and somewhat mathematical account of how such molecular orbitals can be formally derived from atomic orbitals by the method known as the 'linear combination of atomic orbitals' (abbreviated to LCAO).

6.2 Relationship between Atomic and Molecular Orbitals (United Atom Approach)

Let us first consider a simple homonuclear diatomic molecule, i.e. two identical atoms held together by a number of electrons. (Although the atoms are identical it is convenient to distinguish the two nuclei by labelling one A and the other B.) We can now create a hypothetical 'United Atom' by letting the two atomic nuclei A and B coalesce, so that we have a nucleus whose mass and charge are twice that of either A or B singly, together with all the attendant electrons. This united atom can be treated like any other atom and its electrons can be allocated to appropriate united-atom orbitals. The orbitals will be described by the usual three quantum numbers n, ℓ and m, and each may accommodate up to two electrons.

If we take this hypothetical atom and separate the two constituent nuclei, the electron charge clouds (filled orbitals of the united atom) will no longer be concentrated about a single nucleus but will spread out about the axis connecting the two nuclei. This is shown pictorially in *Figure 6.1*, which illustrates naively the effect on s and p charge clouds. It can be seen that we would expect the s orbital charge cloud of the united atom,

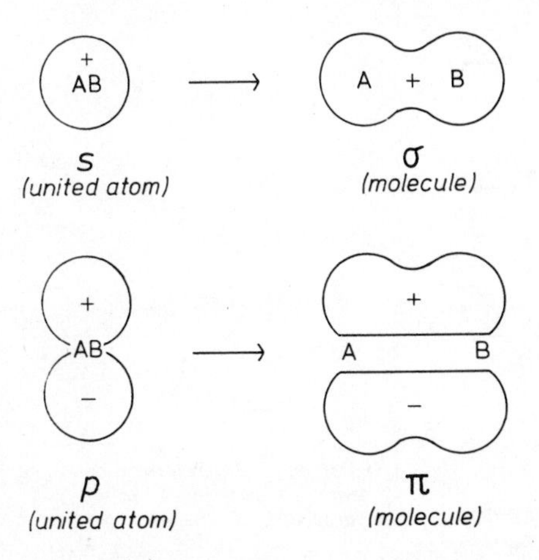

Figure 6.1 United atom. σ and π orbitals

which is spherically symmetrical, to spread out into a charge cloud that is symmetrical about the A–B axis. This new molecular orbital is given the label σ. In the same way a p_y or p_z charge cloud will spread out to form two cylindrical charge clouds above and below the A–B axis (considered as the x axis). This molecular orbital is referred to as π.

Although this approach is very limited and imprecise as far as this discussion goes, it does show the relationship between the atomic s and p orbitals and the molecular σ and π orbitals which have the same symmetry features.

In practice this approach can be taken much further and relationships developed between the atomic and molecular orbitals, the results being expressed in the form of a correlation diagram (cf. *Figure 6.2*). This indicates not only the energies of orbitals for both the united atom and the separated atoms, which will be quite different because of the different nuclear charges, but also the energy levels at intermediate internuclear distances corresponding to molecule formation. These molecular orbitals fall into two categories:

(a) Those levels, or molecular orbitals, corresponding to a decrease in energy as the atoms get closer together, i.e. related to lines sloping downwards from right to left (namely the $(k)z\sigma$, $z\sigma$, $w\pi$, . . . , etc.), are called bonding molecular orbitals.

(b) Those corresponding to an increase in energy as the internuclear distance decreases, and so related to lines sloping upwards from right to left (e.g. $(k)y\sigma$, $y\sigma$, $v\pi$, . . . , etc.), are called antibonding molecular orbitals.

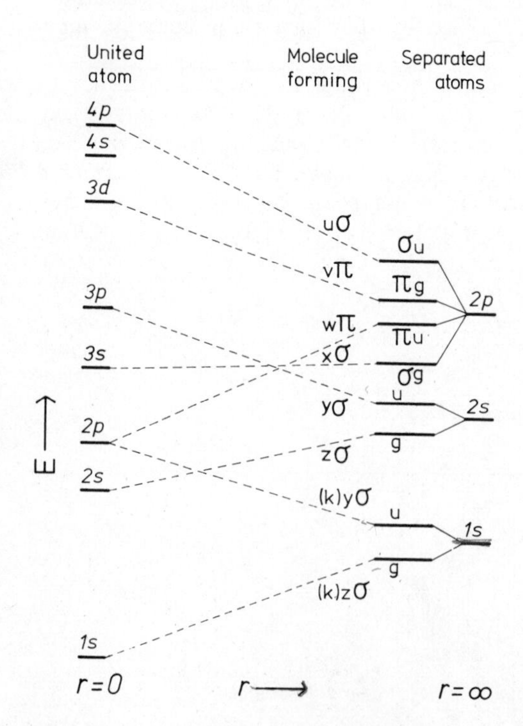

Figure 6.2 Correlation diagram. Energy levels in separated and united atoms

The symbolism used to describe these molecular orbitals will be explained in the next section and the significance of the terms bonding and antibonding will also become clearer when we have discussed the construction of molecular-orbital wave functions by the LCAO method.

6.3 The LCAO Method

Let us consider again the homonuclear diatomic molecule A–B and the behaviour of a single electron under the influence of both of the nuclei and of the other electrons. Now the Schrödinger equation for such a complicated system cannot be solved completely to give us accurate wave functions representing the molecular orbitals referred to in the previous section, but with the help of chemical and mathematical intuition it is possible for us to make shrewd guesses about the form of these functions. Thus if we assume that in the immediate neighbourhood of nucleus A the electron is acted on by only this nucleus and any associated electrons, then it can be described by a simple atomic wave function (atomic orbital) ϕ_A. In a similar way, another wave function ϕ_B will describe the behaviour of the electron when it is close to nucleus B. Clearly ϕ_A and ϕ_B must be solutions to the Schrödinger equation for the molecule, but neither will correspond to that lowest energy situation in which the electron is bonding together both nuclei.

A general mathematical procedure for solving wave equations makes use of the fact that if we have two solutions to a given wave equation, then any linear combination of these two solutions will also be a solution. In the case under discussion we know that ϕ_A and ϕ_B are solutions to the Schrödinger equation, so it is reasonable to take as a 'trial' function the linear combination $C_A\phi_A + C_B\phi_B$, where C_A and C_B are simple numbers. In practice, the expression is written in the form

$$\psi = N\{C_A\phi_A + C_B\phi_B\}$$

where N is included as a normalizing constant, chosen so that $\psi^2 dv$ taken over the whole of space is unity. This means that the chance of finding the electron somewhere in space is unity.

The next step is to find those values for the coefficients C_A and C_B which will give the most satisfactory 'trial' function, i.e. the one which is nearest to the true molecular orbital, and this can be done using a technique called the 'Variation Method'. We can write down a large number of approximate wave functions, $\psi_1, \psi_2, \psi_3, \ldots, \psi_n$, by giving different values to C_A and C_B, and each of these functions will have an associated energy $E_1, E_2, \ldots, E_n$. The energy of any linear combination of ϕ_A and ϕ_B will be lower than that corresponding to ϕ_A or ϕ_B separately, and the Variation Method states that the best approximation to the true wave function will be the combination giving the lowest energy. As applied to the molecule A–B this means that the energy E must be a minimum with respect to the coefficients C_A and C_B, so that in mathematical terms we get two linear equations:

$$\frac{\partial E}{\partial C_A} = 0 \quad \text{and} \quad \frac{\partial E}{\partial C_B} = 0$$

These equations can be solved to give E and the ratio C_A/C_B. The equation from which the energy E can be obtained has two roots, one for which $C_A = C_B$, and one for which $C_A = -C_B$. There are accordingly two linear combinations of ϕ_A and ϕ_B that we must consider:

$$\psi_+ = NC_A(\phi_A + \phi_B) \quad \text{and} \quad \psi_- = NC_A(\phi_A - \phi_B)$$

The molecular wave function ψ_+ corresponds to the situation in which there is a build-up of electron density in the region between the nuclei and a more effective screening of one nucleus from the other. Hence there will be a decrease in the energy of the system by comparison with the separated atoms so that a bond has been formed which is described by the 'bonding' molecular orbital ψ_+. We say that this molecular orbital has been formed by the overlap of the two atomic orbitals represented by ϕ_A and ϕ_B.

The reader may well ask how these particular atomic orbitals are chosen. A detailed examination of the expression for the energy associated with the function ψ_+ shows that ϕ_A and ϕ_B will combine effectively only if

(a) they represent states of similar energy;
(b) they overlap to a considerable extent; and
(c) they have the same symmetry with respect to the molecular axis A–B.

If these conditions are not satisfied, the coefficients C will have very small values, and the corresponding wave functions will make little contribution to the linear combination. Shortly we shall discuss particular molecules and see how the choice of ϕ_A and ϕ_B is made.

The function ψ_- is called an 'antibonding' molecular orbital, since it represents a state of higher energy, in which the electrons are displaced from the internuclear region. The energy of the antibonding molecular orbital is greater than that of either atomic orbital, as *Figure 6.3* indicates, so that if the electron should find itself in the antibonding orbital, it would return if possible to a more stable atomic orbital of lower energy. This point will become increasingly significant when we come to discuss more complex molecules, since then electrons may occupy both bonding and antibonding molecular orbitals; the presence of an electron in an antibonding orbital introduces a factor opposing the formation of a stable molecule.

6.4 Hydrogen Molecule Ion (H_2^+)

The molecular-orbital method is now illustrated by reference to the simplest possible molecule, the hydrogen molecule ion, H_2^+, in which two protons are held together by a single electron. This molecule can be detected spectroscopically when hydrogen gas under reduced pressure is subjected to an electric discharge, and it is found to have a bond length of 106 pm and a dissociation energy of 269.3 kJ mol^{-1}. The atomic orbitals ϕ_A and ϕ_B used in the LCAO procedure will be the hydrogen 1s orbitals (see p.32) because they are the most suitable orbitals on grounds of energy and symmetry. The linear combination of these two atomic orbitals will produce

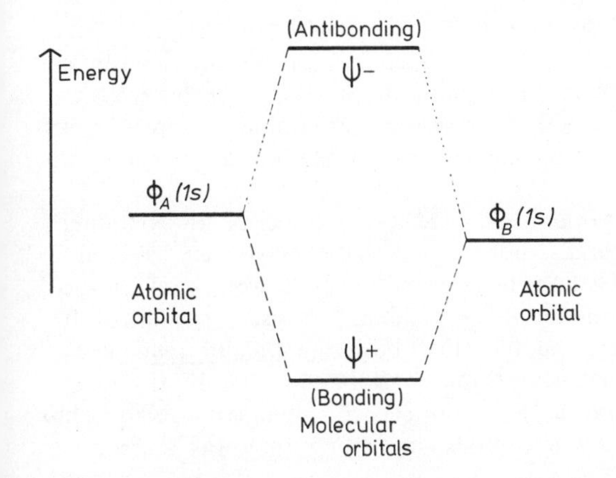

Figure 6.3 The relative energies of molecular orbitals and their constituent atomic orbitals

two molecular orbitals, one bonding and one antibonding, as shown schematically in *Figure 6.3.* (We shall see later on that in general the combination of n atomic orbitals gives n molecular orbitals, half of which are bonding and half antibonding.) In the ground state of H_2^+ the single electron occupies the bonding orbital. This comparatively simple approximation method gives a calculated energy which has a minimum value at an internuclear separation of 132 pm when the dissociation energy is 169.8 kJ mol^{-1}. Although these values are not very close to the experimental ones, it must be appreciated that it is something of an achievement that these results should be of the right order of magnitude. Calculated values for the internuclear distance and dissociation energy which agree exactly with experimental values have been obtained, but only by making very considerable modifications to the trial functions, with a corresponding elaboration of the computations involved.

The bonding and antibonding molecular orbitals, and their pictorial relationship to the atomic 1s orbitals, are illustrated in *Figure 6.4.* The significance of such boundary surfaces for atomic orbitals was discussed in Chapter 3, and it may be worth pointing out yet again that surfaces (or cross-sections of surfaces) merely represent the region in which the electron is most likely to be found. The maximum 'electron density' is normally well inside such surfaces. As we shall see shortly, this representation of atomic orbitals with positive and negative signs is helpful in choosing combinations of the correct symmetry.

The wave functions ψ_+ and ψ_- are usually written ψ_g and ψ_u in the molecular-orbital theory of homonuclear diatomic molecules. The subscripts g and u are abbreviations for the German words 'gerade' (even) and 'ungerade' (odd), and refer to important symmetry properties of the wave

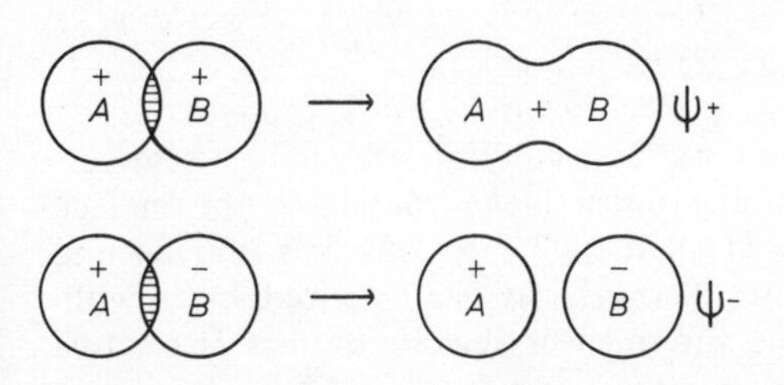

Figure 6.4 Combination of s atomic orbitals to form molecular orbitals

functions. An orbital is said to be symmetrical (gerade) if the wave function is unchanged in sign when the coordinates x, y, z of the electron are replaced by $-x$, $-y$, $-z$, i.e. when any point in the orbital is reflected in the midpoint or centre of symmetry; conversely, an orbital is antisymmetrical (ungerade) if the wave function changes sign when the x, y, z coordinates are replaced by $-x$, $-y$, $-z$.

We have used the general symbols ψ_+ and ψ_- to describe the bonding and antibonding molecular orbitals, but more specific labels can be used, namely $\sigma 1s$ and σ^*1s, to indicate which atomic orbitals were used in the formation of the molecular orbitals. The symbol * is used to mark antibonding orbitals, and σ reflects the fact that both the bonding and antibonding orbitals are symmetrical about the A–B bond axis. In the United Atom approach (Section 6.2) we noticed this symmetry relationship between atomic s and molecular σ orbitals; in neither case was there a nodal plane.

We can now use the LCAO approach to combine together other atomic orbitals of the two nuclei, and so build up a set of molecular orbitals that can be used to describe a whole range of homonuclear diatomic molecules.

6.5 Homonuclear Diatomic Molecules

The combination of two $2s$ atomic orbitals will produce a bonding $\sigma 2s$ and an antibonding σ^*2s molecular orbital, which once again are symmetrical about the A–B bond axis. These $\sigma 2s$ molecular orbitals are higher in energy than the $\sigma 1s$ molecular orbitals, since they are formed from the higher energy $2s$ atomic orbitals.

Some new problems appear when we discuss the combination of p type atomic orbitals to form molecular orbitals. Here we have to remember two facts about p orbitals: first, there are three equivalent p orbitals, p_x, p_y and p_z, for each principal quantum number ($n \geqslant 2$), and these are set at right angles to each other; secondly, the two halves of the p orbital have opposite mathematical signs, one lobe being positive and the other negative. When the p_x orbitals are combined (cf. *Figure 6.5*) the bonding and antibonding molecular orbitals which are formed are similar to the σs and σ^*s orbitals; they are symmetrical about the molecular A–B axis (taken as the x axis), and are denoted σp and σ^*p. The combination of p_y orbitals produces molecular orbitals of very different shapes, however, since in this case there is overlap of both lobes of the atomic orbitals, leading to molecular orbitals that are no longer symmetrical about the molecular axis. We saw that the simple United Atom approach gave a π orbital in which there were two cylindrical charge cloud streamers, one above and the other below the molecular axis. This bonding π molecular orbital, written $\pi_y 2p$, can be visualized somewhat crudely as a 'meringue' of charge clouds in which the cream forms the nodal plane containing the molecular axis.

The combination $\phi_{A(2p_y)} - \phi_{B(2p_y)}$ [see *Figure 6.5(iv)*] produces the antibonding molecular orbital, written $\pi_y{}^*2p$, and this orbital has a more complex shape, since in addition to the nodal plane containing the molecular axis there is also a nodal plane at right angles to this axis and cutting it midway between the nuclei. This π^* orbital is characterized by a withdrawal of electronic charge from the internuclear region, so that there is a

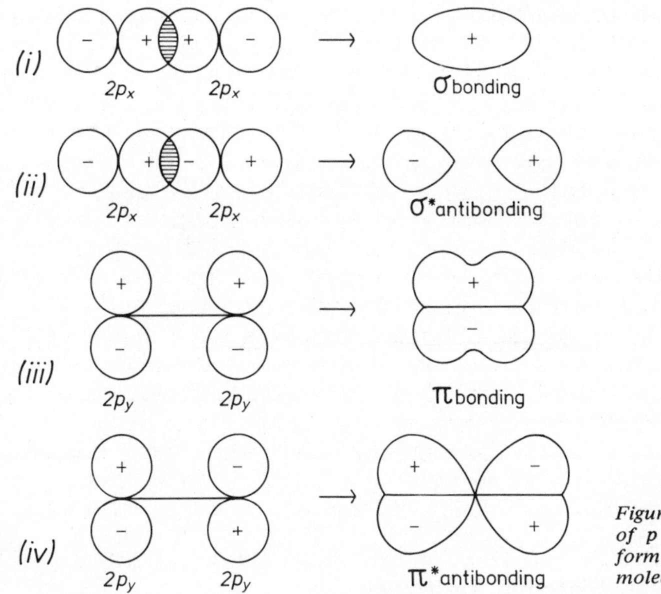

Figure 6.5 Combination of p atomic orbitals to form σ, σ, π and π* molecular orbitals*

higher repulsion between the nuclei; the π^* orbital is accordingly higher in energy than the π orbital (see *Figure 6.5*). The combination of $2p_z$ atomic orbitals gives rise to two analogous molecular orbitals, one bonding ($\pi_z 2p$) and one antibonding ($\pi_z{}^*2p$).

The component atomic orbitals, $2p_y$ and $2p_z$, have the same energy and are said to be 'degenerate', so it is hardly surprising to find that the corresponding π molecular orbitals are equal in energy. We refer to the $\pi 2p$ molecular orbitals as being 'doubly degenerate', since there are two orbitals of equal energy ($\pi_y 2p \equiv \pi_z 2p$); the π^*2p orbitals are also doubly degenerate ($\pi_y{}^*2p \equiv \pi_z{}^*2p$). The energy relationships are shown in *Figure 6.6.*

Reference to *Figure 6.5* shows that the $\pi_y 2p$ orbital is unsymmetrical since when any point is reflected in the centre (i.e. x, y, z become $-x$, $-y$, $-z$) its sign changes; hence ψ is written ψ_u. The $\pi_y{}^*2p$ orbital, on the other hand, is symmetrical with respect to inversion in the centre of the A–B bond, and accordingly ψ is written ψ_g in this case. These distinctions are important because it can be shown that, when electrons in a given molecule change from one orbital to another, they only do so if they go into an orbital of a different symmetry type. Electronic transitions do not normally occur between two *gerade* orbitals or between two *ungerade* orbitals.

It is convenient to assign a new quantum number λ to the molecular orbitals such that the component of the angular momentum about the internuclear axis is $\lambda h/2\pi$, where λ takes the values 0, ±1, ±2, . . . , etc. λ has the same significance for molecules that the quantum number m has for atoms. The σ type molecular orbitals have $\lambda = 0$, and the π type have $\lambda = 1$. Orbitals with $\lambda = 2$ are termed δ orbitals; they can be obtained by a combination of d atomic orbitals and resemble 'double-π' orbitals in that they have two nodal planes intersecting in the molecular axis. Thus the number of nodal planes is the same as the value of λ.

Some readers may be wondering why the combinations have been only of 1*s* with 1*s*, 2*s* with 2*s*, 2*p* with 2*p*, etc., and why we did not consider the combination of a 1*s* atomic orbital of one atom with the 2*s* atomic orbital of the other. However, in homonuclear molecules, where the atoms A and B are identical, the 1*s* and 2*s* atomic orbitals correspond to very different energies, and we have seen that the component atomic orbitals must have similar energies if they are to combine effectively. For the same reason, *s* and *p* orbitals do not combine in homonuclear diatomic molecules. We shall see later in this chapter, however, that such combinations may be possible with heteronuclear diatomic molecules when the atoms A and B are different, as with hydrogen fluoride, for instance.

Before we can discuss the electron arrangements (and apply the 'Aufbau' principle to molecules) in molecules other than the hydrogen molecule ion, we must arrange the molecular orbitals in order of their energies (see *Figure 6.6*). These energies have been determined from spectroscopic observations, and when arranged in increasing order of magnitude, they give the sequence

$$\sigma 1s < \sigma^* 1s < \sigma 2s < \sigma^* 2s < \sigma 2p < \pi_y 2p = \pi_z 2p < \pi_y{}^* 2p = \pi_z{}^* 2p < \sigma^* 2p$$

A similar arrangement exists for the $\sigma 3s$ to $\sigma^* 3p$ molecular orbitals, but these energies are known with less certainty and, moreover, the bonding is now rarely between pure *s*, *p*, *d* and *f* atomic orbitals, but rather between various combinations of them (see p.88).

Although the notation we have used is very clear, it is a little cumbersome, particularly in respect of the π molecular orbitals, and Mulliken has suggested an alternative terminology for the orbitals $\sigma 2s$ to $\sigma^* 2p$ inclusive:

$$z\sigma < y\sigma < x\sigma < w\pi < v\pi < u\sigma$$

This notation can be used also for the one-quantum (*K* shell) and three-quantum (*M* shell) molecular orbitals by writing (*K*) or (*M*) before the Mulliken symbols, i.e. $(K)z\sigma$ and $(K)y\sigma$, and $(M)z\sigma$, $(M)y\sigma$, . . . , $(M)u\sigma$.

The Mulliken notation is much the briefer of the two, and it has two further advantages:

1. It may be applied to heteronuclear diatomic molecules, where the molecular orbital may be formed by combining two atomic orbitals from different quantum shells. We shall discuss this again later, but can clarify the point by referring to the HF molecule, where the bonding molecular orbital is compounded from the H(1*s*) and F($2p_x$) atomic orbitals. This molecular orbital can be simply written as $x\sigma$ in the Mulliken notation, but it would be difficult to describe the orbital in more specific terms.
2. The Mulliken notation does not specify which atomic orbitals were combined to give the molecular orbital, so that $z\sigma$ for instance merely says that this molecular orbital is the one of lowest energy, and does not imply that it is derived solely from 1*s* atomic orbitals.

However, the Mulliken notation suffers from two important disadvantages:

1. Confusion may arise between the z, y and x of the Mulliken notation and Cartesian co-ordinates (x, y, z).
2. The easily visualized relationship between the molecular orbital and the constituent atomic orbitals is lost.

Thus it can be seen that each notation has its merits, and students should be familiar with both. In the following molecules we discuss the electronic distributions in simple homonuclear diatomic molecules by applying the 'Aufbau' principle, and assigning electrons one at a time to the orbitals of lowest energy, the Pauli Principle limiting each orbital to a maximum of two 'paired' electrons. In the course of these assignments we shall use both notations to illustrate their application.

6.5.1 H_2^+

We have already discussed the structure of the hydrogen molecule ion in some detail, and have seen that the electron is in the bonding $\sigma 1s$ molecular orbital. The structure is accordingly written $H_2^+[(\sigma 1s)]$.

6.5.2 H_2

The hydrogen molecule contains one electron more than the hydrogen molecule ion, and in the ground state this electron is also found in the $\sigma 1s$ orbital, giving the configuration $H_2\,[(\sigma 1s)^2]$. Thus the hydrogen molecule contains two electrons in the lowest energy bonding molecular orbital, $\sigma 1s$, and therefore the two-electron bond of H_2 is much stronger than the one-electron bond of H_2^+; the bond dissociation energies are 458.5 and 269.3 kJ mol^{-1}, respectively.

6.5.3 He_2^+

This molecule has been detected spectroscopically. Its formation can be considered as the combination of a helium atom, He, and a helium ion, He^+, i.e. $He(1s^2) + He^+(1s) \rightarrow He_2^+[(\sigma 1s)^2(\sigma^* 1s)]$. The He_2^+ molecule differs from H_2 in that an extra electron is present in the antibonding $\sigma^* 1s$ orbital, and this antibonding electron reduces the bond strength: the 'relaxing' effect of an antibonding electron is rather greater than the binding force resulting from a bonding electron, so that the bond in He_2^+ is even weaker than a simple one-electron bond.

6.5.4 He_2

This molecule does not exist under normal conditions; its configuration would be $(\sigma 1s)^2(\sigma^* 1s)^2$, and the bonding power of $(\sigma 1s)^2$ would be more than cancelled out by the antibonding $(\sigma^*1s)^2$. The non-existence of the He_2 molecule illustrates the general principle that the superposition of equal numbers of doubly-filled bonding and antibonding orbitals leads to either no bonding or even to a slight overall repulsion.

Although the He_2 molecule is less stable than two uncombined helium atoms, it can be detected in discharge tubes. This is because in the discharge tube we are comparing the energies of two excited helium atoms with that of an excited He_2 molecule, of configuration $[(\sigma 1s)^2(\sigma 2s)^2]$, in which there are four bonding electrons. While this molecular configuration corresponds to a higher energy than two uncombined helium atoms (in their ground states), it is more stable than two separated excited helium atoms.

6.5.5 Li_2

$$2\text{Li}(1s^2 2s) \rightarrow \text{Li}_2\,[(\sigma 1s)^2(\sigma^*1s)^2(\sigma 2s)^2]$$

We have already seen that the bonding associated with the $(\sigma 1s)^2$ electrons is cancelled by the antibonding effect of $(\sigma^*1s)^2$, so that the bonding between the two lithium atoms results from the pairing of the Li 2*s* electrons in the bonding $\sigma 2s$ orbital. We can simplify the nomenclature by writing *KK* in place of $(\sigma 1s)^2(\sigma^*1s)^2$, thus giving $Li_2\,[KK(\sigma 2s)^2]$ as the ground state of the lithium molecule. In general we shall treat such inner-shell electrons as non-bonding, even though rather more refined treatments may have to consider inner-shell interactions.

The diatomic molecules formed by other alkali metals in the gas phase have analogous configurations, $Na_2\,[KKLL(\sigma 3s)^2]$, . . . , etc.

6.5.6 N_2

$$2\text{N}(1s^2 2s^2 2p^3) \rightarrow \text{N}_2\,[KK(\sigma 2s)^2(\sigma^*2s)^2(\sigma 2p)^2(\pi_y 2p = \pi_z 2p)^4]$$

or

$$\text{N}_2\,[KK(z\sigma)^2(y\sigma)^2(x\sigma)^2(w\pi)^4]$$

The *K* shell electrons again take no part in the bonding, and we have therefore two 2*s* and three 2*p* electrons from each nitrogen atom to feed into the available molecular orbitals (cf. *Figure 6.6*). The bonding resulting from $(z\sigma)^2$ is effectively cancelled out by the antibonding $(y\sigma)^2$, leaving the $(x\sigma)^2$ and $(w\pi)^4$ to provide the molecular bonding. These six bonding electrons produce a N≡N triple bond, one bond being σ and the other two π in character. This triple bond is symmetrical about the N–N bond axis, the electronic charge cloud forming a cylinder around the bond axis.

The P_2 molecule, which is found at high temperatures in the gas phase, has an analogous configuration but with both the *K* and *L* shells full. The bonding electrons are the $[(m)x\sigma]^2$ and $[(m)w\pi]^4$, or $(\sigma 3p)^2$ and $(\pi_y 3p = \pi_z 3p)^4$.

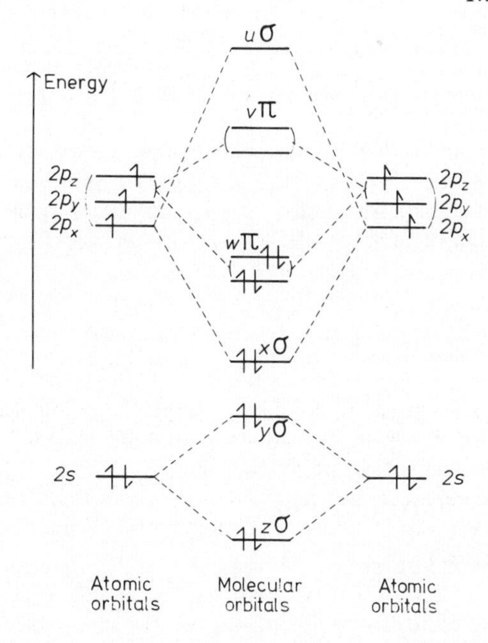

Figure 6.6 The formation of the molecular orbitals for N_2

6.5.7 O_2

$$2O(1s^2 2s^2 2p^4) \rightarrow O_2\,[KK(z\sigma)^2(y\sigma)^2(x\sigma)^2(w\pi)^4(v\pi)^2]$$

The oxygen molecule contains two electrons more than the nitrogen molecule, and these electrons enter the antibonding $v\pi$ orbitals, so that the triple bond formed from $(x\sigma)^2$ and $(w\pi)^4$ is effectively reduced to a double bond by the antibonding character of $(v\pi)^2$. Since, however, there are two $v\pi$ antibonding molecular orbitals, then, by Hund's Rule of Maximum Multiplicity (see p.44), one electron will enter each of these, giving a paramagnetic molecule with two unpaired electrons. This explanation of the well-known paramagnetism of the O_2 molecule, and an analogous explanation for the paramagnetism of S_2, was one of the early successes of the molecular-orbital theory.

6.5.8 F_2

$$2F(1s^2 2s^2 2p^5) \rightarrow F_2\,[KK(z\sigma)^2(y\sigma)^2(x\sigma)^2(w\pi)^4(v\pi)^4]$$

The $(z\sigma)^2$ and $(y\sigma)^2$ cancel each other out, as do the $(w\pi)^4$ and $(v\pi)^4$, so that although all fourteen electrons contribute in theory, the effective bonding is provided by the $(x\sigma)^2$. The F–F bond is therefore a single two-electron bond.

Cl_2 and Br_2 have structures analogous to F_2, Cl_2 having its inner K and L, and Br_2 its inner K, L and M, shells full.

6.5.9 Ne_2

$$2Ne(1s^2 2s^2 2p^6) \rightarrow Ne_2\,[KK(z\sigma)^2 (y\sigma)^2 (x\sigma)^2 (w\pi)^4 (v\pi)^4 (u\sigma)^2]$$

This molecule is not formed under normal conditions because, as we saw in the corresponding example of the He_2 molecule, the number of bonding and antibonding electrons is equal; for similar reasons the other noble gas atoms do not form stable diatomic molecules.

6.6 Heteronuclear Diatomic Molecules

The same general principles may be used in describing simple heteronuclear diatomic molecules, i.e. molecules in which the atoms A and B are no longer identical. We must take particular care in selecting the atomic orbitals of A and B that will combine effectively and recall that three conditions have to be satisfied for the effective combination of atomic orbitals. Thus the atomic orbitals must

(i) have similar energies;
(ii) have charge clouds that overlap as much as possible; and
(iii) have the same symmetry properties with respect to the molecular axis A–B.

We stressed the first condition in our discussion of homonuclear diatomic molecules, where the atomic orbitals being combined were of the same type (i.e. 1*s* with 1*s*, 2*s* with 2*s*, etc.), and where the other two conditions were fulfilled automatically. The application of these principles to heteronuclear diatomic molecules can best be illustrated by a consideration of a number of simple examples.

6.6.1 HF AND THE OTHER HYDROGEN HALIDES

The electronic configurations of hydrogen and fluorine atoms are

$$\text{H} \quad 1s$$

$$\text{F} \quad 1s^2 2s^2 2p^5$$

Now the molecular orbital describing the H–F bond must be compounded from a linear combination of the H(1*s*) atomic orbital with one of the F atomic orbitals. Spectroscopic evidence shows that the energies of the $1s^2$ and $2s^2$ electrons in fluorine are far too low, however, to contribute appreciably to the bonding; this is merely another way of saying that the inner electrons are 'deep-seated' and do not take part in the bonding but remain essentially in atomic orbitals. The 2*p* electrons are the only ones of suitable energy, but of the three 2*p* orbitals the only orbital of the correct symmetry is the one pointing along the H–F bond axis. If we take the

H–F bond axis as the x axis, then only the F $2p_x$ atomic orbital contributes to the bonding. (Readers should note that in many more advanced texts the bond axis is taken as the z axis, and this also applies to molecules such as N_2, so that the π bonding is then based on the interaction of p_x and p_y atomic orbitals. The choice of co-ordinate for the bond axis is arbitrary, and throughout this book we shall take the principal axis – as in a linear molecule – as the x axis, and the plane of a molecule such as BCl_3 (see p.91) as the xy plane.)

Figures 6.7(a) and *6.7(b)* show the effect of combining the hydrogen $1s$ atomic orbital with the fluorine $2p_x$ and $2p_y$ atomic orbitals, respectively. The overlap characteristic of the $2p_z$ atomic orbital is directly comparable with that of the $2p_y$ orbital.

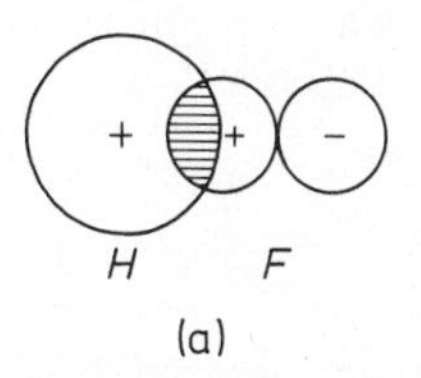

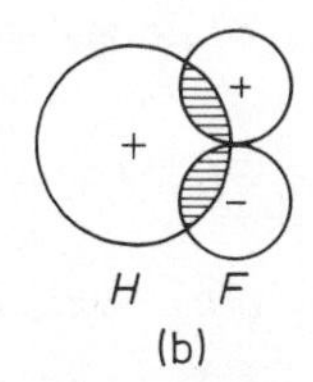

***Figure 6.7** Possible combination of hydrogen 1s and fluorine 2p orbitals (overlap indicated by horizontal shading)*

The $2p_x$ orbital gives a much greater overlap with the hydrogen $1s$ orbital for a given internuclear distance than does the $2p_y$ orbital, and whereas in *Figure 6.7(a)* the overlapping orbitals are of the same sign, in *Figure 6.7(b)* the s orbital is overlapped by both lobes of the p orbital, so that the overlap from the positive lobe of the p orbital is exactly balanced by that of the negative lobe. These overlaps cancel one another out, or in mathematical terms the overlap integral $\int \phi_{(H_{1s})} \phi_{(F_{2p_y})}$ is zero.

Thus the bonding between the hydrogen and fluorine atoms results almost entirely from the H($1s$) and F($2p_x$) electrons and the other electrons of fluorine remain in atomic orbitals. We can write the configuration of HF as

$$\mathrm{H}(1s) + \mathrm{F}(1s^2 2s^2 2p^5) \rightarrow \mathrm{HF}[K(2s)^2 (x\sigma)^2 (2p_y)^2 (2p_z)^2]$$

and the bonding $x\sigma$ orbital is the only molecular orbital. This configuration closely resembles that of the neon atom ($1s^2 2s^2 2p^6$) where all the orbitals are atomic.

Now the $x\sigma$ molecular orbital was obtained by combining the H($1s$) and F($2p_x$) atomic orbitals according to the wave function

$$\psi = c_1 \phi_{\mathrm{H}(1s)} + c_2 \phi_{\mathrm{F}(2p_x)}$$

This is often rewritten in the form

$$\psi = \phi_{\mathrm{H}(1s)} + \lambda \phi_{\mathrm{F}(2p_x)}$$

where λ replaces the coefficients c_1 and c_2. In homonuclear diatomic molecules $c_1 = c_2$, and $\lambda = 1$, but this is no longer true for heteronuclear diatomic molecules such as HF, because the electrons are no longer shared equally between the nuclei. Thus λ is greater than unity for HF, showing that the $F(2p_x)$ atomic orbital is contributing more than the $H(1s)$ orbital to the molecular wave function. λ is said to be a measure of the asymmetry or polarity of the bond.

The other hydrogen halides have similar single bond structures with an uneven sharing of the electron pair between the nuclei. In these HX molecules the chlorine, bromine and iodine atoms contribute respectively $3p_x$, $4p_x$ and $5p_x$ atomic orbitals.

6.6.2 MOLECULES SUCH AS NO AND CO

To a first approximation we can use the energy diagram of *Figure 6.6*. Carbon monoxide then has an identical configuration to the nitrogen molecule, and nitric oxide has one additional electron which is placed in the antibonding $v\pi$ orbital, i.e.

$$\mathrm{CO}[KK(z\sigma)^2(y\sigma)^2(x\sigma)^2(w\pi)^4]$$

$$\mathrm{NO}[KK(z\sigma)^2(y\sigma)^2(x\sigma)^2(w\pi)^4(v\pi)]$$

Because of the single antibonding electron, the nitric oxide molecule is paramagnetic, and the bonding is weaker than in the nitrogen molecule. Nitric oxide is unexpectedly stable for a molecule with an unpaired electron (i.e. a radical), and this stability is attributed to the electron being distributed over both atoms in a molecular orbital rather than localized on one atom.

The configuration of carbon monoxide is exactly the same as that of the nitrogen molecule, which implies a triple bond between the carbon and oxygen atoms. This description is not entirely adequate, however, because upon ionization, CO gives CO^+ and N_2 gives $N_2{}^+$, but whereas the loss of an electron (bonding) from N_2 leads to bond weakening, this is not the case with CO, since the bonding is stronger in CO^+, as inferred by a shorter internuclear distance and a larger vibrational frequency. This suggests that CO contains only a double bond, and that ionization results in more electronic charge being pulled into the internuclear region. A simple interpretation is that the molecular orbitals are considerably distorted towards the more electronegative oxygen atom (much as in HF), and that the ionization reduces this electronegativity difference.

Of course the extension of the bonding diagram for N_2 to the NO and CO molecules ignores the effect of the different atomic numbers, i.e. as we go from C to N to O so the energies of the analogous atomic orbitals are reduced because of the increasing attraction to the nucleus. Hence the energy of the oxygen 2*s* atomic orbital will be appreciably lower than that of the carbon 2*s* orbital. *Figure 6.8* illustrates the sort of diagram that emerges when the correct energies for the atomic orbitals are used, and it may be seen that the bonding $w\pi$ molecular orbitals are now lower in energy than the bonding $x\sigma$ orbital. The carbon and oxygen atomic orbitals no longer contribute equally to the molecular orbitals, the coefficients of the oxygen atomic orbitals being larger for the bonding molecular orbitals and smaller for the antibonding molecular orbitals. This confirms the qualitative picture we gave of the bonding electrons occupying orbitals that are more 'concentrated' on the oxygen atom.

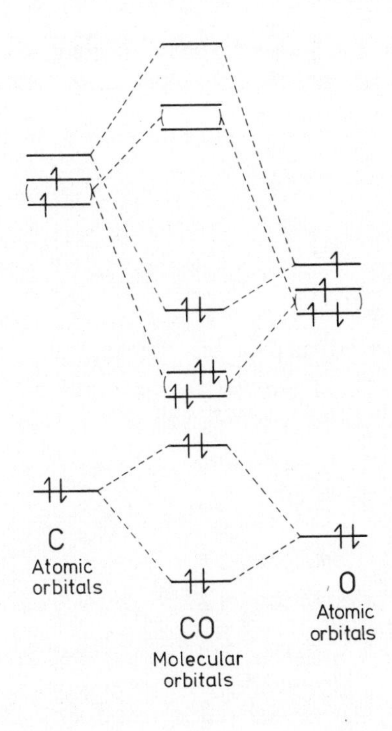

Figure 6.8 Molecular-orbital energy diagram for carbon monoxide

6.7 Bond Strength

The examples so far discussed in this chapter show that stable molecules are formed whenever there is an excess of bonding electrons. The excess of bonding electron pairs over antibonding electron pairs is called the 'bond order'. *Table 6.1* summarizes some experimental data on some of the molecules we have discussed.

We can see that there is a rough correlation between 'bond order' and bond dissociation energy in the two series N_2, NO and O_2, and H_2 and H_2^+.

Similarly, if we compare diatomic molecules involving second-row elements with their first-row analogues, we observe a decrease in bond dissociation energy. Thus the value for SiO is 774 kJ mol^{-1} (cf. 1071 for CO) and

Table 6.1 Relation of excess of bonding electrons to bonding strength for simple diatomic molecules

Molecule	*Pairs of bonding electrons*	*Pairs of antibonding electrons*	*Excess of bonding pairs*	*Bond dissociation energy/* kJ mol^{-1}
CO	4	1	3	1071.0
N_2	4	1	3	942.9
NO	4	1½	2½	627.6
O_2	4	2	2	493.9
H_2	1	0	1	458.5
H_2^+	½	0	½	269.6
He_2	1	1	0	—

for S_2 is 348 (cf. 494 for O_2). The electrons of higher principal quantum number evidently have a smaller bonding power, and this we attribute to their more diffuse nature and poorer overlap.

6.8 Reference

1. GLASSTONE, S., *Theoretical Chemistry*, 275, Van Nostrand, New York (1945)

For further details of the molecular-orbital method see, for example

GRAY, H.B., *Electrons and Chemical Bonding*, Benjamin, New York (1964)

7

THE VALENCE-BOND METHOD

7.1 Introduction

In the preceding chapter we discussed the molecular-orbital method for obtaining approximate solutions to the appropriate molecular Schrödinger equation. We considered the skeleton of the molecule (as given by the atomic nuclei, or nuclei plus inner shell electrons) and evaluated the molecular energy levels that could be occupied by electrons. These molecular energy levels, or orbitals, were readily obtained by taking a suitable linear combination of atomic orbitals. For clarity we restricted our discussion to diatomic molecules.

In this chapter we shall describe an alternative approximation approach, the so-called valence-bond method, which incorporates the idea of the pairing of electrons, with each pair of electrons linking just two nuclei. Initially the discussion will be concerned with σ-bonded homonuclear (H_2^+ and H_2) and heteronuclear (HCl) diatomic molecules, but later in the chapter we shall apply the method in principle to the problem of π electrons in molecules such as benzene, and introduce the concept of resonance. The problem of σ-bonded polyatomic molecules, such as ammonia and water, is deferred until Chapter 8. Here we shall see that the valence-bond approach, which involves the notion of hybridization, gives a very satisfactory framework for a discussion of the shapes of such molecules. At the same time we shall consider further the molecular-orbital approach in which the electrons can be placed in molecular orbitals embracing all the atomic nuclei.

7.2 The Hydrogen Molecule Ion

We will start with this molecule (see also p.54) because it is the simplest and involves only a single electron and two nuclei. It is convenient to consider the system initially with the two nuclei well separated, with the electron associated with only one of the nuclei; in other words we have a hydrogen atom and a proton. As the nuclei are gradually brought together they repel one another, since both are positively charged, but at the same time the electron becomes attracted to both nuclei and acts as a kind of 'molecular glue'. If this attractive force is greater than the internuclear repulsion force then a stable molecule will form.

Figure 7.1 shows how the overall energy of the system changes with internuclear distance. As the nuclei are brought together, so the attractive force becomes apparent and the energy decreases, corresponding to a

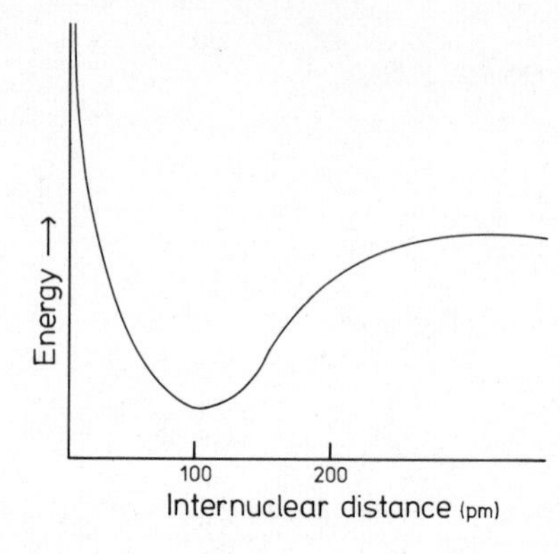

Figure 7.1 The hydrogen molecule ion: variation of energy with internuclear distance

more stable arrangement. As the internuclear distance decreases further, however, the internuclear repulsion becomes more important and eventually dominant, so that the curve passes through an energy minimum and then increases rapidly. The minimum energy in the curve corresponds to the normal internuclear, or bonding, distance (106 pm) in the H_2^+ molecule in the ground state.

The Schrödinger equation for the molecule H_2^+ is

$$\nabla^2\psi + \frac{8\pi^2 m}{h^2}\left(E + \frac{e^2}{r_a} + \frac{e^2}{r_b} - \frac{e^2}{R}\right)\psi = 0$$

This equation differs from that given for the hydrogen atom (p.27) only in that the expression for the potential energy V now contains three terms, $-e^2/r_a$ and $-e^2/r_b$, which arise from the attraction of the electron for the nuclei A and B, and e^2/R, which arises from the internuclear repulses. *Figure 7.2* shows the appropriate distances.

When the two nuclei in this system are very far apart, we have to consider two possible arrangements. In one arrangement the electron is associated entirely with the hydrogen nucleus A, nucleus B being a bare proton (we denote the arrangement, or structure, as $H_AH_B^+$ and describe it by a wave function ψ_I). In the other possible arrangement the electron is associated entirely with nucleus B and we assign a wave function ψ_{II} to describe the structure $H_A^+H_B$. These two structures correspond to states of equal energy, and if the nuclei were at an infinite distance apart, one would give an accurate description of the electronic standing wave associated with either

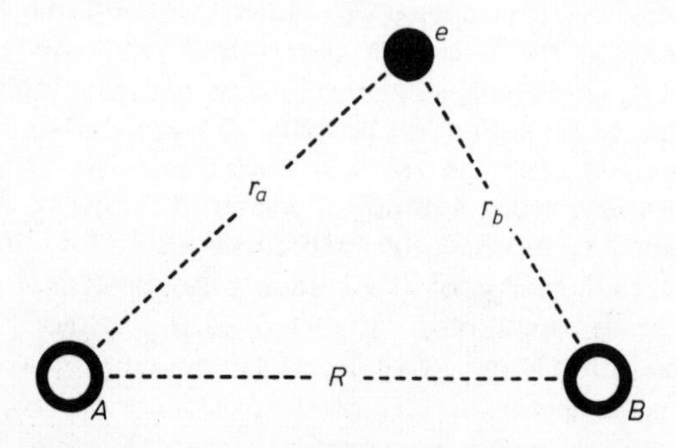

Figure 7.2 The hydrogen molecule ion: A and B represent the two nuclei (protons) and e denotes the single electron

proton A or proton B. When the two nuclei are brought together, however, neither structure adequately describes the situation, and we can obtain a better description by taking a linear combination of ψ_I and ψ_{II}, so that we describe the molecule by the function

$$\psi = C_I\psi_I + C_{II}\psi_{II}$$

Now ψ_I and ψ_{II} represent states of equal energy, so they contribute equally to ψ, C_I being equal to C_{II} or $-C_{II}$. Thus we can write down the two obvious linear combinations

$$\psi = \psi_I + \psi_{II}$$

and

$$\psi_- = \psi_I - \psi_{II}$$

where both C_I and C_{II} are taken as unity. These functions are modified by multiplying them by a normalization constant, $1/\sqrt{2}$ (see p.53), which ensures that for each function the value of $\int\psi^2 dv$ is unity when taken over the whole of space. Hence the two linear combinations may now be written

$$\psi_+ = 1/\sqrt{2} \times (\psi_I + \psi_{II})$$

and

$$\psi_- = 1/\sqrt{2} \times (\psi_I - \psi_{II})$$

Now the charge density is proportional to ψ^2 and hence

$$\psi_+^2 = \tfrac{1}{2}[\psi_I^2 + \psi_{II}^2 + 2\psi_I\psi_{II}]$$

while

$$\psi_-^2 = \tfrac{1}{2}[\psi_I^2 + \psi_{II}^2 - 2\psi_I\psi_{II}]$$

The ψ_+ function thus describes a state in which the charge density is greater than the sum of the separate charge densities, $\tfrac{1}{2}[\psi_I^2 + \psi_{II}^2]$, by an amount $\psi_I\psi_{II}$. These charge density distributions are shown in *Figure 7.3(a)* in which the broken lines represent the density functions ψ_I^2 and ψ_{II}^2, and the full lines ψ_+^2 and ψ_-^2; ψ_+^2 and ψ_-^2 can also be represented by contours – lines of equal electron density – as in *Figure 7.3(b)*. The diagrams for the ψ_+^2 function show clearly that the increased electron charge density associated with this linear combination is concentrated in the region between the nuclei, whereas the charge cloud is pushed away from this region in the ψ_- combination.

Figure 7.4 shows how the energy varies with the internuclear distance for both ψ_+ and ψ_-. The function ψ_+ is evidently to be associated with the stable or 'ground' state of the H_2^+ molecule, since the curve shows an energy minimum at an internuclear separation or 'bond length' of r_0. The curve labelled ψ_- shows a steady increase of energy as the nuclei approach each other and no stable molecule is formed.

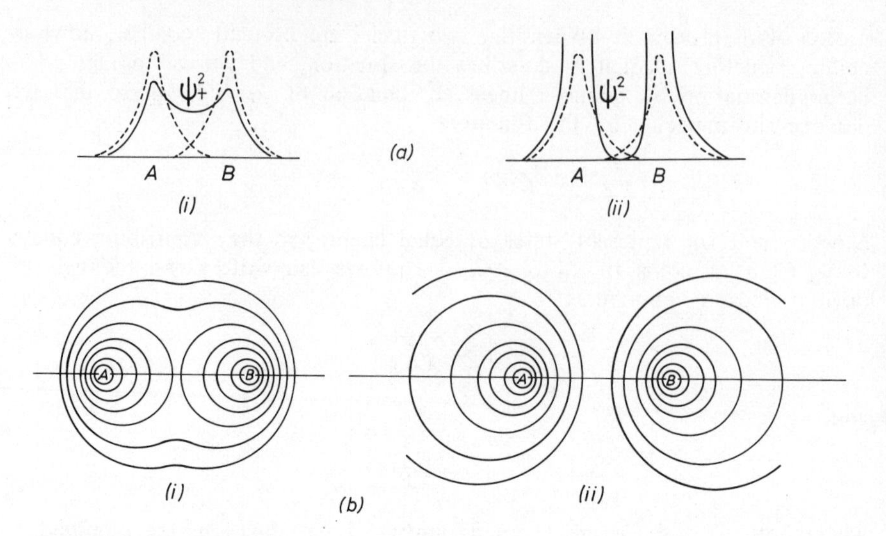

Figure 7.3 The hydrogen molecule ion: (a) electron distribution functions (full lines represent ψ_+^2 or ψ_-^2; broken lines represent ψ_I^2 or ψ_{II}^2); (b) contours of equal electron density for ψ_+^2 and ψ_-^2

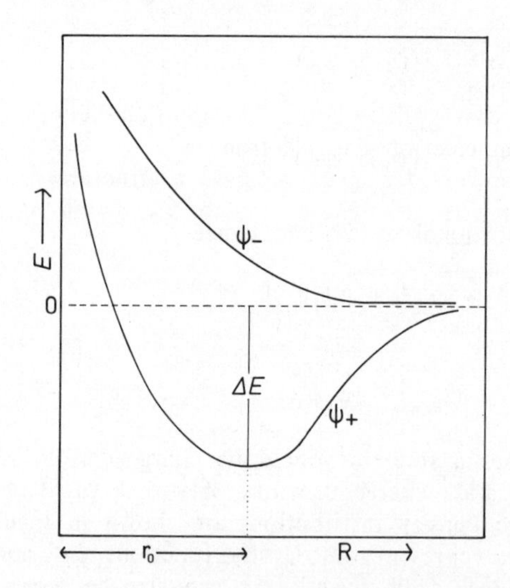

Figure 7.4 The hydrogen molecule ion: variation of energy with internuclear distance

In the case of the hydrogen molecule ion both the simple molecular-orbital and the valence-bond treatments lead to identical values for the energy and the bond length. Since ψ_I represents a structure in which the electron is associated only with nucleus A, it is an atomic orbital of A and identical with one of the component atomic orbitals (ϕ_A) of the LCAO treatment; similarly, ψ_{II} is identical with ϕ_B. Both methods use the linear combination technique, but the valence-bond treatment considers hypothetical 'structures', whereas the emphasis is on the electron in the molecular-orbital approach. The difference between the two methods will become more apparent when we consider more complicated molecules, for then the component wave functions of the linear combinations will no longer be the same.

7.3 The Hydrogen Molecule

The potential energy term V in the Schrödinger equation for the hydrogen molecule is

$$-e^2/r_{a_1} - e^2/r_{b_1} - e^2/r_{a_2} - e^2/r_{b_2} + e^2/R_{AB} + e^2/r_{12}$$

where the distances r_{a_1}, etc., are those shown in *Figure 7.5*. The four negative terms correspond to the attraction of the electrons 1 and 2 for

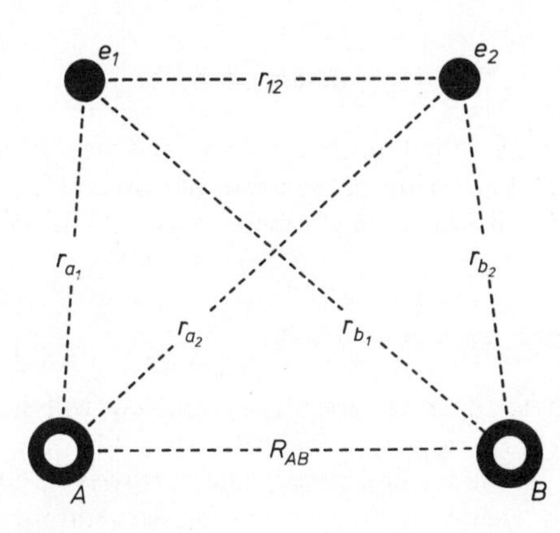

Figure 7.5 The hydrogen molecule, showing distances between the two nuclei (A and B) and the two electrons (e_1 and e_2)

the protons A and B, and the two positive terms to repulsion between proton and proton, and between electron and electron.

When the nuclei A and B are very far apart we have a structure $H_A(1) + H_B(2)$, where $H_A(1)$ represents electron 1 associated with nucleus A, and $H_B(2)$ represents electron 2 associated with nucleus B. We can describe this system of two hydrogen atoms by a wave function ψ_I. An equally good description is afforded by a structure in which electron 2 is associated with nucleus A, and electron 1 with nucleus B, since the two electrons are indistinguishable, and this state, $H_A(2) + H_B(1)$, is described by the wave function ψ_{II}. As the atoms are brought together, their standing waves will interact, and a new wave motion will result which Heitler and London suggested could be described as a linear combination of ψ_I and ψ_{II}, i.e.

$$\psi = N[c_I\psi_I + c_{II}\psi_{II}]$$

Here again, ψ_I and ψ_{II} represent states of the same energy and they will be equally important in the linear combination; in fact $c_I = \pm c_{II}$, giving

$$\psi_+ = Nc_I(\psi_I + \psi_{II})$$

and

$$\psi_- = Nc_I(\psi_I - \psi_{II})$$

A plot of energy against the internuclear distance R_{AB} shows a minimum for the ψ_+ combination, corresponding to molecule formation. The calculated dissociation energy (303 kJ mol^{-1}) and bond length (87 pm) are roughly comparable with the experimentally determined values of 456 kJ mol^{-1} and 74.1 pm.

The structures represented by ψ_I and ψ_{II} are referred to as covalent structures. The Heitler–London method can be improved, however, by considering two additional structures in which both electrons are associated with only one nucleus (either A or B). Thus additional structures are the ionic forms

$$H_A{}^+ + H_B{}^- \quad \{\text{or } H_A + H_B(1,2)\} \text{ described by } \psi_{III}$$

and

$$H_A{}^- + H_B{}^+ \quad \{\text{or } H_A(1,2) + H_B\} \text{ described by } \psi_{IV}$$

$H_A(1,2)$ represents the arrangement in which both electrons are associated solely with nucleus A, and $H_B(1,2)$ that with both electrons associated with nucleus B. Thus we can write down a new general form of linear combination

$$\psi = c_I\psi_I + c_{II}\psi_{II} + c_{III}\psi_{III} + c_{IV}\psi_{IV}$$

Since ψ_I and ψ_{II} represent structures of equal energy, c_I and c_{II} will have the same numerical value; similarly, ψ_{III} and ψ_{IV} describe structures of equal energy and so c_{III} and c_{IV} will have the same value. However, the ionic structures are much higher in energy than the covalent structures, so the value of c_{III} and c_{IV} will be much smaller than that of c_I and c_{II}. An approach of this kind has been used and the variation method applied to determine the 'best' values of the mixing coefficients (c_I, c_{II}, c_{III} and c_{IV}), i.e. the values corresponding to the state of lowest energy. These calculations gave a dissociation energy of 310 kJ mol^{-1} and an equilibrium bond length of 88.4 pm for c_I and c_{II} equal to unity and c_{III} (and c_{IV}) = 0.158.

7.4 Comparison of the Molecular-orbital and Valence-bond Methods

We can now usefully compare the molecular-orbital and the valence-bond descriptions of the hydrogen molecule, and to do this we must examine the component wave functions in more detail. Consider first of all the molecular-orbital method. If we neglect inter-electron repulsion, the probability of finding electron 1 in a volume dv_1(= $\psi_1{}^2 dv_1$) is independent of the probability of finding electron 2 in a volume dv_2(= $\psi_2{}^2 dv_2$), so that the probability that electron 1 is in the volume dv_1 while, simultaneously, electron 2 is in volume dv_2, is the product of the individual probabilities (= $\psi_1{}^2 dv_1 \cdot \psi_2{}^2 dv_2$). The wave function ψ_+, which describes the molecule, is therefore given by the product $\psi_1\psi_2$, i.e. the product of the functions for the individual electrons. Now we have seen (p.53) that in the molecular-orbital treatment, the wave function for a bonding electron in a molecule is given by the sum of atomic orbitals. Thus

$$\psi_1 = \phi_{A(1)} + \phi_{B(1)}$$

where $\phi_{A(1)}$ and $\phi_{B(1)}$ represent systems in which electron 1 is associated with nuclei A and B, respectively. (We are simplifying the expressions by omitting the normalizing constant and the coefficients c_A and c_B.) Similarly for electron 2

$$\psi_2 = \phi_{A(2)} + \phi_{B(2)}$$

The overall molecular wave function is, therefore

$$\begin{aligned}\psi_+ = \psi_1\psi_2 &= (\phi_{A(1)} + \phi_{B(1)})(\phi_{A(2)} + \phi_{B(2)}) \\ &= \phi_{A(1)}\phi_{A(2)} + \phi_{B(1)}\phi_{B(2)} \\ &\quad + \phi_{A(1)}\phi_{B(2)} + \phi_{B(1)}\phi_{A(2)} \qquad (7.1)\end{aligned}$$

Here the term $\phi_{A(1)}\phi_{A(2)}$ represents a structure in which both electrons are on nucleus A, thus producing ions H_A^- and H_B^+, while $\phi_{B(1)}\phi_{B(2)}$ represents the ionic form $H_A^+H_B^-$.

In the valence-bond treatment, each constituent function of the linear combination represents a particular arrangement of protons and electrons, so that ψ_I, for example, refers to the structure in which electrons 1 and 2 are associated with nuclei A and B, respectively. Here again, the probability of finding electron 1 near nucleus A and, simultaneously, electron 2 on nucleus B, is given by the product of the individual probabilities, so that ψ_I can be written $\phi_{A(1)}\phi_{B(2)}$ and ψ_{II} can be written $\phi_{A(2)}\phi_{B(1)}$. Hence, taking the linear combination

$$\psi_+ = \phi_{A(1)}\phi_{B(2)} + \phi_{A(2)}\phi_{B(1)} \qquad (7.2)$$

Equation 7.2 differs from equation 7.1 in the absence of the 'ionic' terms $\phi_{A(1)}\phi_{A(2)}$ and $\phi_{B(1)}\phi_{B(2)}$. The simple (Heitler–London) valence-bond theory assumes, in effect, that these structures will not be important, because the mutual repulsion between electrons would reduce the probability that they would be found simultaneously close together on the same nucleus. The molecular-orbital theory, on the other hand, neglects the effect of interelectronic repulsion, and gives equal weight to the ionic and non-ionic (i.e. covalent) terms. It is clear that neither method will give reliable results unless it is suitably modified, because each describes limiting conditions, and it can indeed be shown that as the two methods are improved they 'converge' and finally become equivalent. The modifications involve the consideration of more elaborate wave functions, with correspondingly more elaborate calculations.

The treatment of the hydrogen molecule by Kolos and Roothaan (1960) uses a function containing 50 terms, and calculations of this kind have only become feasible because of the development of high-speed electronic computers. *Table 7.1* shows values for the internuclear distance and the binding energy for H_2^+ and H_2 as calculated by a number of workers, and as determined experimentally. These values show that the two-electron bond in H_2 is much stronger and shorter than the one-electron bond in H_2^+.

Table 7.1 Bond lengths and binding energies for H_2^+ and H_2

Method	H_2^+		H_2	
	D/kJ mol^{-1}	R/pm	D/kJ mol^{-1}	R/pm
Experimental	269	106	458	74.1
Simple LCAO	170	132	259	85.0
James (1935)	267	–	–	–
Heitler–London (1927)	–	–	303	86.9
Weinbaum (1933)	–	–	310	88.4
Kolos and Roothaan (1960)	–	–	458	74.1

These results also show that complete accuracy is only obtained by abandoning the simple LCAO functions, but, as Coulson has pointed out, this also involves abandoning conventional chemical concepts and simple pictorial quality in the results. The great advantage of the admittedly inaccurate linear combination method is that it can be related to the chemist's familiar pictures of bond diagrams and structural formulae.

7.5 Heteronuclear Diatomic Molecules

In the last chapter we considered the heteronuclear diatomic molecule HF in some detail, and considered the two bonding electrons to be in a molecular orbital ($x\sigma$) obtained by taking a linear combination of the hydrogen $1s$ and fluorine $2p_x$ atomic orbitals. Allowance was made for the difference between the two nuclei by assigning different values for C_1 and C_2 in the molecular wave function

$$\psi = C_1\phi_{H(1s)} + C_2\phi_{F(2p_x)}$$

This was rewritten

$$\psi = \phi_{H(1s)} + \lambda\phi_{F(2p_x)}$$

which contains a single coefficient λ (= C_2/C_1) that is a measure of the polarity of the HF bond. λ takes a value greater than unity, showing that the electronic charge cloud is concentrated most on the fluorine end of the bond.

In the valence-bond description the two nuclei are also held together by two bonding electrons, but the system is described by an overall wave equation that incorporates expressions describing covalent and ionic structures:

$$\psi = \psi_{\text{covalent}} + a\psi_{\text{ionic}}$$

We used a similar type of expression for the hydrogen molecule (p.72), where ψ_{covalent} was ($\psi_{I} + \psi_{II}$) and ψ_{ionic} was ($\psi_{III} + \psi_{IV}$); in this instance there were equal contributions from ψ_{III} and ψ_{IV} which described the two possible structures with both electrons associated with only one nucleus. There are theoretically four similar structures for HF that can contribute, two covalent and two ionic:

H(1)	+ F(2)	described by ψ_{I}
H(2)	+ F(1)	described by ψ_{II}
H	+ F(1,2) (i.e. H^+F^-)	described by ψ_{III}
H(1,2)	+ F (i.e. H^-F^+)	described by ψ_{IV}

Commonsense tells us that the last structure will contribute very little to the description, so we need consider only the first three structures. $\psi_{covalent}$ is thus the sum of ψ_{I} and ψ_{II}, and ψ_{ionic} is ψ_{III}.

The wave functions can be written out more fully so that we have

$$\psi = [\phi_{H(1s)}(1)\phi_{F(2p_x)}(2) + \phi_{H(1s)}(2)\phi_{F(2p_x)}(1)] + a[\phi_{F(2p_x)}(1)\phi_{F(2p_x)}(2)]$$

Because fluorine is much more electronegative than hydrogen (see Chapter 9) the charge cloud of the σ bond will be distorted towards the fluorine end of the bond, and this is allowed for in the overall wave function by the incorporation of the ionic term. The value of a is clearly related to the asymmetry of the charge cloud and hence to the polarity or 'degree of ionicity' of the bond. As the covalent and ionic contributions are weighted in the ratio of $1 : a^2$, the percentage ionic character may be represented by the expression

$$\frac{100a^2}{1 + a^2}$$

Other heteronuclear diatomic molecules can be described in much the same way. With molecules such as CO and NO, of course, we will be concerned with both σ and π bonds, and the wave functions involved in $\psi_{covalent}$ and ψ_{ionic} must take account of this.

7.6 Resonance

So far we have tried to avoid using the word 'resonance', which has long been associated with the valence-bond theory. We have described the hydrogen molecule by a wave function which is a linear combination of two functions ψ_{I} and ψ_{II}, each of which describes a hypothetical state or structure, i.e. $H_A(1) + H_B(2)$ and $H_A(2) + H_B(1)$. The actual state of the molecule is often described as being a resonance hybrid of the states represented by ψ_{I} and ψ_{II} or, more simply, that there is resonance between the structures $H_A(1) + H_B(2)$ and $H_A(2) + H_B(1)$. The danger of using this terminology is that it is almost impossible to avoid giving the impression that the hypothetical structures have a real molecular existence. In the case quoted, the structures $H_A(1) + H_B(2)$ and $H_A(2) + H_B(1)$ correspond to separated atoms, and do therefore have a real existence, but we must not imagine that there is some kind of rapid exchange of electron between the nuclei, or some kind of equilibrium between the structures.

In the more complicated molecules that we shall be discussing shortly, the contributing structures will not, in general, have any real existence. They are, in fact, merely a convenient device for getting an approximate solution of a complicated wave equation, but they also have the great practical advantage in that they enable the chemist to picture molecule formation in terms of conventional and familiar bond diagrams, even though the actual state of the molecule cannot be described directly by a single conventional structural formula.

The selection of hypothetical structures from which the true structure is compounded by taking a linear combination is not an arbitrary matter. With HF we saw that H^+F^- was important, but not H^-F^+. If possible structures have wave functions ψ_I, ψ_{II}, . . . , etc., the important ones will be those for which the coefficient c is large in the combination $\psi = N[c_I\psi_I + c_{II}\psi_{II} + \ldots]$ and a detailed study of the expression for the energy associated with the function ψ shows that effective contributions are made by structures that

(a) have similar energies
(b) have, approximately, the same relative positions of the nuclei
(c) have the same numbers of unpaired electrons

These conditions are, in fact, developed from those already mentioned on p.54 where we are discussing linear combinations in terms of the molecular-orbital theory. We saw there that effective combination only occurs when the component functions represent states of almost equal energy when their orbitals overlap to a considerable extent, and when the orbitals have the same symmetry with respect to a molecular axis.

These points can be illustrated by discussing a specific example – the nitromethane molecule – for which two alternative bond diagrams can be drawn if we regard the nuclei as occupying fixed positions in space:

$$H_3C{-}\overset{+}{N}{\overset{\textstyle O}{\underset{\textstyle O^-}{\diagup\!\!\diagup\atop\diagdown}}}$$ Structure I

$$H_3C{-}\overset{+}{N}{\overset{\textstyle O^-}{\underset{\textstyle O}{\diagup\atop\diagdown\!\!\diagdown}}}$$ Structure II

From a study of a range of nitrogen–oxygen molecules it has been established that a nitrogen–oxygen double bond is much shorter than a single bond (see Chapter 11), but experimental work on nitromethane has shown that both nitrogen–oxygen bonds are of the same length (122 pm). The actual state of the molecule is described in terms of hypothetical structures I and II, and we write the wave function for the molecule as the linear combination

$$\psi = N[c_I\psi_I + c_{II}\psi_{II}]$$

We can picture this as a superposing and blending of the structures I and II. Each structure contributes equally to the combination, since the energies of each are the same, they have the same relative positions of the nuclei, and the same number (zero) of unpaired electrons. In Chapter 9 we shall see how molecular-orbital theory can be used to give a more adequate picture of bonding in this molecule.

Our second example will be carbon dioxide, which has long been given the structural formula O=C=O. This seems to satisfy conventional valency requirements, but the carbon–oxygen bond length is found to be 115 pm, whereas the length of a normal carbon–oxygen double bond (as in ketones) is 122 pm, and that calculated for triple bond is 110 pm. The bond in carbon dioxide thus seems to be intermediate in character between a double and triple bond, and this is explained by describing the structure in terms of three hypothetical forms:

$$\mathrm{O{=}C{=}O} \qquad \psi_{\mathrm{I}}$$

$$\overset{+}{\mathrm{O}}{\equiv}\mathrm{C}{-}\overset{-}{\mathrm{O}} \qquad \psi_{\mathrm{II}}$$

$$\overset{-}{\mathrm{O}}{-}\mathrm{C}{\equiv}\overset{+}{\mathrm{O}} \qquad \psi_{\mathrm{III}}$$

so that $\psi = N[c_{\mathrm{I}}\psi_{\mathrm{I}} + c_{\mathrm{II}}\psi_{\mathrm{II}} + c_{\mathrm{III}}\psi_{\mathrm{III}}]$ describes the actual state of the molecule.

The inadequacy of the O=C=O formulation is also demonstrated by energy considerations. The energy of a carbon–oxygen double bond is known to be 732 kJ $\mathrm{mol^{-1}}$, so that the heat of formation of the molecule O=C=O should be approximately twice this value, i.e. 1464 kJ $\mathrm{mol^{-1}}$; in fact, the measured heat of formation is 1602 kJ $\mathrm{mol^{-1}}$ and the difference between these two values, 138 kJ $\mathrm{mol^{-1}}$, represents the extent to which the actual structure is more stable than that represented by the simple formula O=C=O. This difference in energy between the actual molecule and the most stable of the hypothetical structures is called the resonance energy (see *Figure 7.6*).

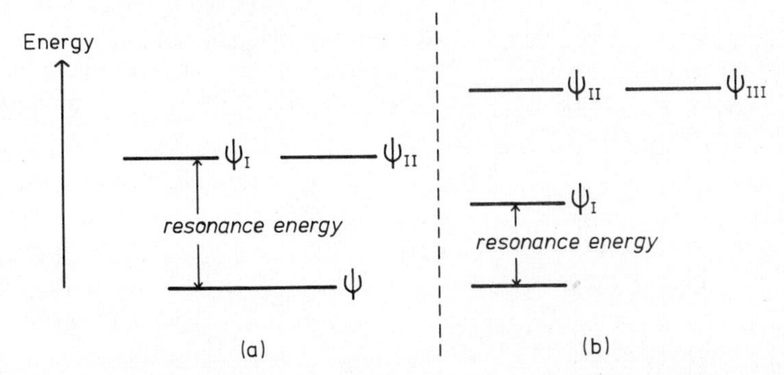

Figure 7.6 Valence-bond energy diagram for (a) hydrogen, and (b) carbon dioxide

The benzene molecule affords perhaps the best example of the way in which the resonance theory can explain some apparently very curious facts – namely, that benzene is extremely stable chemically, although the conventional structural formula introduced by Kekulé includes three double bonds, that the molecule is completely planar, and that the carbon–carbon bonds all have the same length. We can write down five conventional structural formulae for the molecule

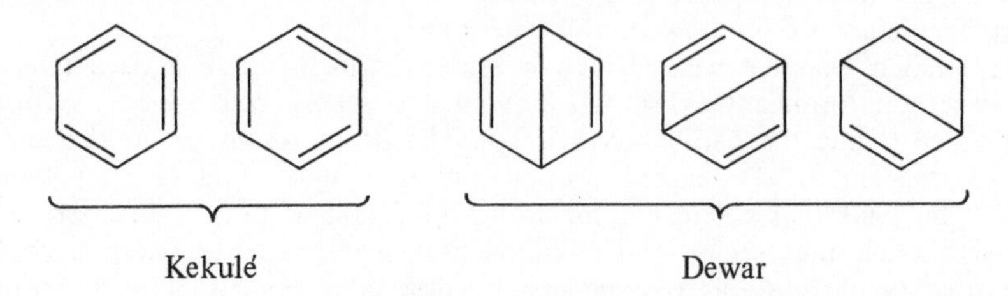

each of which can be represented as before by an appropriate wave function ψ_I, $\psi_{II}, \ldots$, etc. The actual structure of the benzene molecule is then described by a wave function ψ given by

$$\psi = c_I\psi_I + c_{II}\psi_{II} + c_{III}\psi_{III} + c_{IV}\psi_{IV} + c_V\psi_V$$

in which the contributions of the two Kekulé forms will be more important than those of the Dewar forms. (The Dewar structures will be less stable, i.e. of higher energy, because of the weak long bonds across the ring.) The bonds between adjacent carbon atoms in the resulting resonance hybrid are now all equivalent, being neither single nor double but being 'flavoured' with the characteristics of each. In fact, the measured carbon–carbon distance in benzene is 139 pm, whereas the single bond length measured in the saturated hydrocarbons is 154 pm and the double bond length in ethene is 132 pm.

7.7 Resonance: Some Misconceptions and Some Guiding Principles

It is important to stress that the structures which are written down to describe molecules, and which are superposed in the valence-bond treatment, have no real existence. Benzene does not contain equal proportions of molecules with structures

and

nor is there a tautomeric equilibrium

set up, in which electrons oscillate between one position and another, nor can it be said that benzene exists for a certain fraction of time in the form

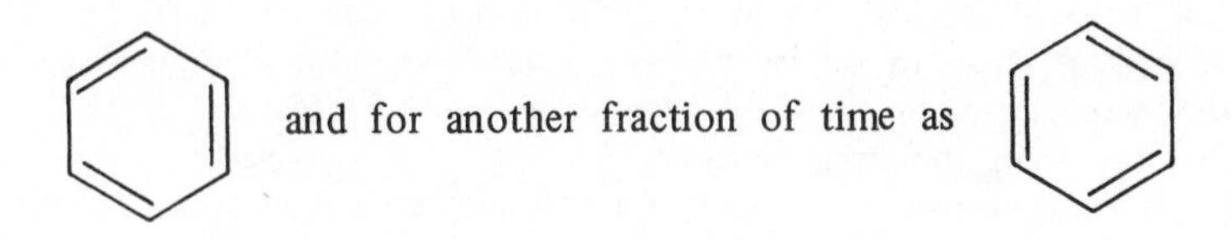

and for another fraction of time as

The molecule itself has a single structure which cannot be described directly by a conventional bond diagram.

It may help clarify the point if we use two analogies. The usual one is to compare the true structure with a mule, which is described in terms of two other beasts, a donkey and a horse. Thus the mule is an animal in its own right but it has the characteristics of both a horse and a donkey; however, it is never either of these. The difficulty of this analogy is that the true 'structure' (the mule) is described in terms of two other *real* 'structures' (the horse and the donkey), whereas a molecule (such as benzene) is usually described in terms of two hypothetical structures (e.g. Kekulé forms) which have no normal existence but which can be readily pictured and represented by simple bond diagrams. Accordingly a better analogy is perhaps one in which the actual structure is compared to a rhinoceros. This is a real animal but it does not live in this country, except in zoos, so if we wished to describe it to a child we could do so in terms of two mythical – or hypothetical – creatures, the unicorn and dragon. The child is familiar with these imaginary creatures and can therefore imagine what a rhinoceros would be like if it is said to have the characteristics of both the unicorn and the dragon.

Resonance theory has been in danger of falling into disrepute because of the uncritical way in which the concept has been applied. It is only too easy to explain away difficulties in interpreting the behaviour of molecules by inventing a large number of improbable structural formulae and saying that the molecule is a hybrid constructed from these forms. The hypothetical structures that are used to form the linear combination must always obey the rules formulated on p.76. Thus the rule that the structures must represent the same relative position of the nuclei illustrates the fundamental difference between resonance and tautomerism. A molecule such as acetone, for instance, exists in the so-called keto and enol forms:

$$(CH_3)_2C{=}O \qquad\qquad CH_3(CH_2{=})C{-}OH$$

keto   enol

which differ not only in the allocation of the bonding electrons but also in the position of the hydrogen atom – bonded to the carbon in the keto form and to the oxygen in the enol form. There is an actual equilibrium between the two tautomeric forms, which are definite entities and can usually (always in theory) be separated from one another. These tautomeric forms do exist and are in equilibrium with one another; the amount of each present depends on its energy content, so with acetone for instance the enol form is present in only very small amounts because it has a much higher energy than the keto form.

The fact that tautomers actually exist, while resonance forms do not, can be further illustrated by a consideration of dipole moments. Thus nitrous oxide may be considered as a hybrid of the two forms

$$\overset{-}{\bar{N}}=\overset{+}{N}=O$$

$$N\equiv\overset{+}{N}-\overset{-}{\bar{O}}$$

The dipoles of these contributing structures have opposed orientation, so the resultant dipole of the actual molecule is very small. If these two forms were actual tautomers in equilibrium, then both structures would align themselves in the field applied during the course of the measurement of the dipole moment, and the resultant dipole would be quite high, being somewhere between the individual tautomer dipoles (according to the relative proportions of the tautomers).

The rule that the contributing structures must be of comparable energy can be illustrated by benzene, where the Dewar forms have a much higher energy, and are therefore less important, than the more stable Kekulé forms in the combination that represents the actual molecule. In the same way, the simple Heitler–London treatment of the hydrogen molecule can be modified to include contributions from ionic structures, such as $H_A(1,2) + H_B$ and $H_A + H_B(1,2)$, but these, being of higher energy, contribute less to the actual structure than the forms $H_A(1) + H_B(2)$ and $H_A(2) + H_B(1)$.

8

DIRECTED VALENCY

8.1 Shapes of Molecules Formed by First-row Elements

Up until now we have been concerned primarily with the nature of bonding in diatomic molecules, although our description of resonance in the last chapter made certain assumptions about molecular shape. We now wish to extend the discussion of bonding to molecules in which one central atom is linked to more than one other atom, and we shall see that the description can be in terms of either the molecular-orbital or valence-bond descriptions. Before these alternative approaches are discussed in detail, however, we can go quite a long way in predicting the shape of covalent molecules if we simply assume that the sets of electron pairs tend to get as far away from one another as possible. Thus in the series of chlorides $BeCl_2$ (gas phase), BCl_3 and CCl_4, we may assume that the chlorine atoms are linked to beryllium, boron or carbon by simple two-electron bonds, so that mutual repulsion between these bonding electron pairs should give the molecules linear, trigonal-planar and tetrahedral shapes, respectively. This is illustrated in *Figure 8.1*.

This simple interpretation of molecular shape on the basis of repulsion between electron pairs has been developed by a number of workers to cover covalent compounds of the first-row elements, and to a more limited extent to compounds of other elements. However, we must take into

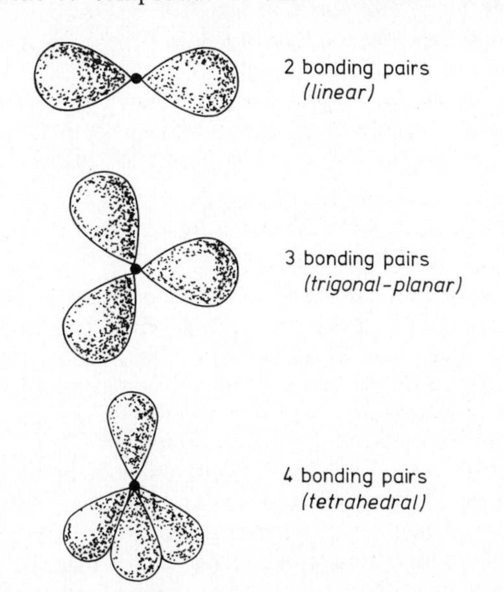

Figure 8.1 Electron-pair repulsion

account all pairs of electrons, and this can be illustrated by reference to the sequence of hydrides CH_4, NH_3 and H_2O. The experimental fact is that the bond angles in these molecules are 109.5°, 107.1° and 104.5°, respectively. Now CH_4 would be expected to have regular tetrahedral HCH angles since in this way the four bonding pairs of electrons would get as far apart as possible. We would expect the ammonium ion, $[NH_4]^+$, to have the same tetrahedral shape, because once again there are four bonding pairs; formally we can regard this ion as based on N^+ which is isoelectronic with the neutral carbon atom, i.e. contains the same number of electrons. The ammonia molecule, NH_3, can be thought of as arising from the $[NH_4]^+$ by the removal of a proton. This will leave the nitrogen atom with three bonding pairs of electrons linking it with the three hydrogen atoms and a lone-pair of electrons (once responsible for holding the fourth hydrogen atom but now restrained only by the nitrogen nucleus). Once again the four pairs of electrons would mutually repel one another, giving a roughly tetrahedral distribution, but since the lone pair of electrons is no longer attracted to two nuclei we would expect it to collapse a little towards the nitrogen nucleus and hence occupy rather more volume than any one of the bonding pairs. The result of the mutual repul sion would be that the bonding pairs would be pushed together somewhat, giving HNH angles rather below the tetrahedral value. It must be emphasized that although we think of the ammonia molecule as 'pseudo-tetrahedral' in terms of all the pairs of electrons, the molecular shape is described as pyramidal because the experimental measurements tell us only where the nuclei are.

The simple approach we have used to account for the shape of the ammonia molecule can be readily extended to the water molecule by a similar sort of argument. Thus NH_3 and H_3O^+ are isoelectronic and should have the same pyramidal shape ('tetrahedral' arrangement of three bonding pairs and one long pair), and we can now 'remove' a proton from H_3O^+ to give H_2O with two bonding pairs and two lone pairs of electrons. Again the lone pairs will occupy more room than bonding pairs with a consequent forcing together of the two O–H bonds.

Summing up, the progressive reduction of bond angles along the sequence CH_4, NH_3, H_2O can be explained on the assumption that repulsion between electron pairs increases in the order

$$\text{bonded pair–bonded pair} < \text{bonded pair–lone pair} < \text{lone pair–lone pair}$$

The argument can be further extended to HF, and the isoelectronic OH^- ion, where there is only one bond but three lone pairs of electrons, so that the distribution of the four electron pairs is again 'tetrahedral'. This description of diatomic species may appear to be rather laboured and artificial, but we shall see in Chapter 10 (p.180) that the approach is of value in understanding the nature of hydrogen bonding in the zig-zag polymers $(HF)_n$.

Of course, we need not restrict our discussion to hydrides, and in Chapter 10 we shall see that all covalent compounds of carbon, nitrogen, oxygen and fluorine which contain only σ bonds can be considered on the basis of the 'tetrahedral' arrangement of bonding and non-bonding (lone) pairs of electrons. *Table 8.1* summarizes the position.

Table 8.1 Arrangements of electron pairs for first row elements*

Linear	*Trigonal planar*	*Tetrahedral*			
		No lone pair	*1 lone pair*	*2 lone pairs*	*3 lone pairs*
—Be—	>B—	Be^{2-}	N	N^-	O^-
—C≡	>C=	B^-	O^+	O	F
=C=	$>N^+=$	C			
		N^+			

*A bonding pair of electrons is represented by a full line, and a lone pair of electrons by a dotted line

Beryllium and boron use all their electrons to form bonds, so that we do not have lone pairs of electrons to worry about. Hence BeX_2 and BX_3 molecules adopt linear and trigonal-planar configurations, respectively. However, there is a considerable tendency for beryllium and boron to achieve the higher tetrahedral symmetry by acquiring additional electron pairs from other molecules. Thus BF_3, for instance, acts as an electron acceptor (Lewis acid) and accepts an electron pair from suitable donors (Lewis bases) such as the ammonia molecule, forming a donor–acceptor or co-ordination compound BF_3,NH_3, in which boron has a tetrahedral configuration and is linked through four bonding pairs (to the nitrogen and the three fluorine atoms). *Figure 8.2* shows how we can picture this bonding. Thus as the lone pair of electrons of the ammonia molecule approaches the planar BF_3 molecule there is repulsion between the lone pair and the three pairs of electrons in the B–F bonds. Hence the planar BF_3 unit turns into a pyramid, and together with the nitrogen atom gives an approximately tetrahedral bond distribution about the boron atom. Other 'tetrahedral' molecules arise through the co-ordination of two donor molecules to simple beryllium compounds; typical examples include $BeCl_2,2O(C_2H_5)_2$ and $[BeCl_4]^{2-}$.

Up until now this discussion has been deliberately restricted to molecules containing only σ bonds (and lone pairs of electrons), but many molecules have both σ and π bonds, although we shall see that π bonds play no part in determining the gross stereochemistry. This can be illustrated especially well by reference to the compounds of carbon. Thus

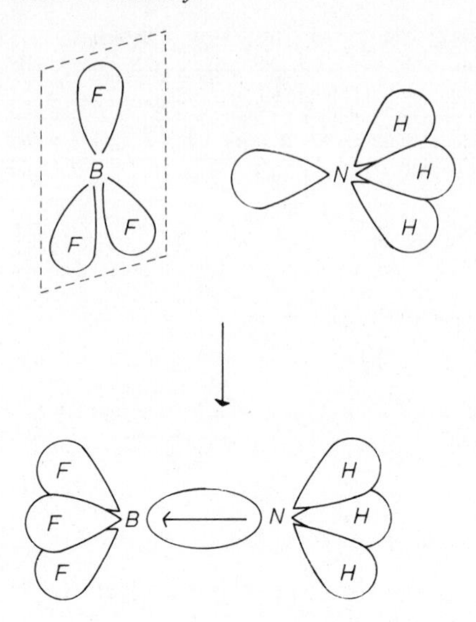

Figure 8.2 Orbital interpretation of the formation of BF_3,NH_3

although a tetrahedral bonding arrangement is found for all saturated compounds of carbon, there is a lower symmetry when carbon is bonded to fewer than four atoms. Bonding to three atoms only, as in say $COCl_2$ or C_2H_4, results in a trigonal-planar arrangement about the carbon atom, i.e.

```
Cl                H       H
  \                \     /
   C=O              C=C
  /                /     \
Cl                H       H
```

While this multiple bonding does not change the planar shape of the molecule, we shall see later (Chapter 10) that the bond angles may differ appreciably from 120° because the electrons in the multiple bond will take up more room than electrons in a single bond. As an example, the Cl–C–Cl bond in $COCl_2$ is reduced to 111°.

When carbon is bonded to only two atoms, it has a linear configuration, with two σ bonds and two π bonds. In such circumstances there may be either =C= or –C≡ skeletons, as typified by such molecules as CO_2 and HCN, respectively:

O=C=O H–C≡N

Whereas first-row elements cannot form more than four covalent bonds, this can increase to six and even more bonds with elements of subsequent rows, so that with these heavier elements we must expect other configurations to be involved besides linear, trigonal-planar and tetrahedral. In particular we shall see that a trigonal-bipyramidal arrangement is likely with five electron pairs and an octahedral arrangement with six electron pairs. These and other more complex configurations will be discussed in some detail in Chapters 10 and 11.

Now that we have seen how the shapes of many simple molecules can be rationalized on the basis of the mutual repulsion of electron pairs, we can take a more sophisticated look at the various ways in which the bonding can be described.

8.2 The Criterion of Maximum Overlapping

In our discussion of the hydrogen molecule by the molecular-orbital and valence-bond methods, we saw that bond formation was described in terms of the overlapping of atomic orbitals from the two hydrogen atoms. This concept is of great importance in discussing directed valency and we must now consider it more fully. Both approximation methods give an expression for the energy of the stable bond which involves a definite integral of the form $\int \phi_A \phi_B dv$, where the integration is taken over the whole of space. When ϕ_A and ϕ_B are small, the integral will be very small; it only becomes significant in regions where ϕ_A and ϕ_B have appreciable values at the same instant – that is when the atomic orbitals occupy the same region of space.

For this reason, $\int \phi_A \phi_B dv$ is called the overlap integral. We also saw in Chapter 6, however, that overlapping of atomic orbitals only results in stable bond formation if certain other conditions are satisfied, viz. that the atomic orbitals are of similar energy and appropriate symmetry. The latter limitations arise because certain orbitals (e.g. *p*, *d*) have lobes of different sign separated by nodal planes. An example of this is discussed on p.63 of Chapter 6. These symmetry limitations must always be borne in mind when the overlap of orbitals is being considered.

Now we can get a very useful qualitative explanation of directed valency if we assume that the strength of a covalent bond is approximately proportional to the amount of overlap. Let us consider some overlapping orbitals from this point of view.

1. *Overlap of* s-*type orbitals*
   The orbitals are spherically symmetrical so that they overlap to the same extent in all directions.

2. *Overlap of* p-*type orbitals*
   We recall from our treatment on p.61 of Chapter 6 that the bond in the F_2 molecule can be ascribed to the overlap of the $2p_x$ atomic orbitals.

In this case, the electron density is greatest along the *x* axis, and so we get maximum overlap for a given internuclear distance if the bond is formed in this direction. Since a large overlap is associated with the formation of a strong bond, we may perhaps relate bond strength to the type of orbital from which the bond is formed. It was pointed out by Pauling that, since the *p* orbitals are concentrated along particular directions, they should overlap more effectively for a given internuclear distance than should *s* orbitals of the same principal quantum number. If the radius of a $3s$ orbital is r, the length of the axis of a $3p$ orbital, from nucleus to boundary, can be shown to be $r\sqrt{3}$; a bond formed by the overlap of a p orbital should be approximately $\sqrt{3}$ times as strong as one formed by the corresponding s orbital. Too much significance must not be attached to the numerical factor, but the general principle remains a useful one.

These ideas can now be used to discuss the structure of polyatomic molecules. The oxygen atom, for instance, has the configuration

$$(1s)^2(2s)^2(2p_x)^2(2p_y)(2p_z)$$

with its two unpaired electrons in orbitals which are at right angles to one another. Let us picture the approach of two hydrogen atoms to the oxygen atom to form a molecule of water. As one hydrogen atom approaches (say along the z axis), the $2p_z$ orbital becomes distorted and finally overlaps the s orbital of the hydrogen atom. The s orbital of the second hydrogen atom overlaps the oxygen $2p_y$ orbital in the same way; it must moreover approach at an angle of 90° to the first OH bond if an appreciable overlap is to result. In other words, the oxygen atom forms two bonds at right angles to one another, since greatest overlap – and hence the strongest bonds – are then obtained. In practice, we find that the HOH angle is not the predicted 90° but 104.5°; this difference has been ascribed to the mutual repulsion of the hydrogen atoms. This is grossly oversimplified, of course, and the reader should not infer that the O–H bonds can be adequately described by an overlap involving oxygen pure p atomic orbitals. A more complete description will be given in Chapter 10. When we consider the related molecules H_2S and H_2Se, however, we find the bond angles are 93° and 91°, respectively, i.e. almost the predicted behaviour for simple p orbital overlap, since the bond lengths are greater and the mutual repulsion of the hydrogen atoms correspondingly diminished.

In a similar way, we get three mutually perpendicular bonds in the NH_3, PH_3, AsH_3 and SbH_3 molecules as a result of the overlap of the s orbitals of the respective hydrogen atoms with the three p orbitals of nitrogen, phosphorus, arsenic and antimony. Here again, the bond angles are not exactly 90°, being 107.1° for NH_3, 93° for PH_3, 92° for AsH_3 and 91° for SbH_3; as the bonds get longer the repulsions between the hydrogen atoms diminish and the bond angle approaches 90°. Once again a fuller description of the structures of these molecules is given in Chapter 10.

8.3 Polyatomic Molecules: VB and MO Descriptions

If we try to apply the ideas of the previous section to a discussion of bonding in compounds of elements such as beryllium, boron and carbon, we at once run into difficulties. Beryllium, with a ground state $(1s)^2(2s)^2$, might be expected to behave as a noble gas such as neon and form no bonds at all. If, however, the beryllium atom receives sufficient energy, one of its $2s$ electrons may be promoted into a $2p$ orbital, giving the excited state $(1s)^2(2s)(2p_x)$, with two unpaired electrons. These could then pair with electrons from two atoms (X), giving rise to a BeX_2 molecule. Normally, we should expect the energy of the two Be–X bonds to be greater than the promotion energy. We can use either the valence-bond or molecular-orbital methods to describe such molecules, and these are now discussed in some detail, first for linear triatomic molecules (e.g. BeX_2), then for both trigonal-planar and tetrahedral molecules.

8.3.1 LINEAR TRIATOMIC MOLECULES

8.3.1.1 *Valence-bond approach*

If we take the specific case of a $BeCl_2$ molecule in the gas phase*, then at a very simple level we can visualize the bonding in two stages, firstly by the $2p_x$ orbital and then by the $2s$ orbital (*Figure 8.3*). The beryllium $2p_x$ orbital overlaps with a chlorine $3p_x$ orbital to form a strong well-defined σ bond, the overlap being best when the p orbitals of the beryllium and chlorine atoms are collinear. The second Be–Cl bond is less clearly defined, however, because the Be $2s$ orbital is spherically symmetrical and accordingly overlaps equally well in any direction. Mutual repulsion between the two chlorine atoms and between the two bonding pairs of electrons will undoubtedly tend to give a large Cl–Be–Cl angle, but no precise value can be predicted on grounds of maximum overlap. Our naive picture of the formation of two Be–Cl bonds from dissimilar beryllium

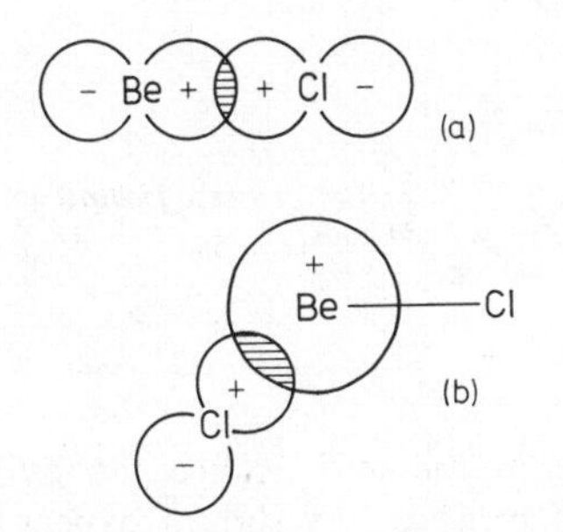

Figure 8.3 Hypothetical bonding in monomeric $BeCl_2$ in terms of Be using its 2s and 2p orbitals for bond formation

orbitals thus describes a structure with ill-defined bond angles and bonds of unequal strength.

It is well established, however, that both bonds in such simple compounds of bivalent beryllium are collinear. It seems necessary, therefore, to describe this equality of bonding in terms of two entirely equivalent orbitals used by the beryllium atoms if we wish to describe each bond in terms of a pair of electrons linking two nuclei. Thus we suppose that the beryllium atom does not use simple $2s$ and $2p_x$ orbitals but a combination of these. When we describe the electronic states of the beryllium atom, we do so, in theory, by solving the appropriate wave equation and obtaining suitable solutions ϕ_{2s} and ϕ_{2p_x}, each of which describes an energy state or orbital capable of accommodating two electrons with opposed spins. Now valid solutions

*The molecule is monomeric in the gas phase, but solid $BeCl_2$ is a polymeric substance (see p.138)

to the wave equation are also given by linear combinations of these wave functions, and we can describe other possible orbitals by combined or 'hybridized' wave functions. If the two solutions, ϕ_{2s} and ϕ_{2p_x}, have equal weight, we get two new equivalent functions which describe two linearly-directed orbitals:

$$\psi_{sp(i)} = \sqrt{1/2} \times (\phi_{2s} + \phi_{2p_x})$$

$$\psi_{sp(ii)} = \sqrt{1/2} \times (\phi_{2s} - \phi_{2p_x})$$

In these combinations, similar to those described in Chapter 7, $\sqrt{1/2}$ is the normalization constant.

In wave-mechanical language we say that two *sp* hybrid orbitals have been formed from one *s* and one *p* orbital. These new hybrid orbitals have strong directional characteristics, and each protrudes further along the axis than the original contributing *p* orbital.

Figure 8.4 shows how these two hybrids, which point in diametrically opposite directions, arise from the combinations shown in the above equations. Thus $\psi_{sp(i)}$ has a large lobe with + sign on the right because the + signs of the *s* and *p* orbitals coincide here; for $\psi_{sp(ii)}$ the combination involves $-\phi_{2p_x}$, which inverts the signs on the $2p_x$ orbital lobes, so that

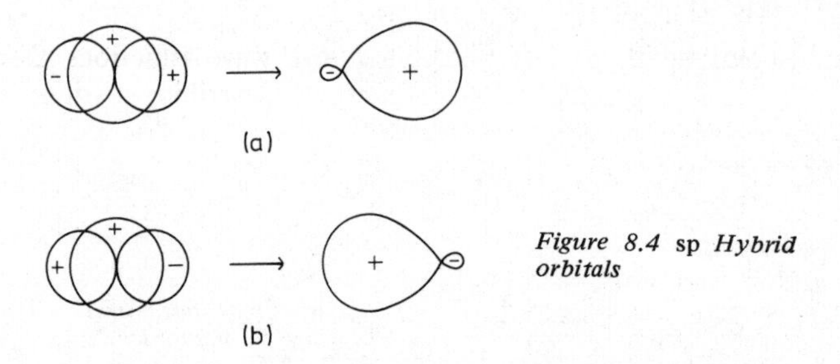

Figure 8.4 sp *Hybrid orbitals*

the large lobe with the + sign is now on the left. Since the hybrid orbitals protrude further than the *s* and *p* orbitals, we should expect them to provide more effective overlapping and stronger bonds.

In a colloquial way, we can describe the hybridization by saying that the electron probability clouds of the pure *s* and *p* orbitals have interacted to produce new standing waves by the merging and reforming of the charge clouds. The charge cloud of the *p* orbital is concentrated along a particular axis, and the merger with the *s* orbital charge cloud extends this axial concentration. The important point is that the two electrons 'occupy' a certain region of space and that we can describe this space either as two *sp* hybrid orbitals or as an *s* and a *p* orbital; the total charge cloud remains symmetrical about the axis.

The formation of the $BeCl_2$ molecule can now be visualized as resulting from the overlap of the *sp* hybrid orbitals of beryllium with the $3p_x$ orbitals of the chlorine atoms. For the most effective overlap the two Be–Cl bonds must now be collinear and equivalent (cf. *Figure 8.5*). This description of the arrangement of the two bonding electron pairs about the beryllium atom in terms of hybrid orbitals is another example of the working of the exclusion principle, which requires electrons of similar spin to be as far away from each other as possible (see p.40).

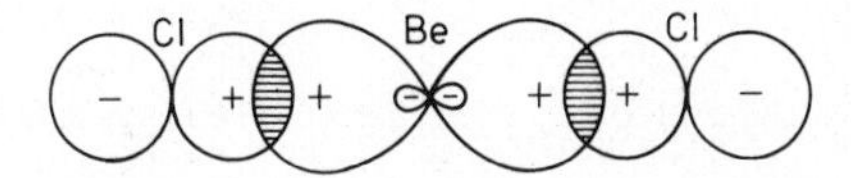

Figure 8.5 Formation of $BeCl_2$; using sp *hybrid orbitals of Be*

At this point we should perhaps make some reservations about colloquial phrases of the type 'two entirely equivalent orbitals used by the beryllium atom . . . ' which have already appeared in this chapter. There is the danger of ascribing real physical significance to the orbitals, as though they were electron 'containers' hooked on to the atom. One should say, rather, that the charge-cloud distribution around the beryllium atom in a molecule such as $BeCl_2$ is such that it can be described mathematically in terms of collinear, equivalent hybrid orbitals. Having made this reservation we shall, however, for the sake of brevity continue to use the colloquial form of words. Readers should note that throughout this discussion of the linear $BeCl_2$ molecule we have implicitly taken the molecular axis as the x axis, and accordingly ascribed the bonding to p_x orbitals.

8.3.1.2 *Molecular-orbital approach*

In the valence-bond description we combined the wave functions describing the $2s$ and $2p_x$ atomic orbitals to give functions describing two *sp* hybridized atomic orbitals. Overlapping of these hybrid orbitals with the $3p_x$ orbitals from the chlorine atoms gives two localized two-electron bonds; each bond links beryllium with just one chlorine atom.

An alternative way of describing the bonding is to use the concepts of molecular orbitals that were developed in Chapter 6. We recall that in the description of the hydrogen molecule the combination of two a.o.'s* (one from each atom) yielded two m.o.'s*, one bonding and one antibonding. In general, the combination of n a.o.'s (n even) yields n m.o.'s, $n/2$ bonding and $n/2$ antibonding. Thus in a linear triatomic molecule ($BeCl_2$), the beryllium atom has two suitable a.o.'s ($2s$ and $2p_x$) and each chlorine atom has one ($3p_x$); the combination of these *four* a.o.'s will accordingly give *four* m.o.'s, two bonding and two antibonding. The combinations, which are expressed mathematically below, are shown pictorially in *Figure 8.6* and energetically in *Figure 8.7.*

$$Cl_a{-}Be{-}Cl_b$$

(i) $\sigma_s = \phi_{Be(2s)} + \phi_{Cl_a} + \phi_{Cl_b}$

(ii) $\sigma_s^* = \phi_{Be(2s)} - \phi_{Cl_a} - \phi_{Cl_b}$

(iii) $\sigma_p = \phi_{Be(2p_x)} - \phi_{Cl_a} + \phi_{Cl_b}$

(iv) $\sigma_p^* = \phi_{Be(2p_x)} + \phi_{Cl_a} - \phi_{Cl_b}$

*For convenience we write a.o. and m.o. for atomic orbital and molecular orbital, respectively

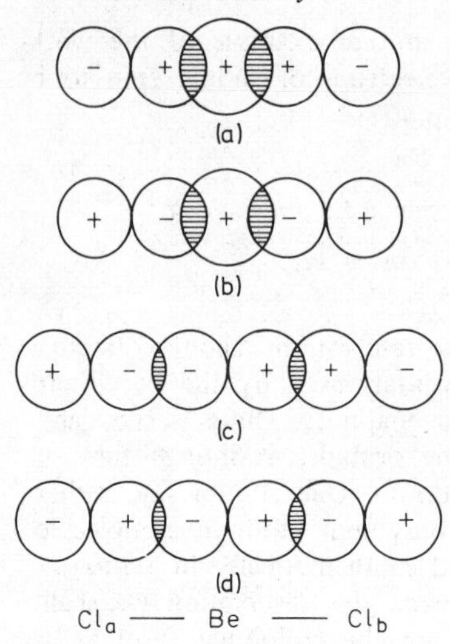

Figure 8.6 Molecular orbitals for $BeCl_2$

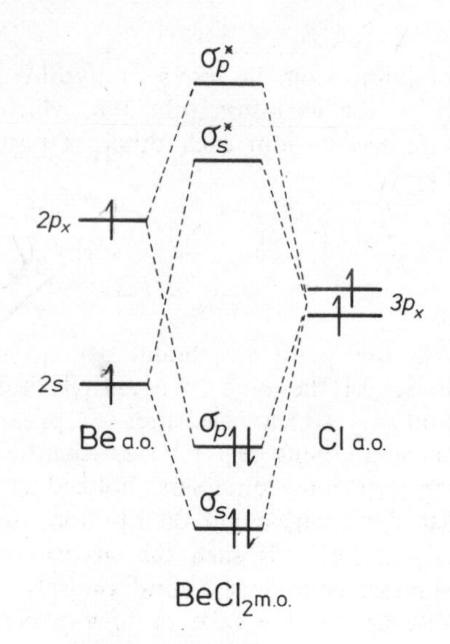

Figure 8.7 Energy level diagram for molecular orbitals in $BeCl_2$

In the pictorial diagram we use the convention that the p orbital of beryllium has a + sign for its right-hand lobe; the chlorine p orbitals have their lobes with a + sign pointing towards the beryllium. In the energy diagram we have assumed for simplicity that the a.o.'s of the chlorine atoms will be of similar energy to those of the $2s$ and $2p_x$ a.o.'s of beryllium, and this is reflected in the wave functions (i)–(iv) where the a.o.'s are mixed equally. Of course, in the general case of MX_2 molecules the energies of the M and X orbitals may be somewhat different, and the molecular-orbital wave functions will then incorporate appropriate mixing coefficients, just as they did for hydrogen fluoride (see p.63). There will, however, still be two bonding m.o.'s, each delocalized and embracing the three nuclei, and each containing two electrons. Thus, in the molecular-orbital approach we describe the bonding in a linear triatomic molecule as consisting of two delocalized two-electron bonds.

If we compare the valence-bond and molecular-orbital approaches for such molecules, we see that the electron density pattern is the same in each case and has a symmetrical distribution about the x axis. We can use either method to describe this density pattern, since both tell us that the three atoms are collinear, bonded by four electrons.

8.3.2 TRIGONAL-PLANAR MOLECULES

In molecules such as boron trichloride we have to account for the experimental observation that the three bonds are coplanar with bond angles of 120°. If we consider the boron atom, with configuration $(1s)^2(2s)^2(2p_x)$, we see that promotion of one $2s$ electron gives the excited state $(1s)^2(2s)(2p_x)(2p_y)$ with three unpaired electrons; these three unpaired electrons give rise to the three bonds, by pairing with an unpaired electron from each of the three chlorine atoms. As with the linear molecules (e.g. $BeCl_2$), there are two convenient ways of discussing this bonding.

8.3.2.1 *Valence-bond approach*

The $2s$, $2p_x$ and $2p_y$ orbitals are mixed or hybridized to give three equivalent sp^2 orbitals. The mathematical treatment shows that the state of lowest energy corresponds to three orbitals which are coplanar and oriented at 120° to each other:

$$\psi_{sp^2\,(i)} = \sqrt{\tfrac{1}{3}}\,\phi_{2s} + \sqrt{\tfrac{2}{3}}\,\phi_{2p_x}$$

$$\psi_{sp^2\,(ii)} = \sqrt{\tfrac{1}{3}}\,\phi_{2s} - \sqrt{\tfrac{1}{6}}\,\phi_{2p_x} + \sqrt{\tfrac{1}{2}}\,\phi_{2p_y}$$

$$\psi_{sp^2\,(iii)} = \sqrt{\tfrac{1}{3}}\,\phi_{2s} - \sqrt{\tfrac{1}{6}}\,\phi_{2p_x} - \sqrt{\tfrac{1}{2}}\,\phi_{2p_y}$$

It should be noted that ϕ_{2p_y} makes no contribution to $\psi_{sp^2\,(i)}$ because we have chosen the x axis as axis for this hybrid orbital. The coefficients differ for the contribution of ϕ_{2p_x} and ϕ_{2p_y} to the other two hybrids, since the angles made by the x and y axes to the hybrid orbital axes are 60° and 30°, respectively. *Figure 8.8* illustrates the formation of the

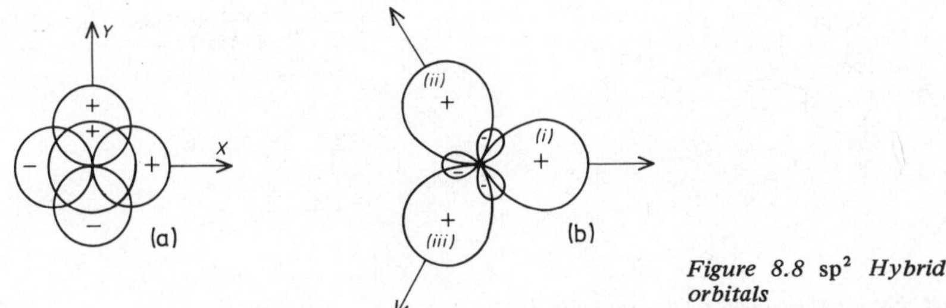

Figure 8.8 sp² *Hybrid orbitals*

hybrid orbitals from the component orbitals. The process can be visualized as the merging and reforming of the s and p charge clouds to give the sp^2 hybrid charge clouds. This concentration of the charge clouds in a plane is as might be expected, since the original $2p_x$ and $2p_y$ orbitals were coplanar. The boron–chlorine bonds are σ bonds and can be described as the overlap of the sp^2 hybrid orbitals of the boron atom with the $3p_x$ atomic orbitals of the chlorine atoms.

8.3.2.2 *Molecular-orbital approach*

We use the same method as for the linear triatomic molecules, except that we now have four atoms and six a.o.'s ($2s$, $2p_x$ and $2p_y$ for B, and $3p_x$ for each of the three Cl atoms), with accordingly six m.o.'s, three bonding (σ_s, σ_x, σ_y) and three antibonding (σ_s*, σ_x*, σ_y*). The energy diagram is

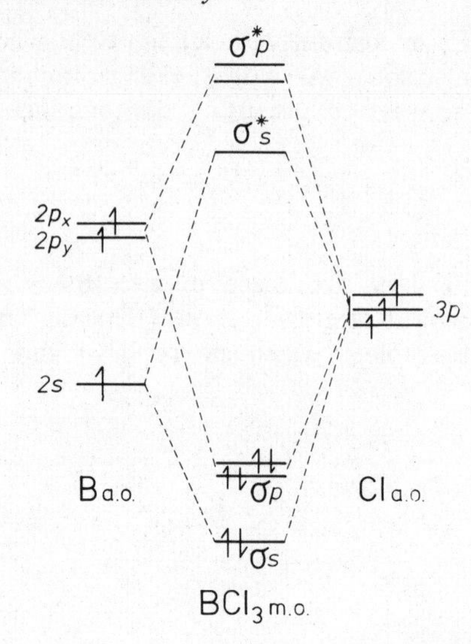

Figure 8.9 Energy level diagram for molecular orbitals in BCl_3

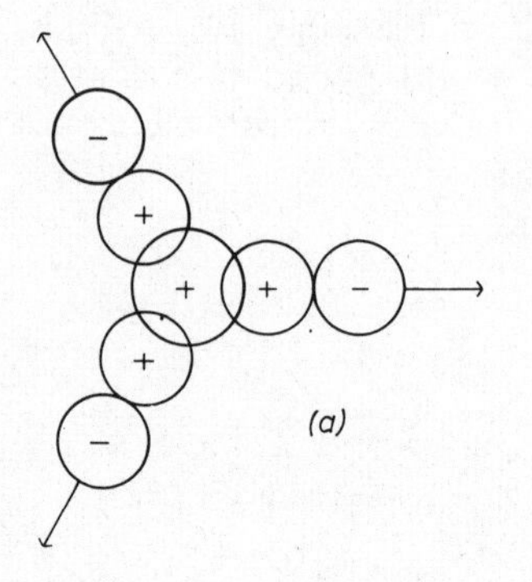

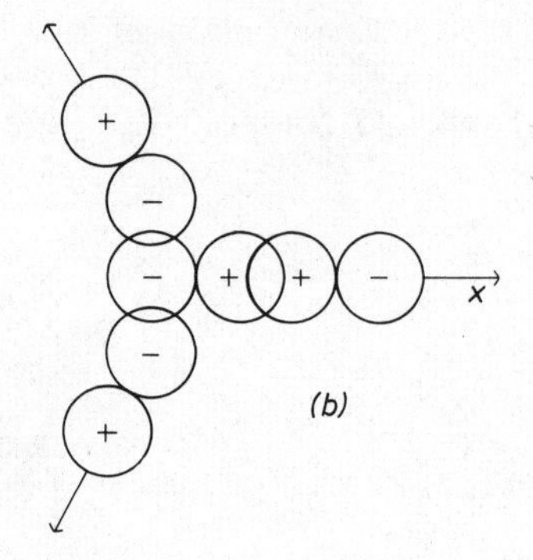

Figure 8.10 Molecular orbitals for BCl_3

shown in *Figure 8.9*, where for simplicity we assume that the energy of the chlorine $3p$ orbitals is between that of the $2s$ and $2p$ orbitals of boron. *Figure 8.10* shows how the σ_s m.o. is obtained by the combination of the boron $2s$ orbital with the $3p$ orbital of each of the three chlorine atoms. It also shows the formation of one of the σ_p m.o.'s from the boron $2p_x$ orbital; here it can be seen that overlap of the $2p_x$ orbital is most effective with the p orbital of chlorine atom 1, and this would be reflected by giving a greater weighting to this chlorine a.o. in the molecular-orbital function. Thus, in the function

$$\sigma_{p_x} = \phi_{B(2p_x)} + C_1\phi_{Cl_1} - C_2(\phi_{Cl_2} + \phi_{Cl_3})$$

C_1 is greater than C_2. There are three bonding m.o.'s, each embracing four nuclei and containing two electrons. Hence in this molecular-orbital description we consider the bonding to consist of three two-electron four-centred bonds. In both of these approaches the total bonding involves six electrons, and the electron-density charge-cloud patterns would in either case have a planar symmetry.

8.3.3 TETRAHEDRAL MOLECULES

We may use methane as a typical example of a molecule with a tetrahedral configuration. As with beryllium and boron, we acquire the necessary number of unpaired electrons by promoting a $2s$ electron, giving carbon the configuration

$$(1s)^2(2s)(2p_x)(2p_y)(2p_z)$$

in which the electron charge cloud is spherically symmetrical. The four C–H bonds can be described by the valence-bond or molecular-orbital technique.

8.3.3.1 *Valence-bond approach*

The concept of four localized two-electron bonds is readily realized by describing the four unpaired electrons on the carbon atom in terms of four sp^3 hybrid orbitals which point towards the corners of a tetrahedron. This is expressed mathematically by taking linear combinations of the $2s$ and $2p$ orbitals:

$$\psi_{sp^3(i)} = \tfrac{1}{2}(\phi_{2s} + \phi_{2p_x} + \phi_{2p_y} + \phi_{2p_z})$$

$$\psi_{sp^3(ii)} = \tfrac{1}{2}(\phi_{2s} + \phi_{2p_x} - \phi_{2p_y} - \phi_{2p_z})$$

$$\psi_{sp^3(iii)} = \tfrac{1}{2}(\phi_{2s} - \phi_{2p_x} + \phi_{2p_y} - \phi_{2p_z})$$

$$\psi_{sp^3(iv)} = \tfrac{1}{2}(\phi_{2s} - \phi_{2p_x} - \phi_{2p_y} + \phi_{2p_z})$$

We then describe the methane molecule in terms of the overlap of the s orbitals of the hydrogen atoms with the sp^3 hybrid orbitals of the carbon atom. Inasmuch as the hybrid orbitals are largely concentrated along the tetrahedral axes, the C–H bonds will also be formed along these axes in order to get the maximum overlap for a given internuclear distance.

Ethane is described in much the same way, with each carbon atom using four sp^3 hybrid orbitals; the C–C bond is described as the overlap of an sp^3 hybrid orbital from each carbon atom and the C–H bonds as the overlap of the carbon sp^3 and hydrogen s orbitals.

The structures of many simple compounds of nitrogen and oxygen can also be discussed in terms of hybrid orbitals, and this treatment has some advantages over that described on p.86. We can, for instance, regard the five outer electrons of nitrogen ($2s^2 2p^3$) as occupying four approximately tetrahedral hybrid orbitals, three orbitals being singly occupied and the other doubly occupied (giving a so-called 'lone pair'). Overlap of the singly-filled orbitals with hydrogen 1*s* orbitals gives the ammonia molecule, in which the bond angles are approximately tetrahedral, i.e. 109.5°. Thus there are three two-electron localized bonds and one lone-pair of (non-bonding) electrons.

Similarly, the six outer electrons of oxygen ($2s^2 2p^4$) can be described by four tetrahedral hybrid orbitals, two being singly occupied and two containing a pair of electrons, and the water molecule formed by the overlap of hydrogen 1*s* orbitals with the singly-occupied oxygen orbitals should then yield a bond angle close to the predicted tetrahedral value.

As we saw earlier, the mutual repulsion of electron pairs in both NH_3 and H_2O would result in bond angles rather less than the tetrahedral one. We shall discuss these ideas in considerably more detail in Chapter 10.

8.3.3.2 *Molecular-orbital approach*

We use the same general principles already applied to linear and trigonal-planar molecules, except that we now have to consider four a.o.'s for the central atom, one *s* and three *p*, together with an a.o. from each of the four univalent atoms. Hence the combination of eight a.o.'s gives eight m.o.'s, four bonding and four antibonding. *Figure 8.11* shows a

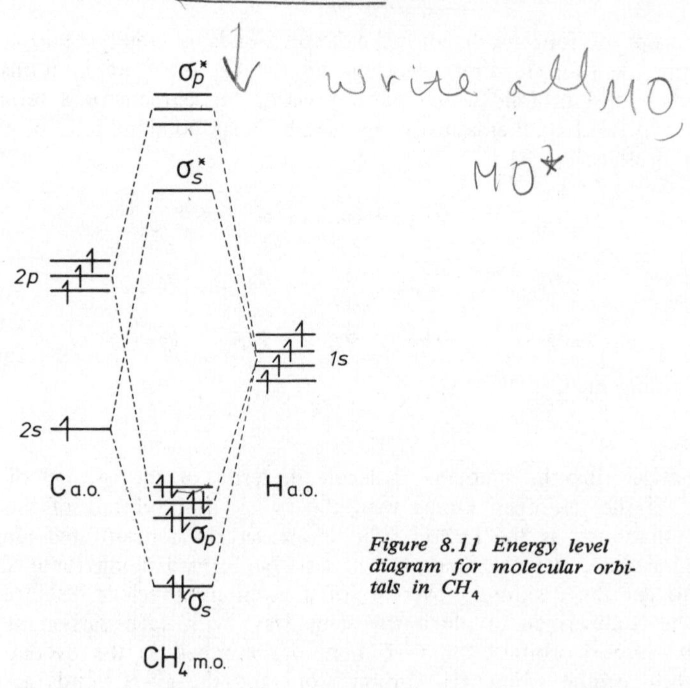

Figure 8.11 Energy level diagram for molecular orbitals in CH_4

typical energy level diagram scheme for CH_4, where once again we have assumed that all the a.o.'s are of similar energy. The eight electrons, four from carbon and one from each hydrogen, are placed in the four bonding m.o.'s. If, for simplicity, we take all the weighting coefficients as unity – that is, regard all the a.o.'s as being of the same energy – then the linear combinations are the following:

$$\sigma_s = \phi_{C_{2s}} + (\phi_{Ha_{1s}} + \phi_{Hb_{1s}} + \phi_{Hc_{1s}} + \phi_{Hd_{1s}})$$

$$\sigma_s^* = \phi_{C_{2s}} - (\phi_{Ha_{1s}} + \phi_{Hb_{1s}} + \phi_{Hc_{1s}} + \phi_{Hd_{1s}})$$

$$\sigma_{px} = \phi_{C_{2p_x}} + (\phi_{Ha_{1s}} + \phi_{Hb_{1s}} - \phi_{Hc_{1s}} - \phi_{Hd_{1s}})$$

$$\sigma_{px}^* = \phi_{C_{2p_x}} - (\phi_{Ha_{1s}} + \phi_{Hb_{1s}} - \phi_{Hc_{1s}} - \phi_{Hd_{1s}})$$

$$\sigma_{py} = \phi_{C_{2p_y}} + (\phi_{Ha_{1s}} - \phi_{Hb_{1s}} + \phi_{Hc_{1s}} - \phi_{Hd_{1s}})$$

$$\sigma_{py}^* = \phi_{C_{2p_y}} - (\phi_{Ha_{1s}} - \phi_{Hb_{1s}} + \phi_{Hc_{1s}} - \phi_{Hd_{1s}})$$

$$\sigma_{pz} = \phi_{C_{2p_z}} + (\phi_{Ha_{1s}} - \phi_{Hb_{1s}} - \phi_{Hc_{1s}} + \phi_{Hd_{1s}})$$

$$\sigma_{pz}^* = \phi_{C_{2p_z}} - (\phi_{Ha_{1s}} - \phi_{Hb_{1s}} - \phi_{Hc_{1s}} + \phi_{Hd_{1s}})$$

8.3.4 MOLECULES USING *d* ORBITALS FOR BONDING

Elements in the second row of the Periodic Table and beyond, where the electrons begin to occupy *d* levels, may form compounds showing symmetries other than linear, trigonal-planar or tetrahedral. By far the commonest configuration is the octahedral one, found in such compounds as SF_6 and $[Co(NH_3)_6]^{3+}$. Five co-ordination, which is observed in simple compounds such as PCl_5 and in co-ordination compounds such as $VCl_3,2NMe_3$, leads to two basic arrangements, trigonal-bipyramidal or square-pyramidal; there are several alternative stereochemical arrangements for seven and eight co-ordination, and these will be discussed in some detail in Chapters 10 and 11. The latter chapter will also give an account of the square planar arrangement that is sometimes found in preference to the tetrahedral one when the central atom forming the four bonds is a transition element.

Once again it is possible to use either a valence-bond or molecular-orbital approach to describe the bonds in any of these configurations. The concept of hybridization, with localized two electron σ bonds, is the simplest picture, but the more sophisticated molecular-orbital approach is generally necessary for the consideration of properties such as spectra and magnetic susceptibility. In Chapter 11 we shall see that there is a third description, the ligand-field one, which is related to the molecular-orbital description but is somewhat simpler; it is applicable to transition-metal compounds where the *d* electrons play a vital role.

For the moment, however, we are more concerned with the simple stereochemistry of compounds, and for this purpose the valence-bond approach is adequate. The octahedral arrangement is described in terms of sp^3d^2 hybrid orbitals, obtained by taking linear combinations of the wave functions describing the s, p_x, p_y, p_z, $d_{x^2-y^2}$ and d_{z^2} orbitals:

$$\psi_{(i)} = \sqrt{\tfrac{1}{6}}\,\phi_s + \sqrt{\tfrac{1}{2}}\,\phi_{p_z} + \sqrt{\tfrac{1}{3}}\,\phi_{d_{z^2}}$$

$$\psi_{(ii)} = \sqrt{\tfrac{1}{6}}\,\phi_s - \sqrt{\tfrac{1}{2}}\,\phi_{p_z} + \sqrt{\tfrac{1}{3}}\,\phi_{d_{z^2}}$$

$$\psi_{(iii)} = \sqrt{\tfrac{1}{6}}\,\phi_s + \sqrt{\tfrac{1}{12}}\,\phi_{d_{z^2}} + \tfrac{1}{2}\,\phi_{d_{x^2-y^2}} + \sqrt{\tfrac{1}{2}}\,\phi_{p_x}$$

$$\psi_{(iv)} = \sqrt{\tfrac{1}{6}}\,\phi_s + \sqrt{\tfrac{1}{12}}\,\phi_{d_{z^2}} + \tfrac{1}{2}\,\phi_{d_{x^2-y^2}} - \sqrt{\tfrac{1}{2}}\,\phi_{p_x}$$

$$\psi_{(v)} = \sqrt{\tfrac{1}{6}}\,\phi_s + \sqrt{\tfrac{1}{12}}\,\phi_{d_{z^2}} - \tfrac{1}{2}\,\phi_{d_{x^2-y^2}} + \sqrt{\tfrac{1}{2}}\,\phi_{p_y}$$

$$\psi_{(vi)} = \sqrt{\tfrac{1}{6}}\,\phi_s + \sqrt{\tfrac{1}{12}}\,\phi_{d_{z^2}} - \tfrac{1}{2}\,\phi_{d_{x^2-y^2}} - \sqrt{\tfrac{1}{2}}\,\phi_{p_y}$$

It should be noted that the six hybrid orbitals point along the x, y and z axes, as might be expected since the atomic orbitals used in their construction also did this; thus we used the d_{z^2} and $d_{x^2-y^2}$ orbitals rather than the d_{xy}, d_{xz} and d_{yz}. *Figure 8.12* shows the combination producing

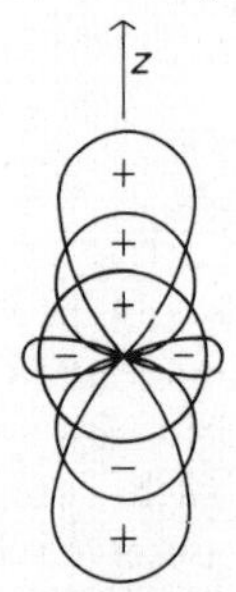

Figure 8.12 Combination of s, p_z and d_{z^2} orbitals

$\psi_{(i)}$; $\psi_{(ii)}$ merely involves changing signs on the p orbital and gives a hybrid orbital of the same shape but pointing in the opposite direction. The reader can see for himself how the functions $\psi_{(iii)}$ – $\psi_{(vi)}$ give orbitals pointing along the x and y axes. The d orbitals used in the hybridization can originate from the penultimate shell, i.e. $(n-1)d^2$, ns, np^3, as in $[Co(NH_3)_6]^{3+}$, or from the same shell as the s and p orbitals, i.e. ns, np^3, nd^2, as in SF_6; the hybrid orbitals have the same shape in either case. To avoid confusion in later discussion involving the use of these orbitals in complex formation we will designate them d^2sp^3 and sp^3d^2, respectively.

Table 8.2 gives a summary of the commoner hybrid orbitals together with typical examples of molecules in which they occur. They will be discussed fully in Chapters 10 and 11.

8.4 Non-equivalent Orbitals

So far, when we have referred to molecules having certain shapes such as trigonal-planar or tetrahedral, the examples we have chosen have been generally those in which the multivalent atom is bonded with a number of identical atoms. Thus boron trichloride is trigonal-planar with Cl–B–Cl angles of 120°, and methane is tetrahedral with all the H–C–H angles the same. When the bonded atoms are not the same, however, the bond angles are slightly different, e.g.

Table 8.2 Hybrid orbitals

Simple			*Hybridized*			
Type	*Degree of protrusion*	*Examples*	*Type*	*Degree of protrusion*	*Shape*	*Examples*
s	1.0	H_2	*sp*	1.93	linear	$HgCl_2$
p	1.73	Cl_2	sp^2	1.99	trigonal-planar	BCl_3
d	not used alone		sp^3	2.00	tetrahedral	CH_4
			dsp^2	2.69	square-planar	K_2PtCl_4
			d^2sp^3	2.93	octahedral	Rb_2TiCl_6
			dsp^3	2.80	trigonal-bipyramidal	PCl_5
			d^4sp^3	–	dodecahedral	$K_4Mo(CN)_8$

CH_4 all H–C–H angles = 109.5°

CCl_4 all Cl–C–Cl angles = 109.5°

CH_3Cl all H–C–H angles = 110.9°

$CHCl_3$ all Cl–C–Cl angles = 110.9°

and although these variations are small, they are nevertheless significant. In these cases some of the hybrid orbitals have a little more *p* character than others, i.e. they are not all exactly equivalent. Many more examples of this type of variation from an 'ideal' bond angle will be met with in Chapter 11.

8.5 π Molecular Orbitals

So far in this chapter we have restricted our more detailed discussion of molecular shape to molecules containing σ bonds only, and we have pointed out that while the valence-bond description involves the appropriate number of localized two-electron bonds, the molecular-orbital approach describes the behaviour of electrons by means of orbitals that are localized. We saw in Chapter 6 that when more than two electrons hold two nuclei together, then the orbitals may be divided into σ and π type depending upon their symmetry. In the N_2 molecule, for instance, bonding involves six electrons in one σ and two π orbitals (cf. p.60).

If we now return to a consideration of the simple compounds of carbon, we see that in a discussion of the structure of ethene, both σ- and π-type bonds appear. Thus it is observed experimentally that all six atoms are coplanar with bond angles close to 120°, and it is convenient to consider that carbon makes use of two of its three 2*p* orbitals and the 2*s* orbital to form three co-planar sp^2 hybrid bonds; the remaining $2p_z$ orbital, which is not hybridized, has its axis perpendicular to the plane (*xy*) of the hybrid orbitals, one lobe being above the plane, the other below. A σ-type carbon–carbon bond is formed by the overlap of two sp^2 hybrid orbitals, one from each atom, and the four carbon–hydrogen bonds are formed by the overlap of the remaining sp^2 hybrid orbitals with hydrogen s orbitals; all the nuclei are therefore co-planar. There is also the possibility of lateral overlap of the two unhybridized p_z orbitals, and while this is not as great as the end-on type of overlap, it will be appreciable because the axes of these p orbitals are parallel. The second bond between the carbon atoms is thus a π bond. It is weaker than the σ bond and its presence accounts for the reactivity of ethene and of other compounds containing similar localized π bonds. Any attempt to twist the molecule about the C–C axis would reduce the overlap of the $2p_z$ orbitals and would in effect break the π bond. The absence of free rotation of the methylene groups

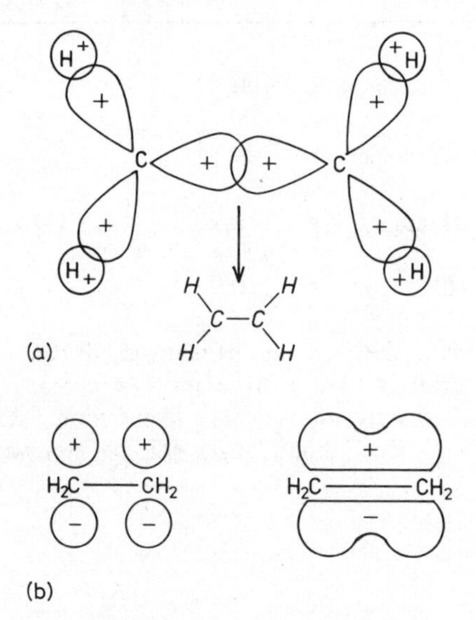

Figure 8.13 Molecular orbitals for ethene

about the C=C bond is thus explained. *Figure 8.13* illustrates this σ and π bonding in ethene.

It is important to realize that this double bond has two regions of charge density, one above and one below the plane of the molecule, and that these two charge clouds belong to the same orbital and cannot be considered independently, i.e. the plane of the molecule is the nodal plane of the π orbital.

With the ethyne molecule (cf. *Figure 8.14*), the valence-bond description takes carbon sp hybrid orbitals as the basis for the formation of the H–C–C–H skeleton, leaving two of the $2p$ orbitals ($2p_y$ and $2p_z$) unchanged. Lateral overlap of these p orbitals gives two π bonds between the carbon atoms, the total charge distribution (for the four electrons) having cylindrical symmetry about the C–C axis [cf. *Figure 8.14(c)*].

The π bonds we have just mentioned for ethene and ethyne are localized between two carbon atoms, but there are many molecules in which the π orbitals may be delocalized and embrace more than two nuclei. Indeed, it is just such molecules that are so difficult to represent on paper

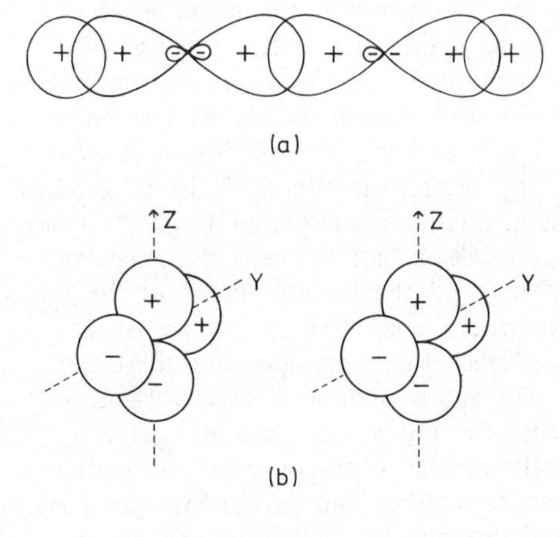

Figure 8.14 Molecular orbitals for ethyne

by conventional structural formulae. Amongst many examples may be mentioned benzene, where we had to resort to the use of at least two structural forms to emphasize that all the bonds were equal. We know from experiment that benzene is a regular hexagon with bond angles of 120°, so we may use a valence-bond description for the framework σ bonds [cf. *Figure 8.15(a)*] in which carbon is sp^2 hybridized, as in ethene. We now

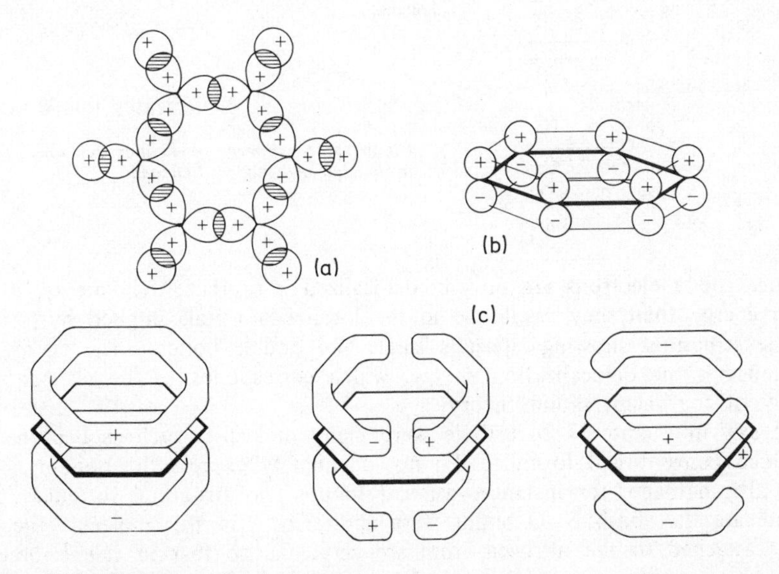

Figure 8.15 Orbital overlap in benzene

have to consider the six p orbitals perpendicular to the plane of the carbon hexagon. Lateral overlap of these p orbitals, taken in pairs, will give three localized π bonds, corresponding to one or other of the Kekulé structures, but there is no reason why the overlap should be limited in this way. Thus the $2p_z$ orbital on any carbon atom in *Figure 8.15(b)* can overlap equally well with the $2p_z$ orbital on either of the neighbouring carbon atoms. The π bonding is therefore considered to result from mutual overlapping of all the $2p_z$ orbitals, giving delocalized molecular orbitals [*Figure 8.15(a)*] associated with all the carbon nuclei in the molecule. In this molecular-orbital description, combinations of the six $2p_z$ atomic orbitals give six molecular orbitals, three bonding and three antibonding, and the bonding orbitals will accommodate the six π electrons. *Figure 8.16* shows the energy scheme for the π electrons in benzene. The lowest energy bonding π orbital provides a charge cloud consisting of two hexagonal shaped 'streamers', one above and one below the plane of the benzene ring. The other two bonding π orbitals, which are of higher energy, have an additional node and may be described colloquially as 'split hexagonal streamers'.

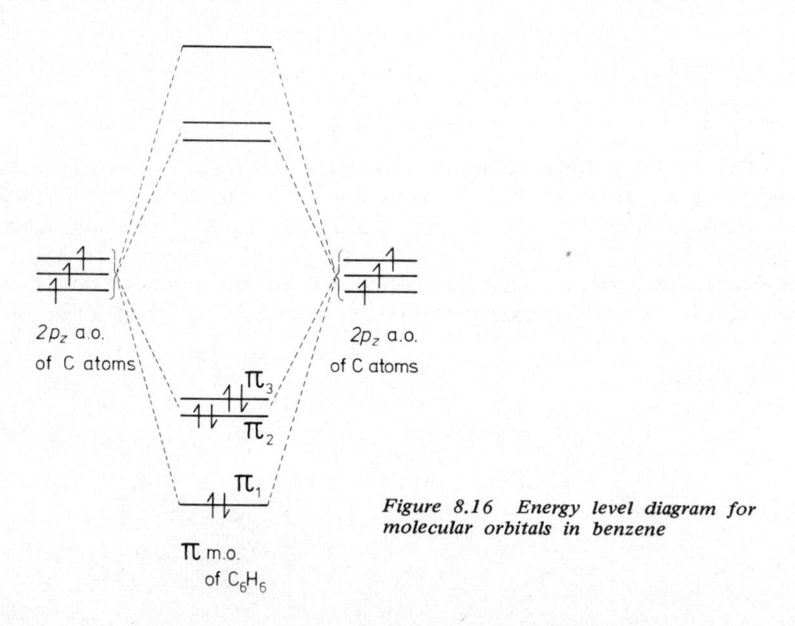

Figure 8.16 Energy level diagram for molecular orbitals in benzene

Since the π electrons are now in delocalized π orbitals, they are of lower energy than they would be in the localized orbitals implied by Kekulé structures showing alternate single and double bonds. The energy difference is the delocalization energy, which corresponds to the resonance energy of the valence-bond theory.

We saw in Chapter 7 that there were many molecules such as benzene in which conventional formulae did not adequately express the bonding. With nitromethane, for instance, we had to use two structural formulae to indicate that both N–O bonds were equivalent. In this molecule the bonds attached to the nitrogen atom are co-planar, so that in valence-bond terminology we can regard the σ bonding as arising from nitrogen sp^2

hybrids. As we saw at the beginning of the chapter, the basic shapes of molecules are determined by their σ bonds and, therefore, may be described simply in terms of hybridization. Any delocalized π bonding may then be expressed in molecular-orbital terms. Thus in nitromethane the nitrogen atom and the two oxygen atoms have their $2p_z$ orbitals perpendicular to the plane of the nitro group, so that the combination of the three a.o.'s will give rise to three m.o.'s. If to a first approximation we consider the nitrogen and oxygen $2p_z$ a.o.'s to have the same energy, then we should expect one bonding, one non-bonding and one antibonding m.o. *Figure 8.17* shows the energy diagram for this π bonding, corresponding to the following three wave functions:

$$\psi_{(i)} = \phi_N(\pi) + \phi_{O_1}(\pi) + \phi_{O_2}(\pi)$$

$$\psi_{(ii)} = \phi_{O_1}(\pi) + \phi_{O_2}(\pi)$$

$$\psi_{(iii)} = \phi_N(\pi) - \phi_{O_1}(\pi) - \phi_{O_2}(\pi)$$

The four π electrons will occupy $\psi_{(i)}$ and $\psi_{(ii)}$ so that the essential bonding is a three-centre two-electron π bond. The two non-bonding electrons will not contribute to the stability of the molecule; they correspond to

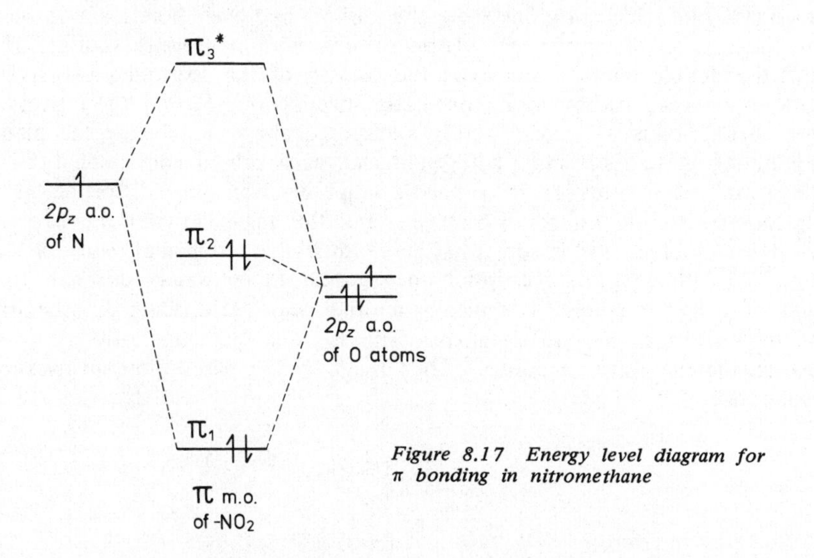

Figure 8.17 Energy level diagram for π bonding in nitromethane

the resonance description of the electrons being in the $2p_z$ a.o. of either oxygen atom. *Figure 8.18* shows diagrammatically how the $\psi_{(i)}$ m.o. arises.

It is, of course, perfectly possible to put down the full m.o. diagram for the $-NO_2$ grouping, which is merely the superposition of the σ energy diagram for a trigonal-planar molecule (cf. *Figure 8.9*) and the π diagram of *Figure 8.17*. In principle, we should always consider this full diagram, but we shall find (see next Chapter) that it is more convenient in many cases to consider the π bonding separately on an m.o. diagram and use the hybridization approach for σ bonds.

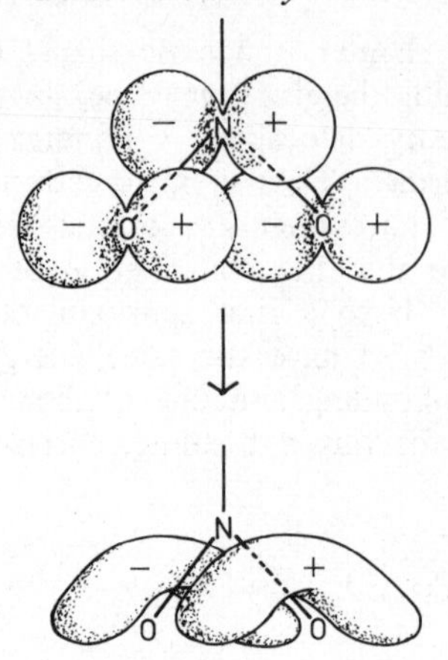

Figure 8.18 Formation of delocalized π molecular orbital in $-NO_2$ *group*

8.6 The Equivalent-orbital Description

Up till now we have described multiple bonds, as in ethene and ethyne, in terms of σ and π orbitals, but it is perfectly possible to discuss these structures in terms of hybrid orbitals that are, approximately, tetrahedrally arranged. In ethene, for instance, the carbon–hydrogen bonds are formed by the overlap of carbon sp^3 hybrid orbitals with hydrogen $1s$ orbitals, and the double bond results from the overlap of the remaining two hybrid orbitals on each carbon atom; this latter overlap (cf. *Figure 8.19*) gives two 'bent' bonds or 'banana' orbitals disposed above and below the plane containing the carbon and hydrogen nuclei, each orbital containing two electrons. The formation of a double bond involves some distortion of the original hybrid orbital distribution, and the angle between the two C–H bonds in ethene is somewhat more than the tetrahedral value of 109.5°. Although this description may appear to be widely different from that of σ and π bonds, it is merely another way of 'dividing up' the four bonding electrons into two regions. This is done mathematically by rearranging the wave functions. Thus the σ and π description involves two functions:

$$\psi_\sigma = \frac{1}{\sqrt{2}}(\phi_{C_{1(\sigma)}} + \phi_{C_{2(\sigma)}})$$

$$\psi_\pi = \frac{1}{\sqrt{2}}(\phi_{C_{1(\pi)}} + \phi_{C_{2(\pi)}})$$

where the σ a.o.'s are '*sp*' hybrids and the π a.o.'s are the $2p_z$ orbitals. If we take linear combinations of ψ_σ and ψ_π we get two equivalent orbitals:

$$\psi_{(i)} = \frac{1}{\sqrt{2}}(\psi_\sigma + \psi_\pi)$$

$$\psi_{(ii)} = \frac{1}{\sqrt{2}}(\psi_\sigma - \psi_\pi)$$

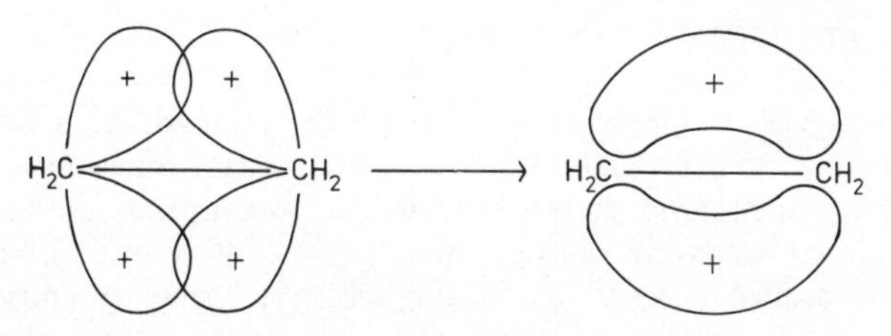

***Figure 8.19** Molecular orbitals for ethene: equivalent orbital description of double bond*

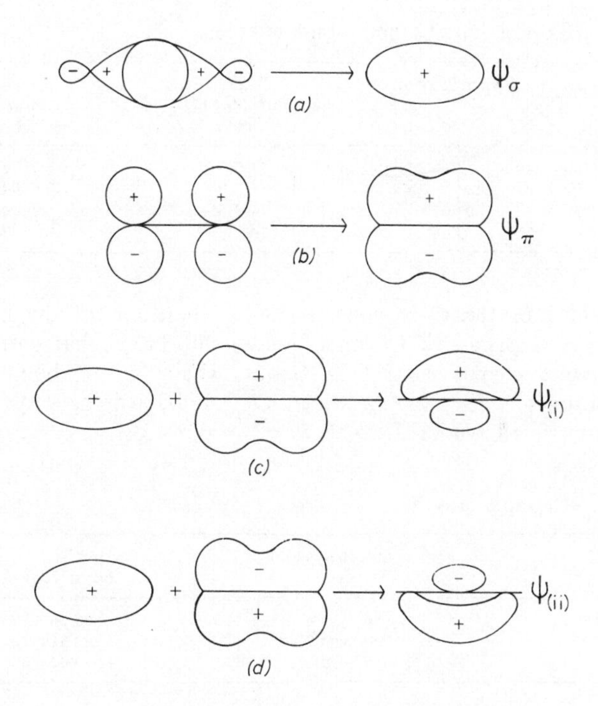

***Figure 8.20** Formation of equivalent orbitals in ethene: (a) σ molecular orbital; (b) π molecular orbital; (c) and (d) equivalent molecular orbitals*

This is illustrated diagrammatically in *Figure 8.20.* We shall refer back to this approach in Chapter 13 when discussing the bonding in diborane.

The triple bond between the carbon atoms in ethyne can also be represented by 'banana' or 'bent' bonds, in this case three, each containing two electrons. The reader will, no doubt, realize that this representation for two-carbon atom linking by one, two or three bonds is the orbital presentation of the old idea of tetrahedral linking through a point, a side or a face.

8.7 Bond Energies

At the beginning of this chapter we related the strength of a bond to the extent of the overlap of the orbitals involved in its formation. Unfortunately this idea, which gives an easily visualized description of bond formation, breaks down when we consider the *sp* type of hybrid orbital. Thus if we give the radius of an *s* orbital the arbitrary value of unity, the axes of the hybrid orbitals, measured from the nucleus, have the following lengths: sp = 1.93, sp^2 = 1.99 and sp^3 = 2.00 pm. We should, therefore, expect the bond strengths to increase slightly in this order, but Walsh has shown that the order is, in fact, the reverse of this. *Table 8.3* gives value for the C–H bond length and C–H bond energy in suitable molecules.

Table 8.3 C–H bond lengths and bond energies

Molecule	*Hybridization*	*C–H bond length/* pm	*C–H bond energy/* kJ mol^{-1}
Ethyne	sp	105.7	∿506
Ethene	sp^2	107.9	∿444
Methane	sp^3	109.4	∿431
CH radical	(p)	112	335

The shortening of the C–H bond as the '*s* character' of the hybrid orbital increases is paralleled in other bonds, and Brown has correlated bond length and hybrid character for C–X bonds, where X = carbon, halogens, oxygen or nitrogen. The way in which C–C bond lengths vary with changes in hybridization is shown in *Table 8.4.*

Table 8.4 C–C bond lengths

Molecule	*Hybridization*	*C–C bond length*/pm
Ethane	sp^3–sp^3	154
Propene	sp^3–sp^2	151
Methylethyne	sp^3–sp^2	146

Coulson has pointed out that while Pauling's comparison of bond length and overlap is very useful in a qualitative sense, it is not adequate for the quantitative consideration of bond energies in the sp^3, sp^2 and sp hybrids, and it would in fact suggest zero energy for π bonds.

A better estimate of the bond strength is given in terms of the overlap integral. The concept of maximum overlap does, however, remain a most useful one in establishing the directions in which bonds are formed, and hybrid orbitals of the type d^2sp^3, dsp^2, etc., which do protrude extensively along their axes, do certainly form very stable bonds.

9

IONIC, HYDROGEN AND METALLIC BONDS

9.1 Introduction

So far we have been concerned with covalent bonding in small molecules usually in the gas phase. Many substances are, however, crystalline solids at room temperature. It may eventually be possible to describe all molecules by a single, comprehensive theory of bonding but, at present, most chemists find it convenient to classify bonds into a number of types – covalent, ionic, molecular, hydrogen, metallic, for example. The bonding in some solids is essentially all of one type, e.g. covalent in diamond, ionic in sodium chloride. Other substances may exhibit more than one type of bonding, e.g. in the rhombic sulphur structure we can recognize S_8 rings in which the S–S bonds are essentially covalent, the S_8 rings themselves being held in a regular three-dimensional array by molecular (van der Waals) bonds. Large molecules such as proteins may have structures which use covalent, ionic, molecular and hydrogen bonds. *Table 9.1* shows a possible classification of structures with a summary of the characteristic features of each structure type.

9.2 Ionic bonds

The Kossel, Lewis and Langmuir theories explained electrovalency, and the formation of electrovalent (i.e. ionic) bonds in terms of the electrostatic attraction between ions of opposite charge, these ions being formed by a complete transfer of electrons between atoms. The formation of ionic bonds is thus related to the ease with which ions can be formed from neutral atoms, and to the way in which the ions are packed together in the crystal structure.

The formation of positive ions from neutral atoms has already been discussed (p.42) in terms of ionization energy. *Figure 5.4* (p.44) shows that the alkali metals will be the most likely elements to lose one electron each and form stable cations, but even in these cases the amount of energy needed is quite large: ca. 500 kJ mol^{-1}. The formation of negatively charged ions is usually discussed in terms of 'electron affinity', the energy liberated when a neutral atom acquires an electron and forms a stable negative ion. Electron affinity can also be regarded as the ionization energy

Table 9.1 Classification of crystal structures

Type	*Structural unit*	*Bonding*	*Characteristics*	*Examples*
Ionic	Cations and anions	Electrostatic, non-directional	Strong, hard crystals of high m.p. Moderate insulators. Melts contain ions and are conductors. Some are soluble in liquids of high dielectric constant. Optical and magnetic properties are largely those of the constituent ions	Alkali halides
Covalent	Atoms	Covalent: limited number of electron-pair bonds, spatially directed	Strong, hard crystals of high m.p. Insulators	Diamond
Molecular	Molecules	Mainly covalent between atoms in molecule. van der Waals (or hydrogen) bonding between molecules	Soft crystals of low m.p. and large coefficient of expansion. Insulators	Iodine; ice; crystalline organic compounds
Metallic	Metal ions	'Metallic'; delocalized valence electron orbitals. Non-directional	Single crystals are soft; strength depends on structural defects and grain; good conductors; variable m.p.	Iron

of a negative ion, i.e. the energy needed to remove one electron from the singly charged ion

$$Cl^- + \text{'electron affinity'} \rightarrow Cl + 1e^-$$

Electron affinity is not easy to measure experimentally. Some values for halogens, oxygen ($O \rightarrow O^-$) and sulphur are quoted in *Table 9.2.*

Table 9.2 Electron affinities (ΔU)

Atom	F	Cl	Br	I	O	S
Electron affinity/ kJ mol^{-1}	333±8	357±6	337±4	308±4	141±40	208±12

A decrease in value with increasing ionic radius would be expected in a series such as the halide ions, but there is as yet no adequate explanation of the anomalous low value for fluorine.

Energy is liberated when the halogen atoms form halide ions with single negative charges. These ions will have the stable noble gas configuration, but the singly charged O^- ion would have to acquire a second electron to achieve this stable configuration. However, in the formation of the stable oxide ion, O^{2-}, more energy is absorbed in overcoming the repulsion between the second electron and the O^- ion than is supplied by the $O \rightarrow O^-$ conversion; there is, therefore, a net absorption of energy (702 kJ mol^{-1}) in the change from oxygen atom to oxide ion. We observe that the numerical values of electron affinities are, in general, smaller than those of the ionization potentials; thus the energy released in the formation of a negative ion may not be sufficient to remove an electron from another atom to form a positive ion. In the formation of an electrovalent compound, A^+B^-, from two atoms A and B, only part of the ionization energy of A is supplied by the electron affinity of B; however, more than sufficient energy is supplied by the electrostatic attraction between the ions when they are brought close together. This electrostatic attraction may be very large in ionic crystals, which consist of a regular arrangement of positive and negative ions in a crystal lattice. (Crystals are built up from small units packed together side by side; the points defining the corners of these units produce what is known as a simple crystal lattice.)

9.3 Crystal Lattice Energy*

M. Born and his collaborators developed a theory of ionic crystals from 1918 onwards, and the applications of this work have proved to be extremely useful for chemistry in general and inorganic chemistry in particular.

Lattice energy is defined as the energy liberated when one mole of gaseous cations and one mole of gaseous anions are brought together from

*This section is reproduced with permission from E. Cartmell, *Principles of Crystal Chemistry*, Monograph for Teachers No.18, Royal Institute of Chemistry, London (1971)

infinite distance apart to their equilibrium position in the crystal lattice at 0 K, i.e. the change in internal energy, ΔU, for the reaction

$$M^{z+}(g) + X^{z-}(g) \rightarrow MX(solid)$$

(Some authors define lattice energy as the increase in internal energy resulting from the separation of ions from their position in the crystal to infinite separation: this will give a different algebraic sign. The equation defining ΔU should always be quoted, otherwise arithmetical errors in crystal energy calculations are easily made.) Lattice energy can be calculated for structures consisting of spherical ions. The interaction energy between two ions of charge z^+e and z^-e separated by a distance R is according to the Coulomb law, $-z^+z^-e^2/R$ (in cgs units; or $-z^+z^-e^2/4\pi\epsilon R$ in SI units). *Figure 9.1*

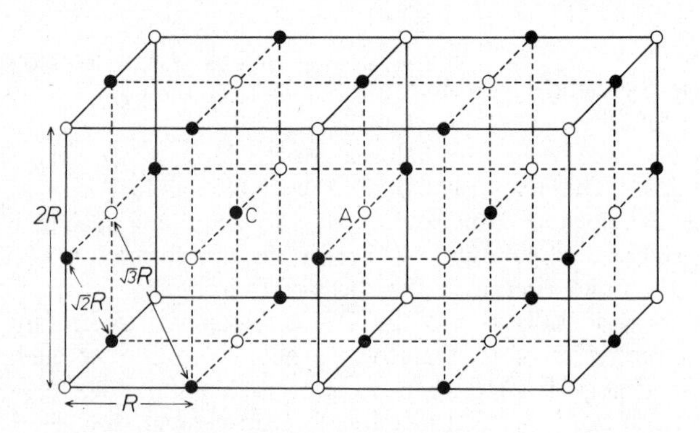

Figure 9.1 The NaCl structure: Na^+ black circles; Cl^- open circles

shows part of a sodium chloride structure. Consider the interaction energy between a particular cation C and all other ions in one mole of crystalline NaCl. In this structure C has

6 nearest neighbours (Cl^-) at a distance R
12 second neighbours (Na^+) at a distance $\sqrt{2}R$
8 third neighbours (Cl^-) at a distance $\sqrt{3}R$
6 fourth neighbours (Na^+) at a distance $\sqrt{4}R$, etc.

so that the Coulombic interaction energy at C becomes

$$-(6z^+z^-e^2/R) + [12(z^+)^2e^2/\sqrt{2}R] - (8z^+z^-e^2/\sqrt{3}R) + [6(z^+)^2e^2/\sqrt{4}R] \ldots$$

or

$$-z^+z^-e^2/R \times [6 - (12z^+/(\sqrt{2})z^-) + (8/\sqrt{3}) - (6z^+/2z^-) + \ldots]$$

The ratio of the ionic charges is constant for a given structure (e.g. $z^+/z^- = 1$ for NaCl, 2 for TiO_2), so that the Coulombic energy term can be written $U_C = -z^+z^-e^2A/R$, where A, called the Madelung constant, is written for the series enclosed within the square bracket above.

We can now repeat the analysis to get the Coulombic interaction at a particular anion A in the sodium chloride structure. This will also be $-z^+z^-e^2A/R$ since cations and anions have the same arrangement of neighbours in the sodium chloride structure. The total Coulomb energy for a lattice of N cations and N anions, where N is the Avogadro number, is then given by

$$U = \frac{1}{2}(U_C + U_A) = \frac{-z^+z^-e^2AN}{R}$$

The factor ½ is needed because $U_C + U_A$ includes the interaction between each pair.

Madelung constants have been calculated for a large number of structures (not without difficulty) and *Table 9.3* lists some of them. The Madelung constant is a measure of the additional interaction energy resulting from a three-dimensional lattice of ions. It may be compared with the interaction

Table 9.3 Madelung constants

Caesium chloride structure	1.763
Sodium chloride structure	1.748
Zinc blende structure	1.638
Wurtzite structure	1.641
Rutile structure	2.408
Fluorite structure	2.519

between a single pair of ions – there is a 75% increase in Coulombic energy when we go from the ion-pair $(Na^+)(Cl^-)$ to crystalline NaCl.

The Coulombic interaction energy, which represents a net attraction between the ions, must be opposed by a repulsion energy since the ions maintain an equilibrium separation in the crystal structure. A number of expressions has been suggested for this repulsion, which arises partly from the effects of overlapping charge-clouds and partly as a consequence of the Heisenberg uncertainty principle. Born and Landé first suggested the expression BN/R^n so that the lattice energy becomes

$$U = -z^+z^-e^2AN/R + BN/R^n$$

The constant B can be eliminated if we make use of the fact that, at $R = R_e$, the equilibrium separation of the ions in the crystal, $(\delta U/\delta R) = 0$. This gives

$$B = \frac{z^+z^-e^2A}{n}R_e^{n-1}$$

whence

$$U = \frac{-z^+z^-e^2AN}{R_e}\left(1 - \frac{1}{n}\right) \quad \text{(cgs units)}$$

$$= \frac{-z^+z^-e^2AN}{4\pi\epsilon R_e}\left(1 - \frac{1}{n}\right) \quad \text{(SI units)}$$

The integer n, which depends upon the nature of the ions, can be estimated from compressibility measurements. A value of $n = 9$ is obtained for alkali-metal halide structures. Later, when the development of quantum mechanics showed that electron wave functions decreased exponentially with increasing distance from the nucleus, Born and Mayer used the expression $B'N \exp(-R/\rho)$ for the repulsion term. The constant B' can again be eliminated by putting $(\delta U/\delta R)_{R = R_e} = 0$ and the expression for the lattice energy now becomes

$$U = \frac{-z^+z^-e^2AN}{R_e}(1 - \rho/R_e)$$

The constant ρ, for ions with inert-gas electron configuration, is about 0.35×10^{-10} m. A typical value of R_e is about 3×10^{-10} m, so that the repulsion term amounts to about 12% of the Coulombic interaction.

In very precise calculations of lattice energy, allowance has to be made for additional interactions. Polarization of one ion by its neighbours introduces a 'dipole–induced dipole' attraction NC/R^6 (where C is an empirical constant) and the residual (or zero-point) energy – vibrational energy retained by the lattice at 0 K – must also be included. These terms are, however, small (from 4 to 40 kJ mol^{-1}) compared with the total lattice energy, which may be several hundreds of kilojoules per mole.

Unfortunately, Madelung constants become increasingly difficult to calculate as the complexity of the structures increases, and in recent years a semi-empirical expression for lattice energies, due to Kapustinskii, has often been used. This is

$$U = -287.2\frac{\nu z^+z^-}{r_+ + r_-}(1 - \frac{0.345}{r_+ + r_-})$$

where ν is the number of ions in the formula of the substance considered (e.g. $\nu = 2$ for NaCl, 3 for CaF_2, etc.) and r_+ and r_- are the ionic radii. Values for the lattice energies of a number of crystals are quoted in *Table 9.4*, where they are compared with values obtained indirectly from thermochemical data, using the Born–Haber cycle, which we must now discuss.

9.4 The Born–Haber Cycle

There is no experimental method for the direct determination of lattice energy. A value can be obtained indirectly, however, using a method developed by Born and Haber which we will discuss using sodium chloride as an example. The enthalpy of formation of one mole of crystalline sodium chloride, NaCl(s), can be measured experimentally:

$$Na(s) + \tfrac{1}{2}Cl_2(g) \rightarrow NaCl(s); \quad \Delta H = -410 \text{ kJ mol}^{-1} \text{ (298 K)}$$

This process can be broken down into a number of stages:

1. $Na(s) \rightarrow Na(g); \quad \Delta H_1 = 108 \text{ kJ mol}^{-1}$ (298 K)
   This is the enthalpy of sublimation of sodium

2. $\frac{1}{2}Cl_2(g) \rightarrow Cl(g)$; $\Delta H_2 = 121$ kJ mol^{-1} (298 K)
   Here $\Delta H_2 = \frac{1}{2}$[enthalpy of dissociation of $Cl_2(g)$]

3. $Na(g) \rightarrow Na^+(g)$; $\Delta H_3 = 489$ kJ mol^{-1} (298 K)
   ΔH_3 is the ionization energy of sodium*

4. $Cl(g) \rightarrow Cl^-(g)$; $\Delta H_4 = -365$ kJ mol^{-1} (298 K) (i.e. exothermic)
   ΔH_4 is the electron affinity of chlorine*

5. $Na^+(g) + Cl^-(g) \rightarrow NaCl(s)$

ΔH_5 for this reaction† at 298 K can now be obtained by using Hess's law since the enthalpy change for the total process, ΔH, is independent of the route by which NaCl(s) is formed from Na(s) and Cl(g). Perhaps the safest way to apply Hess's law is to plot enthalpies on a vertical scale in a diagram of the type shown in *Figure 9.2*, where an arrow pointing

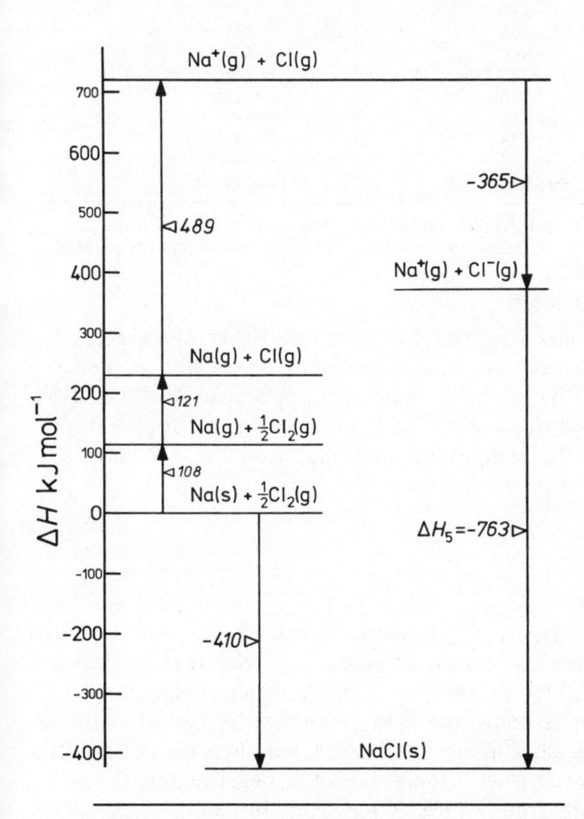

Figure 9.2 The Born–Haber diagram for NaCl

*Ionization energies and electron affinities are usually recorded in volts at 0 K. They are internal energy changes and when inserted as stages in the Born–Haber cycle the values have to be corrected to enthalpy changes (in kJ mol^{-1}) at 298 K

†The heat needed to vaporize crystalline NaCl is considerably less than the lattice energy since there is an appreciable concentration of ion-pairs, Na^+Cl^-, at temperatures just above the boiling point of NaCl

upwards indicates an increase in enthalpy (ΔH positive) for the particular stage. The vertical distances on the two sides of the diagram [i.e. from the NaCl(s) datum to the $Na^+(g) + Cl(g)$ datum] must be numerically equal so that

$$410 + 108 + 121 + 489 = -(365 + \Delta H_5)$$

whence $\Delta H_5 = -763$ kJ mol^{-1} at 298 K. This enthalpy change is often called the lattice energy in Born–Haber calculations. However, it should be corrected to give the true lattice energy, which is ΔU at 0 K for $Na^+(g) + Cl^-(g) \rightarrow NaCl(s)$. This correction ($\Delta H = \Delta U - 2RT$) is about 8–12 kJ mol^{-1} for simple compounds and so can usually be neglected.

Table 9.4 Lattice energies*/kJ mol^{-1}

Compound	*Calculated value*	*Born–Haber value*
LiCl	−825	−817
NaCl	−764	−764
KCl	−686	−679
KI	−617	−606
CdI_2	−1996	−2410
MgF_2	−2915	−2908
CaF_2	−2584	−2611
CaO	−3485	−3464
Al_2O_3	−15514	−15326

*The very high values for CaO and Al_2O_3 reflect the importance of the ionic charges; the product z^+z^- in the expression for the calculated lattice energy is 4 for CaO and 6 for Al_2O_3

In *Table 9.4* lattice energies obtained by the Born–Haber cycle are compared with those calculated assuming the solid is essentially an ionic structure. It will be seen that there is very good agreement for the alkali halides, reasonable agreement for halides such as MgF_2 and CaF_2, but no agreement for CdI_2 where the assumption that the structure is ionic is clearly not valid.

9.5 Ionic Radii

The size of ions is determined by the attractive force exerted on the outer electrons by the effective nuclear (positive) charge, i.e. the true nuclear charge diminished by the effect of the 'inner' or 'screening' electrons. Thus, when a neutral atom is converted into a positive ion we should expect the size to decrease since there has been a net increase in the effective nuclear charge. Conversely, we should expect a negative ion to be larger than the neutral atom from which it has been formed.

The distance between neighbouring positive and negative ions in crystals can be measured by X-ray diffraction methods; details of the method are outside the scope of the present work. Some values of the interionic distances in a series of alkali metal halides are given in *Table 9.5*. If we subtract the interionic distance in a sodium halide from the interionic distance in the corresponding potassium compound, we get an almost constant

Table 9.5 Interionic distances

Interionic distance/pm		Δ(KF − NaF, *etc.*)
KF	266	35
NaF	231	
KCl	314	33
NaCl	281	
KBr	329	31
NaBr	298	
KI	353	30
NaI	323	

value for the difference, Δ, as shown in the table. This is most simply explained if we assume that each ion acts as a sphere of constant radius; the measured interionic distance R (*Figure 9.3*) is then the sum of the radii of two spheres in contact, $r_+ + r_-$. It may seem surprising that this assumption works so well, for we emphasized in earlier chapters that, although the electron charge cloud is concentrated in shells close to the

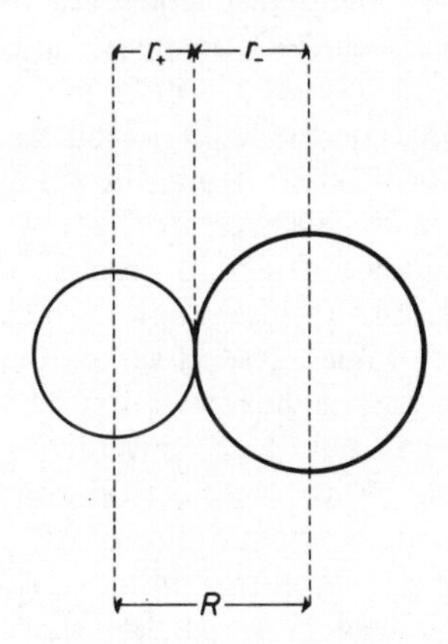

Figure 9.3 Interionic distance and ionic radii

nucleus, there is still a finite probability of finding an electron at a considerable distance from the nucleus. However, reference to *Figure 5.1*, p.39, which illustrates the radial distribution of charge density in the sodium ion, shows that this density decreases very rapidly as r increases, so that there is little error in assuming that the ion has a finite radius. We can, therefore, use experimentally determined values of interionic distances to obtain the radii of all other ions if the radius of one ion is already known. Thus, from *Figure 9.3*, if R and r_+ are known, r_- is given by:

$$r_- = R - r_+$$

The construction of tables of ionic radii now becomes possible. For many years, crystal chemists and geochemists have used tables prepared by V. Goldschmidt, which in turn were based on values of individual radii calculated by J.A. Wasastjerna using a theory that related the polarizability of an ion to its volume. Goldschmidt had himself measured interionic distances in a large number of crystals by the X-ray diffraction method and he made slight adjustments to the Wasastjerna values to get the best fit between observed interionic distances and the sums of radii.

The theoretical basis of Wasastjerna's approach is not now generally accepted, and the tables devised by Linus Pauling are normally used. Pauling emphasized that the radii would be a function of co-ordination number, the ratio of the radii and the charge on the ions. He considered isoelectronic compounds such as K^+Cl^- and divided the interionic distance in the inverse ratio of the effective nuclear charges of these ions. He obtained values for $r_+(K^+)$ = 133 pm and $r_-(Cl^-)$ = 181 pm. This enabled him to assemble a table of values of univalent radii and, subsequently, values for the radii of multivalent ions (called crystal radii) in structures of octahedral co-ordinations. Here again values were adjusted to get the best possible fit between observed interionic distances and sums of radii.

In recent years very precise X-ray diffraction studies have provided an experimental measurement of electron densities in crystals such as sodium chloride. The charge density contours appear in two-dimensional section as concentric circles centred on a regular array of sodium and chlorine nuclei. The electron density goes through a minimum along the line joining a sodium nucleus to its nearest-neighbour chlorine nucleus, and the position of this minimum should indicate the point at which the spherical cation and anion are in contact. Unfortunately the radii so obtained from the results of W. Witte *et al.* are strikingly different from the Pauling or Goldschmidt values. A new set of ionic radii has been devised by B.S. Gourary and F.J. Adrian based on these X-ray diffraction results. They are shown in *Table 9.6* with the corresponding Goldschmidt and Pauling values. It will be seen that the diffraction values give larger cations and smaller anions. *Table 9.6* also includes a column headed atomic radii. These are not concerned with ionic structures but it is convenient to include them here for comparison purposes. They refer to neutral atoms, and the most recent values are obtained from computations of the radius of maximum radial charge density $[4\pi r^2 R^2(r)]$ in the outermost shell of the atom. The ionic radii of cations are considerably smaller than their atomic radii, whereas the anion radii are greater than the corresponding atomic radii.

Table 9.6 Ionic and atomic radii/pm

Ion	*Goldschmidt*	*Pauling*	*Diffraction*	*Atomic*
Li^+	78	60	94	145
Na^+	98	95	117	180
K^+	133	133	149	220
Rb^+	149	148	163	235
Cs^+	165	169	186	260
Mg^{2+}	78	65	–	150
Ca^{2+}	106	99	–	180
Sr^{2+}	127	113	–	200
Ba^{2+}	143	135	–	215
F^-	133	136	116	50
Cl^-	181	181	164	100
Br^-	196	195	180	115
I^-	220	216	205	140
O^{2-}	132	140	–	60
S^{2-}	174	184	–	100

It is interesting to note that, in the early days of structure analysis, W.L. Bragg developed and used a set of radii characterized by large metal ions and small anions. Later, the theories of ion polarization seemed to demand small cations and large anions. Now we have electron-density determinations from precise X-ray diffraction studies which indicate a return to large cations and small anions, although the numerical values for the radii are not the same as those adopted by Bragg. However, it is too soon to consider a general replacement of Pauling or Goldschmidt values by a set based on diffraction results. In the first place only a very small number of electron-density determinations by high precisions methods has so far been completed, and the Fourier method used has serious limitations when applied to heavy atoms. The position of the electron-density minimum is also sensitive to the nature and arrangement of the surrounding ions. Thus, the minimum electron density along the calcium–fluorine line in the CaF_2 (fluorite) structure gives a Ca^{2+} radius of 96 pm and a F^- radius of 140 pm, which does not agree at all with the F^- radius (116 pm) obtained from the electron density in alkali-metal fluorides such as KF. The traditional Goldschmidt or Pauling values have been of immense assistance in working out some very complex structures. The important point is that any consistent set of radii chosen to get the best fit between the sum of the radii and the measured interionic distance will be helpful in discussing solid structures. The danger lies in the uncritical and 'quantitative' use of these same radii in other situations, as, for example, in discussions of ion hydration.

A discussion of trends in chemical properties in relation to the periodic classification can often usefully be linked with trends in ionic radii. As we go down Group IA, from lithium to rubidium, there is a steady increase in ionic radius since the effect of increasing nuclear charge is more than counteracted by increasing principal quantum number of the outer electrons and increased screening by the inner electrons. Negative ions also increase in size as we go down a group (cf. F^-, Cl^-, Br^- and I^-). It should be noted that negative ions are larger than isoelectronic positive ions (cf. F^- with Na^+, Cl^- with K^+) since the increased effective nuclear charge in the positive ions pulls the electrons closer to the nucleus. Again, if we compare the radii of Na^+ and Mg^{2+} we observe a marked decrease in radius, the net increase in positive charge pulling the electron closer to the nucleus.

9.6 Ionic Structures

Since ions with noble gas electronic structures are spherical, their attraction for ions of opposite charge is exerted to the same extent in every direction, and simple electrostatic considerations show that in an ionic structure containing spherical A^+ and B^- ions, the most stable arrangement is one in which the ions are in contact with each other and arranged in a symmetrical way. The structures are determined by two factors: the relative sizes of the ions and the requirement that the structure, as a whole, must be electrically neutral. The size factor largely determines the geometry of the structure, and we can discuss this in terms of co-ordination numbers. The most stable arrangement for a co-ordination number of 2 is a linear one, B–A–B, because this minimizes the repulsion between the negatively charged B^- ions, and the stable arrangements for three, four, six and eight co-ordinations are coplanar, tetrahedral, octahedral and cubical, respectively. Now a large positive ion can be in contact with a large number of negative ions, but a small positive ion can only accommodate a small number of large negative ions around and in contact with it. Let us consider a particular example, where the co-ordination number of the A^+ ion is six.

Figure 9.4(a) shows the immediate environment of an A^+ ion in the crystal structure; the A^+ ion is in contact with four coplanar B^- ions; there are, in addition, two more B^- ions in contact with A^+, one above and one below the plane of the diagram – these are omitted for clarity from *Figure 9.4*. *Figure 9.4(b)* shows the corresponding environment of A^+ in another compound AC, where the radius of the C^- ion is greater

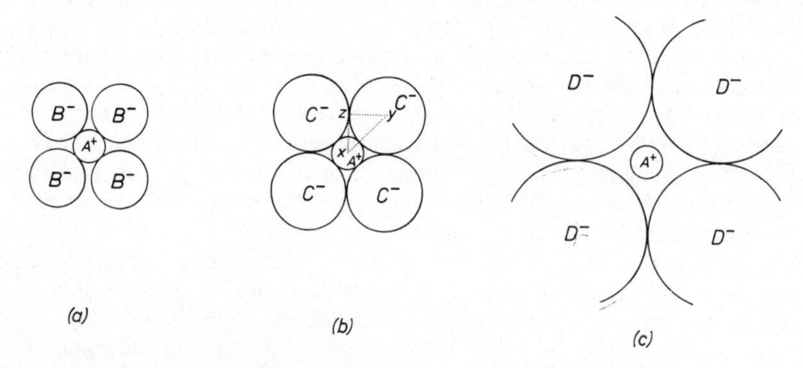

Figure 9.4 Ionic co-ordination and the limiting value of the radius ratio

than that of B^-; the negative ions are now in contact with each other as well as with the ion A^+. This represents a limiting condition for octahedral co-ordination.

If a third substance AD were considered, containing a still larger negative ion D^-, it would be impossible to get a stable structure with six co-ordination, because the negative ions would then be in contact with each other but not with the positive ion (*Figure 9.4(c)*). In this case, a more stable structure (with positive and negative ions in contact) would be obtained with fourfold co-ordination. The limiting condition, illustrated in *Figure 9.4(b)*, is defined by the relative sizes of the ions; there is a limiting value of the radius ratio, r_+/r_-, which can easily be calculated from the geometry of *Figure 9.4(b)*:

$$zy = xy \cos 45°$$

i.e.

$$r_- = (r_+ + r_-)/\sqrt{2}$$

whence

$$\frac{r_+}{r_-} = \sqrt{2} - 1 = 0.414$$

Table 9.7 Limiting values for the radius ratio

Co-ordination	*Limiting value* $\frac{r_+}{r_-}$
8 (Cubic)	0.732
6 (Octahedral)	0.414
4 (Square-coplanar)	0.414
4 (Tetrahedral)	0.225
3 (Trigonal-coplanar)	0.155

The radius ratio for the structure of the compound AD shown in *Figure 9.4(c)* will be less than 0.414 since D^- is larger than C^-. Thus octahedral co-ordination is only possible if the radius ratio is greater than the limiting value. The limiting values of the radius ratio for other co-ordination numbers can easily be calculated and are given in *Table 9.7. Figure 9.5* is a three-dimensional representation of (*a*) anions in contact with each other but not with the cation, and (*b*) anions in contact with the cation.

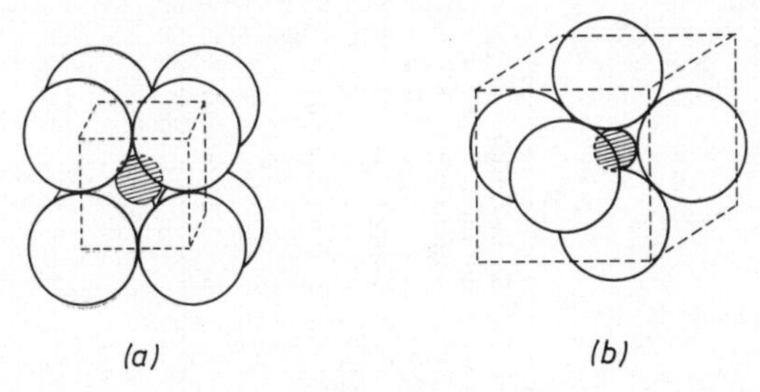

Figure 9.5 (a) Anions in contact with each other but not with the cation; (b) rock salt structure, with six anions in contact with the cation

A knowledge of ionic radii may be a guide to the structure of an ionic crystal. Thus a radius ratio greater than 0.732 implies that, from a geometrical point of view, cubical co-ordination should be possible, whereas a ratio between 0.732 and 0.414 would require the co-ordination number to fall to 6 (octahedral). Unfortunately, the radius ratio is only a rough guide. Sodium fluoride has a radius ratio of 0.74, and potassium fluoride one of 1.0, but both these compounds crystallize in the rock salt structure with co-ordination numbers of 6. Although twelve fluoride ions can pack around a positive potassium ion, geometry does not permit a packing of twelve potassium ions around each fluoride ion in the crystal structure. Geometry would allow cubical co-ordination, however, giving a caesium chloride structure. It seems likely that in these cases 'second-order' effects such as van der Waals' forces may be decisive in determining the co-ordination, since there is only a small difference in crystal energy between a caesium chloride and a sodium chloride structure.

The co-ordination number of the positive ion is not necessarily the same as that of the negative ion. It will be the same in the case of structures of general formula AB, where the requirement of electrical neutrality means that there must be equal numbers of A^+ and B^- ions in the structure, but in a structure of formula AB_2, containing A^{2+} and B^- ions, there must be twice as many negative as positive ions to preserve overall neutrality. The co-ordination of the positive ion must hence be double that of the negative ion.

Table 9.8 Ionic structures

Type of structure	*Examples*	*Radius ratio*	*Co-ordination number of positive ion*	*Co-ordination number of negative ion*
Caesium chloride	CsCl, CsBr, CsI	>0.732	8 (Cubic)	8 (Cubic)
Rock salt	NaCl, NaBr, NaI MgO, CaO, MnO	<0.732 >0.414	6 (Octahedral)	6 (Octahedral)
Fluorite	CaF_2, SrF_2, ThO_2	>0.732	8 (Cubic)	4 (Tetrahedral)
Rutile	TiO_2, SnO_2, PbO_2	>0.414	6 (Octahedral)	3 (Trigonal)

Table 9.8 summarizes the essential structural features of a number of simple inorganic substances. *Figure 9.6* illustrates the 'unit cell' of the caesium chloride structure and the way in which the units stack together to build up the crystal. The caesium ions form a simple cubic lattice interlocking with, but displaced from, a similar lattice of chloride ions.

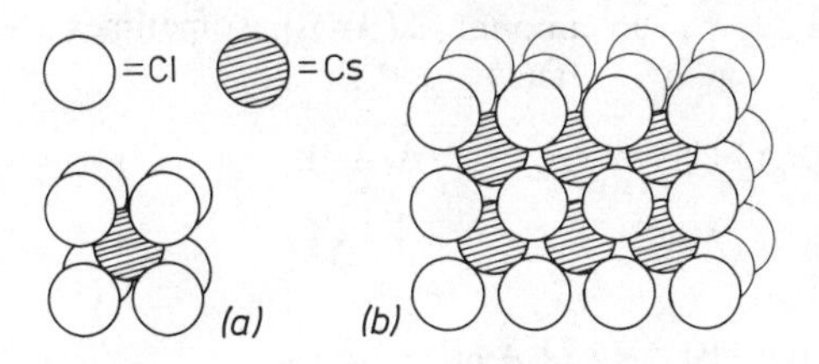

Figure 9.6 Caesium chloride: (a) unit cell of crystal; (b) structure

Rock salt has a structure in which the sodium ions form a lattice with points at the corners of a cubic unit and also at the centres of the cube faces (see *Figure 9.1*). The chloride ions form an interlocking face-centred cubic lattice.

9.7 Electronegativity

Although it is convenient to distinguish between covalent and ionic bonds, many bonds of a type intermediate between the purely ionic and the purely covalent are known. Thus, in the series of isoelectronic molecules CH_4, NH_3, OH_2 and FH, the electronic charge distribution in the C–H, N–H, O–H and F–H bonds is increasingly concentrated in a direction away from the hydrogen atom, so that these essentially covalent bonds have increasing 'ionic character' as we go along the series from C–H to F–H. The C, N, O and F atoms are arranged in order of increasing 'electronegativity' – defined as the power of an atom in a molecule to attract electrons to itself. This definition may be extended as follows.

Consider a diatomic molecule A–B: then, using valence-bond theory, we can describe this molecule by a wave function which is a linear combination of wave functions corresponding to structures such as A–B, A^+B^-, A^-B^+, i.e.

$$\psi = c_I\psi_I(A\text{–}B) + c_{II}\psi_{II}(A^+B^-) + c_{III}\psi_{III}(A^-B^+)$$

If, in this expression, $c_{III} > c_{II}$, we say that A is more electronegative than B, whereas if $c_{III} = c_{II}$ the atoms have the same electronegativity.

There have been many attempts to assign quantitative electronegativity values, and we shall only discuss the more important ones.

9.7.1. METHOD I (Pauling)

If, in the expression for the wave function of the diatomic molecule A–B given above, $c_{II} = c_{III} = 0$, the bond is said to be purely covalent, and the bond energy, $D(\text{A–B})$, will be equal to E_{cov}, the pure covalent bond energy. E_{cov} cannot be measured directly but its value is usually taken as the arithmetic mean of the bond energies of the homonuclear diatomic covalent molecules A–A and B–B. If A and B have different electronegativities, however, the bond A–B will no longer be purely covalent, and the bond energy, $D(\text{A–B})$, will be greater than that of the purely covalent bond, E_{cov}, by an amount $\Delta(\text{A–B})$, sometimes called the 'ionic–covalent resonance energy'. Thus

$$D(\text{A–B}) = E_{cov} + \Delta(\text{A–B})$$

or

$$\Delta(\text{A–B}) = D(\text{A–B}) - E_{cov}$$

Now the greater the difference between the electronegativities of A and B, the greater will be the value of $\Delta(\text{A–B})$, and Pauling makes use of the following empirical expression to relate $\Delta(\text{A–B})$ with electronegativity differences:

$$0.208\sqrt{\Delta} = |x_A - x_B|$$

where x_A represents the electronegativity of atom A. (The factor 0.208 arises from the conversion of $\Delta(\text{A–B})$ measured in kcal mol^{-1} into electron volt energy units.)

Pauling selected an arbitrary x value of 2.1 for hydrogen – this gave a suitable range of numerical values for the elements C to F in the first row of the Periodic Table; other values can then be assigned if the $\Delta(\text{A–B})$ values are known. Unfortunately, however, the determination of $\Delta(\text{A–B})$ is subject to considerable errors, and numerical values for electronegativities have to be scrutinized with some care.

9.7.2 METHOD II (Mulliken)

Mulliken defines the electronegativity, M_A, of an atom A as the arithmetic mean of its electron affinity, E_A, and its ionization energy, I_A, i.e.

$$M_A = \tfrac{1}{2}[E_A + I_A]$$

The reason for choosing this function can be explained qualitatively if we realize that, if A is more electronegative than B, then the energy required to form an ionic bond A^-B^+ from the neutral atoms will be less than that needed to form an ionic bond A^+B^-, since it is clearly easier to transfer the electron from atom B to atom A than vice versa, if atom A has the greater tendency to attract the electron. Now the energy which is required to produce A^- and B^+ from A and B is given by $I_B - E_A$; similarly, the energy needed to produce A^+ and B^- is given by $I_A - E_B$. Thus if

$$I_A - E_B = I_B - E_A$$

then atoms A and B must be equally electronegative. On the other hand, A will be more electronegative than B if

$$I_A - E_B > I_B - E_A$$

i.e. if

$$I_A + E_A > I_B + E_B$$

The sum of the ionization potential and electron affinity is thus a measure of the electronegativity. It is important to note, however, that the appropriate values of ionization energy and electron affinity are not the same as those of an isolated atom. Electronegativity relates to an atom combined with other atoms, so that in beryllium, for example, where the atom uses *sp* hybrid orbitals, we can obtain two different ionization energies, corresponding to the removal of the *s* or the *p* electron. We saw on p.40 that *s* electrons are more firmly held than *p* electrons of the same principal quantum number, so that the value of ionization energy used in electronegativity determinations will depend upon the nature of the bond hybridization (see also p.122). The Mulliken approach is also limited by the difficulty of getting reliable electron affinity values. A plot of Mulliken electronegativities against Pauling values gives a straight line going through the origin, and the slope of this line gives the equation

$$x_{\text{Pauling}} = x_{\text{Mulliken}}/3.15$$

relating the two electronegativity scales.

9.7.3 METHOD III (Sanderson)

R.T. Sanderson suggests that electronegativity is related to the 'compactness' of the electron charge cloud of an atom relative to that of a hypothetical isoelectronic inert atom, since if an atom is very electronegative, i.e. strongly attracts other electrons, its own electrons will be held close together. He therefore represents electronegativities as a Stability Ratio, *SR*, defined as the ratio of the average electron density of the atom to that of a hypothetical isoelectronic inert atom. The Sanderson and Pauling values are related by the expression

$$\sqrt{x_{\text{Pauling}}} = 0.21SR + 0.77$$

The two scales agree quite closely except for germanium, arsenic and antimony, where the Sanderson values are appreciably higher than the corresponding Pauling electronegativities. A high value for Ge is also obtained in an electronegativity scale devised by Allred and Rochow, and based on calculations of the electrostatic force exerted by a nucleus on an electron in a bonded atom. Sanderson, and Allred and Rochow, claim that there is a great deal of chemical evidence in favour of the high value for germanium; Drago, however, has pointed out that these chemical properties

Table 9.9 Electronegativity values

H 2.1						
Li 1.0	Be 1.5	B 2.0	C 2.5	N 3.0	O 3.5	F 3.9
Na 0.9	Mg 1.2	Al 1.5	Si 1.8	P 2.1	S 2.5	Cl 3.0
K 0.8	Ca 1.0	—	Ge 1.8	As 2.0	Se 2.4	Br 2.8
Rb 0.8	Sr 1.0	—	Sn 1.8	Sb 1.9	Te 2.1	I 2.5
Cs 0.7	Ba 0.9					

can equally be explained, while retaining the Pauling values for electronegativity, as a consequence of the presence of *d* electrons in the electron configuration of the germanium atom. *Table 9.9* lists a selection of electronegativity values.

This concept of electronegativity is also of importance to theoretical organic chemistry, where chemical reactivity can be correlated with differences in the electron charge density around particular atoms. The ionization energy of an *s* electron is greater than that of a *p* electron, since the *s* electron is, on the average, more under the influence of the nucleus. This means that in a hybrid (*sp*) orbital, the greater the *s* character of the hybrid the greater will be the effective electronegativity of the atom giving rise to the hybrid orbital. The electronegativity of the carbon atom in ethyne (*sp* hybrids) is thus greater than it is in methane, where carbon is using sp^3 hybrid orbitals. It is this fact that accounts for the acidic properties of ethyne, e.g. the ease with which one of its hydrogen atoms can be replaced by sodium.

9.8 The Hydrogen Bond

The covalent and the ionic bonds that we have been discussing are usually very strong, with bond energies in the range ca. 100–500 kJ mol^{-1}. We now discuss a very much weaker link, the so-called 'hydrogen bond', where the bond energies are in the 10–40 kJ mol^{-1} range. A hydrogen bond between two atoms A and B can be written A–H· · ·B; this implies that A and B are sufficiently close together for a bond to exist, and that the hydrogen atom involved in the bond is attached to atom A by a covalent bond. Hydrogen bonding is only important when the atoms A and B are strongly electronegative, e.g. nitrogen, oxygen and fluorine, and *Table 9.10* lists the lengths of some typical hydrogen bonds in selected compounds of these elements.

9.8.1 HYDROGEN BONDING AND CRYSTAL STRUCTURE

The structures of many crystalline substances are determined by hydrogen bonding, and we can conveniently discuss these as structures in which

Table 9.10 Hydrogen bond lengths

A	B	*Example*	$R_{A-H...B}$/pm
F	F	HF_2^- in KHF_2	226
O	O	HCO_3^- in $KHCO_3$	261
		H_2BO_3	270–273
		Ice (cubic)	276
N	O	Proteins	267–307

there are (a) discrete complex ions containing hydrogen bonds, (b) ions linked through hydrogen bonds into infinite chains, (c) infinite two-dimensional layers, and (d) three-dimensional macromolecules.

9.8.1.1 *Discrete ions*

A detailed analysis of KHF_2 by neutron diffraction reveals an ionic crystal containing K^+ cations and HF_2^- anions. The anion is linear, the two fluorine atoms being linked through a hydrogen atom mid-way between them, i.e. F· · · H· · · F.

9.8.1.2 *Chain structures*

The planar carbonate ions in $KHCO_3$ are linked through hydrogen bonds into infinite chains (1). Hydrogen bonding is also an essential feature of protein structures, where the molecules are built up by the linking of peptide (amide) units (2) to form long chains. There is now considerable

(1)

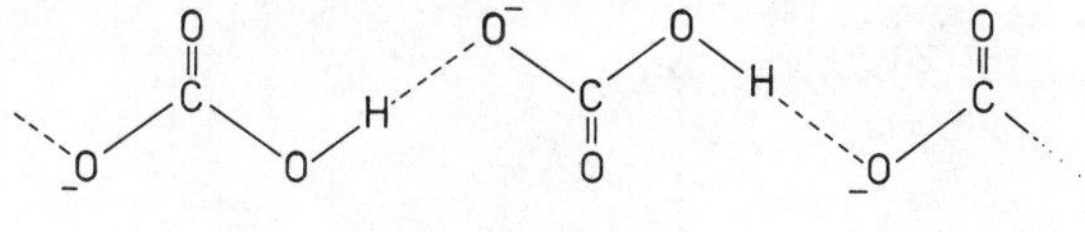

(2)

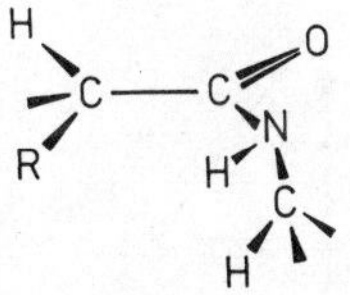

evidence that in many crystalline proteins these polypeptide chains are coiled into a spiral structure, the α-helix, held together by hydrogen bonds linking the carbonyl oxygen of one peptide residue with a nitrogen atom in the third unit along the chain.

9.8.1.3 *Infinite layers*

Boric acid is a good example of the way in which a planar group of atoms such as BO_3^{3-} can be linked into a planar network through hydrogen bonding. In the structure shown in (3), dashed lines indicate the position of the hydrogen bonds.

(3)

9.8.1.4 *Three-dimensional macromolecules*

Ice can exist in a number of crystalline forms. Thus a cubic (diamond) structure can be obtained below −80 °C, while at 0 °C ice has a hexagonal structure. In the cubic form, each oxygen atom is surrounded tetrahedrally by four other oxygen atoms at a distance of 276 pm, giving a very open structure which owes its stability to hydrogen bonding. The hydrogen atom are located between the oxygen atoms in such a way that, at any instant, one oxygen atom has two hydrogen atoms attached at distances corresponding with the length of a covalent O–H bond (99 pm) and two hydrogen atoms at much greater distances (177 pm) held by hydrogen bonding (*Figure 9.7(a)*).

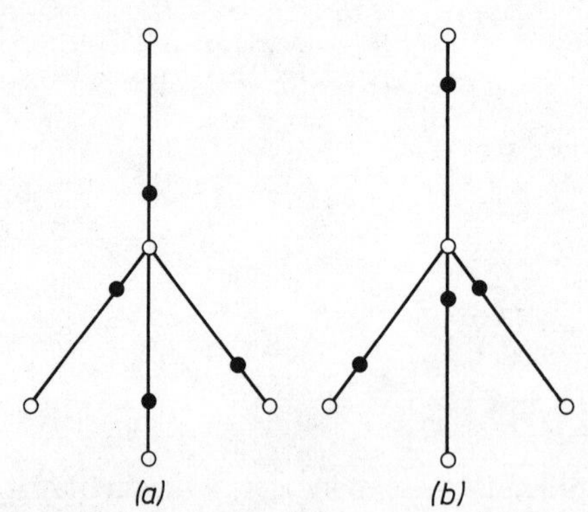

Figure 9.7 The structure of ice: (a) and (b) show alternative arrangements of the hydrogen atoms (represented by black circles)

At another instant the hydrogen atoms may have the configuration indicated in *Figure 9.7(b)*. A hydrogen atom does not therefore have a precise location between two oxygen atoms in the ice crystal; it effectively oscillates between two extreme positions.

9.8.2 THEORIES OF HYDROGEN BONDING

A hydrogen atom, of configuration $1s^1$, can form only one covalent bond, since the $2s$ and $2p$ orbitals have energies too high to be involved in the formation of additional bonds. We have, therefore, to seek another explanation of the way in which hydrogen can link two atoms together in the bond A–H· · · B.

9.8.2.1 *The electrostatic approach*

The fact that hydrogen bonding occurs only between very electronegative elements suggests that electrostatic interaction between polar groups may be a sufficient explanation.

Let us consider the covalent bond between hydrogen and some very electronegative atom A. The electron charge cloud between the two atoms will be considerably distorted to give a greater concentration of electron density in the vicinity of A and a strong dipole will be produced. A second electronegative atom B attached to another atom or molecule will also form the negative end of a dipole. If the two dipoles approach one another along the line A–H· · ·B, then the electrostatic attraction between the positive end of the dipole on A–H and the negative charge on B will be greater than the repulsive forces between like charges, and the closer that B can approach to H the stronger will be this electrostatic link between them. The strength of this electrostatic attraction can be calculated approximately, assuming a model in which orbitals are represented by point charges at their centroids. Lone-pair electrons play an important part in determining the strength and direction of hydrogen bonds, and Schneider's calculations on the water molecule indicate that the strongest bond is formed when the O–H axis of one molecule is collinear with the lone-pair axis on the oxygen atom of a neighbouring water molecule. Calculations of this kind usually give a reasonable hydrogen bond energy of about 25 kJ mol^{-1}, but there are some objections to this theory, and it seems that in many cases the hydrogen bond may have some covalent character.

9.8.2.2 *Valence-bond treatment*

Coulson and Danielsson have attempted to determine the covalent contribution to the hydrogen bond by considering structures such as

A–H	covalent A–H bond
$A^-H^+\cdots B$	ionic A–H bond
A^-H——B^+	covalent H–B bond

and a later treatment by Tsubomura considers also the structure

$$A^+H^-\cdots B$$

The calculations are necessarily approximate, and they involve a considerable number of assumptions, but they indicate that the covalent contribution may be appreciable, especially in 'short' hydrogen bonds (ca. 250 pm).

9.8.2.3 *Molecular-orbital treatment*

Pimentel has discussed hydrogen bonding in the HF_2^- ion using molecular-orbital theory. He takes linear combinations of the hydrogen s orbital with fluorine p orbitals directed along the bond. Thus if the p orbitals of the two fluorine atoms are denoted p_A and p_B, the symmetrical F—H· · ·F bond is described by

$$\psi_1\text{(bonding)} = p_A + p_B + a_1 s$$

$$\psi_2\text{(non-bonding)} = p_A + p_B$$

$$\psi_3\text{(anti-bonding)} = p_A + p_B - a_3 s$$

where a_1 and a_3 are mixing coefficients. The two electrons involved in bond formation are contained in a molecular orbital (ψ_1) which spreads out axially on either side of the hydrogen atom, thus forming two equivalent weak bonds, and there are also two electrons concentrated mainly on the fluorine atoms in the non-bonding orbital ψ_2. See also I_3^-, p.188.

9.8.3 THE IMPORTANCE OF HYDROGEN BONDING

The comparatively small value of the hydrogen bond energy has some important consequences, because chemical reactions which involve the breaking of weak bonds may take place readily at room temperature; this is especially the case in many biochemical systems.

Hydrogen bonding is also of importance in explaining the properties of certain liquids, e.g. the high dielectric constants, high boiling points and low vapour pressures of solvents such as water and the alcohols, the so-called polar solvents, are ascribed to intermolecular association through hydrogen bonds. Intramolecular (or internal) bonding may also occur; thus in *o*-nitrophenol the hydrogen of the phenolic group and an oxygen atom of the nitro group are sufficiently near to each other for appreciable electrostatic interaction to occur, and a six-membered ring system (4)

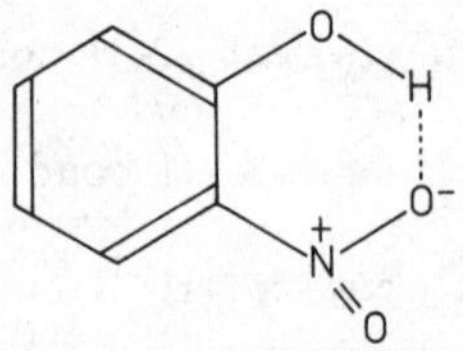

(4)

is formed. This explains why the *o*-isomer is more volatile than the *m*- and *p*-nitrophenols; the latter show intermolecular association and have low vapour pressures, whereas the *o*-isomer is not associated since its hydrogen bond has been formed internally, and the isomers can be readily separated from each other.

9.9 Metallic Bonds

Metal structures have some very characteristic properties. Each atom in the crystal has a very high co-ordination number (frequently twelve but sometimes eight), and the structure has high electrical and thermal conductivity. The metal atoms are packed close together in the crystal; this implies that there is extensive overlap of the outer electron orbitals and that the valency electrons can no longer be associated with a particular nucleus, but are completely delocalized over all the atoms in the structure. A given metal can thus be regarded as an assembly of positive ions which, in general, are spheres of identical radius (manganese and uranium are exceptions) packed together to fill space as completely as possible. There are two ways of 'close-packing' identical spheres: one of these has hexagonal and the other cubic symmetry, but in each case the co-ordination number is twelve. The 'body-centred' cubic structure of, e.g. the alkali metals, is less closely packed, for here the co-ordination number is eight, and each ion has eight nearest-neighbours at the corners of a surrounding cube.

9.9.1 VALENCE-BOND THEORY OF METALS

The bonds between one atom in a metal and its nearest neighbours cannot be electron-pair links, since there are not enough electrons available – an alkali metal atom, for example, has only one electron to share with eight neighbours. However, we can write down a large number of structures of the type

```
Na–Na   Na–Na   and   Na–Na   Na    Na
                               |     |
Na–Na   Na–Na         Na–Na   Na    Na
```

and compound them to give a structure which is a resonance hybrid of these forms. The bond will be weak, however, since one electron pair has to link eight atoms. A considerably stronger bond will be obtained if, as Pauling suggests, we also consider structures such as

```
Na+ Na–Na Na
           |
Na–Na−–Na Na
```

in which there are negative ions forming two covalent bonds by making use of *sp* hybrid orbitals. These structures only become important in the linear combination if there are unoccupied orbitals of suitable energy available for hybridization. If these orbitals are available, there will be a large number of these structures, and the bond will be correspondingly stronger. Thus in sodium the energy of the 3*p* orbitals, normally unoccupied in isolated sodium atoms, is not much greater than that of the 3*s* orbital containing the valency electron when the atoms are close together as they are in the metal, and *sp* hybrid orbitals can be formed. Carbon, on the other hand, is a non-metal, since, in the diamond structure, the four tetrahedrally arranged sp^3 hybrid orbitals are all doubly occupied, and the energies of the 3*s* and 3*p* orbitals are too great to be available for the formation of new hybrid orbitals.

The strength of the bond in these structures will depend upon the number of valency electrons that each atom can contribute. Thus, as we go along the first long row of the Periodic Table, the K, Ca, Sc, Ti, V and Cr atoms can contribute 1, 2, 3, 4, 5 and 6 electrons, respectively, to the structure, and the corresponding increase in bond strength from K to Cr is made evident by the steady increase in melting points and hardness, and a decrease in inter-atomic distances. These physical properties remain roughly constant from Mn to Ni, and Pauling assigns a 'metallic' valency of 6 to these elements also. His values of 5½ and 4½ for the 'metallic' valencies of copper and zinc, respectively, are based on magnetic properties of these elements. The fractional values represent situations in which, at a given instant, some atoms are in one valency state (e.g. 6) and some in another (e.g. 3 or 4). These values have, however, been criticized.

9.9.2 HYBRID ORBITALS IN METAL STRUCTURES

Altmann, Coulson and Hume-Rothery have likewise discussed metal structures in terms of the directional character of hybrid orbitals. Just as carbon atoms in the diamond are arranged in a structure described by the tetrahedral disposition of sp^3 hybrid orbitals, so, in a metal, hybrid orbitals may also describe the structure. In metals, however, the hybrid orbitals are only 'partially occupied'. This means that we can write down a large number of possible structures in some of which a particular orbital may be occupied by electrons, whereas this same orbital may be unoccupied in other structures. The linear combination of wave functions corresponding to these structures then describes a state in which the hybrid orbitals are 'partially occupied'.

The details of their discussion lie outside the scope of this book but, briefly, they consider that body-centred metal structures are determined by the geometry of sd^3 hybrid orbitals, cubic close-packed structures by p^3d^3 hybrids and hexagonal close-packed structures by sd^2, pd^5 and spd^4 hybrid orbitals. These hybrids have differing *d* 'character' or 'weight' and the structure adopted by a transitional metal, for example, is determined by the number of *d* electrons it can contribute to these hybrid orbitals.

9.9.3 MOLECULAR-ORBITAL THEORY OF METALS

The delocalization of the 'free' electron orbitals over all the atoms of a metal structure is, essentially, a molecular-orbital approach to metallic bonding. Now we saw in Chapter 6 that when two atoms combine to form a diatomic molecule, two molecular orbitals are formed by the overlap of atomic orbitals of the combining atoms. When a third atom is added to the diatomic molecule, three molecular orbitals are obtained, and, in general, when there is overlap of orbitals on N atoms in a solid structure, N molecular orbitals will be obtained. Now each orbital has

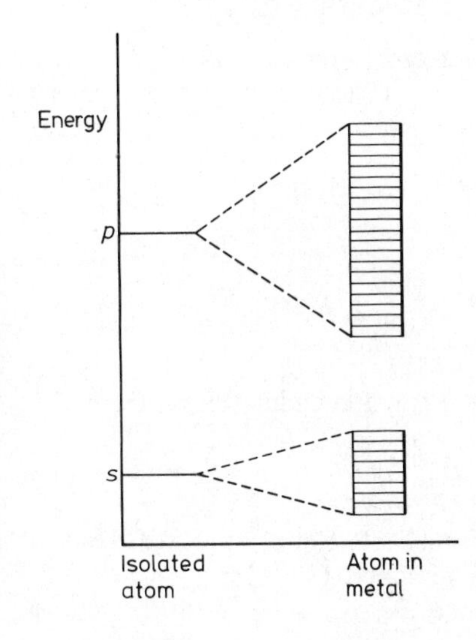

Figure 9.8 Energy levels and energy bands in isolated atoms and in model structures

an associated energy value and *Figure 9.8* illustrates diagrammatically the difference between the energy levels of two orbitals of an isolated atom, and the large number of closely spaced levels which result from the overlapping of orbitals in a metal structure. The levels become so close together in the metal that they effectively form a 'band' of energy values. These bands may be separated from each other, as in *Figure 9.8*, or they may overlap. Each level can accommodate two electrons, so that, if there are N 'free' electrons, the first $N/2$ states will be doubly occupied.

Now electrical conduction depends upon the movement of electrons through the structure under the influence of an applied field, and this movement can only occur if the electrons can accept energy and move to higher, unoccupied levels. Metals, therefore, are structures in which accessible, unoccupied levels are available, whereas in non-metallic substances, which are usually insulators, all the accessible levels are already occupied. The so-called 'semi-conductors' are substances in which there is an unoccupied energy band not too far removed from a band which is completely occupied. At low temperatures the electrons in the occupied band have not got sufficient energy to transfer to the unoccupied or 'conduction' band, but at high temperatures they may acquire sufficient energy to make this transition; the increase in conductivity with temperature, characteristic of semi-conductors, is thus explained.

9.10 The Molecular or van der Waals' Bond

The 'non-ideal' behaviour of gases on compression was ascribed by van der Waals to the existence of weak forces of attraction between atoms or molecules in the gaseous state. These forces are also present in the liquid and solid state; thus crystals of the noble gases have structures in which the monatomic molecules are held in a close-packed arrangement by the van der Waals' forces. The same force produces the so-called 'molecular bond' linking, for example, discrete diatomic molecules in solid halogen structures.

Many organic crystals, e.g. naphthalene, have structures in which discrete molecules are linked by molecular bonds, and the low melting points of such crystals show that the link is very weak compared with covalent or electrovalent bonds – about 42 kJ mol^{-1}.

A quantum theory of this bond has been worked out by London but an account of it is outside the scope of this book.

9.11 Bibliography

1. WELLS, A.F., *Structural Inorganic Chemistry*, Clarendon Press, Oxford 4th Edition (1975)
2. PIMENTEL, G.C. and McCLELLAN, A.L., *The Hydrogen Bond,* Freeman, San Francisco and London (1960)
3. CARTMELL, E., *Principles of Crystal Chemistry*, Monographs for Teachers, No.18, Royal Institute of Chemistry, London (1971)
4. GREENWOOD, N.N., *Ionic Crystals, Lattice Defects and Nonstoichiometry,* Butterworths, London (1968)

10

THE STRUCTURES OF SOME SIMPLE INORGANIC COMPOUNDS

10.1 Bond Lengths

A number of bond lengths have been quoted in previous chapters, and it has been pointed out that the bond length (for a given pair of linked atoms) decreases as the bond order increases. Thus single, double and triple carbon–carbon bonds are 154, 133 and 120 pm, respectively. We have further seen that bond lengths vary somewhat even when there is no multiple bond character, and depend upon the precise nature of the orbitals used, i.e. the relative amounts of *s* and *p* character. Thus the C–H bond length ranges from 105.7 in ethyne to 109.4 pm in methane. These latter deviations are relatively small, however, and it is useful to list average bond lengths for pairs of bonded atoms. *Table 10.1* gives a set of values for quite a few single and multiple covalent bonds, the data being based on the Chemical Society's *Table of Interatomic Distances* (1958,

Table 10.1 Selected bond lengths

Bond	*Bond length/* pm	*Molecule*	*Bond*	*Bond length/* pm	*Molecule*
B–Br	187	BBr_3	N–C	147	$N(CH_3)_3$
B–C	156	$B(CH_3)_3$	N–F	137	NF_3
B–Cl	175	BCl_3	N–H	101	NH_3
B–F	130	BF_3	P–C	184	$P(CH_3)_3$
B–O	136	$B(OH)_3$	P–Cl	204	PCl_3
C–Br	194	CBr_4	P–F	155	PF_3
C–C	154	C_2H_6	P–H	142	PH_3
C=C	133	C_2H_4	P–O	160	P_4O_{10}
C≡C	120	C_2H_2	P=O	145	$POCl_3$
C–Cl	177	CCl_4	P=S	185	$PSCl_3$
C–F	132	CF_4			
C–H	110	CH_4	As–C	196	$As(CH_3)_3$
C–N	147	$N(CH_3)_3$	As–Cl	216	$AsCl_3$
C≡N	115	HCN	As–F	171	AsF_3
C–O	141	$(CH_3)_2O$	As–H	152	AsH_3
C=O	122	$(CH_3)_2CO$	O–Cl	170	OCl_2
C–S	180	$(CH_3)_2S$	O–F	141	OF_2
Si–C	189	$Si(CH_3)_4$	O–H	96	OH_2
Si–Cl	208	$SiCl_4$	S–Cl	199	SCl_2
Si–H	148	SiH_4	S–F	158	SF_6
			S–H	132	SH_2
Ge–C	198	$Ge(CH_3)_4$	S=O	141	SO_2F_2
Ge–Cl	208	$GeCl_4$			
Ge–H	153	GeH_4	Se–H	146	SeH_2
Sn–C	218	$Sn(CH_3)_4$			
Sn–Cl	231	$SnCl_4$			

1965)[1], and Gordy and Cook's compilation of microwave data[2]; a few of the values are taken from more recent research papers. It should be emphasized, however, that the bond lengths quoted are for bonds in the molecules listed, and may be slightly different in other molecules.

10.2 Calculated Bond Lengths

From time to time we shall refer to calculated values of particular bond lengths. These values are obtained by adding together the 'covalent radii' of the two bonded atoms, where the covalent radius of a given atom A is obtained by taking one half of the measured A–A bond length. Thus the measured C–C bond length is 154 pm so the covalent radius is considered to be 77 pm. In the same way the chloride covalent radius (one half of the Cl–Cl bond length) is 99 pm, so that a C–Cl bond is calculated to be the sum of 77 and 99, which is 176 pm; the C–Cl bond length in carbon tetrachloride is found experimentally to be 177 pm, which is in very good agreement with the calculated value.

In general, though, there is usually a bigger discrepancy than this, and the calculated bond length values are normally significantly larger than those determined experimentally. Readers can see this for themselves by calculating bond lengths from the values of single bond covalent radii (*Table 10.2*) or multiple bond covalent radii (*Table 10.3*) and comparing them with the corresponding experimental values given in *Table 10.1*. As one illustration we can look at, say, the Sn–Cl bond length, which is found to be 231 pm in $SnCl_4$, whereas the sum of the Sn and Cl single covalent radii (141 + 99 pm) is 240 pm. The difference largely arises because the theoretical values of covalent radii are based on bond lengths found in homonuclear molecules, whereas the bonds in molecules such as $SnCl_4$ are between atoms of differing electronegativity. This difference in electronegativity can be corrected empirically by means of the Schomaker–Stevenson relationship:

$$r_{A-B} = r_A + r_B - 0.09\Delta$$

where r_{A-B} is the required bond length, r_A and r_B the covalent radii of A and B, respectively, and Δ is the difference between the electronegativities of A and B.

It must be emphasized, however, that such calculations give only approximate bond lengths.

10.3 Stereochemistry of Compounds Formed by Main-group Elements

In Chapter 8 we saw that covalent compounds formed by first-row elements adopted one of the basic shapes: linear, trigonal-planar, tetrahedral, pyramidal and angular. The latter two shapes were considered as arising from a tetrahedral distribution of pairs of electrons, with one and two lone pairs of electrons, respectively. We can now extend this discussion to a general review of molecules formed by main-group elements, and we

Table 10.2 Single bond covalent radii/pm

H 32						
Li 123	Be 89	B 81	C 77	N 74	O 74	F 72
		Al 125	Si 117	P 110	S 104	Cl 99
			Ge 122	As 121	Se 117	Br 114
			Sn 140	Sb 141	Te 137	I 133

Table 10.3 Multiple bond covalent radii/pm

	C	N	O	S	Se
Double bond	67	62	60	94	107
Triple bond	60	55	55	87	–

shall see that the gross stereochemistry of these molecules can be rationalized on this basis of repulsion between electron pairs.

The elements of second and subsequent rows also form compounds with most of these basic shapes, but since the elements can form more than four covalent bonds, the stereochemistry generally has more possibilities. Thus five and six electron pairs give trigonal-bipyramidal and octahedral arrangements, respectively, while seven pairs tend to give a pentagonal-bipyramidal stereochemistry.

To describe the bonding in these molecules we can use either the molecular-orbital or valence-bond methods. The molecular-orbital approach will be used for diatomic molecules, but where the main interest is in the qualitative aspects of the stereochemistry we shall use the valence-bond (hybridization) description as it is simpler and easier to relate to molecular shape.

10.4 Lithium and the Alkali Metals

Each of these elements has a single electron in an outer *s* orbital, and this can be used to form a covalent bond, as in the simple diatomic molecules Li_2, Na_2, . . . , etc. This was discussed in Chapter 6, p.60, where the molecular-orbital description of Li_2 was given as

$$KK(\sigma 2s)^2$$

These diatomic alkali metal molecules are found in small amounts (~1%) in the metal vapour; their dissociation energies (Li_2 = 105, Na_2 = 72, K_2 = 49, Rb_2 = 45, Cs_2 = 44 kJ mol^{-1}) steadily decrease, showing that the bond becomes weaker as the atoms become larger.

The alkali metals do not usually form covalent bonds, however. We have already seen in Chapter 9 that the bonding in many alkali metal compounds can be satisfactorily discussed in terms of the *ionic* bond. *Table 10.4* gives ionization energies and atomic radii for both the alkali metals and the alkaline earth elements. These energies are smaller than those for other elements, showing that the outer *s* electrons are relatively easily removed.

Table 10.4 Radii (atomic and ionic) and first ionization energies for alkali metal and alkaline earth elements

Element	*Radii*/pm		*First ionization energies*	
	Atomic (covalent)	*Ionic**	kJ mol^{-1}	eV $atom^{-1}$
Li	123	60	520	5.39
Na	157	95	496	5.14
K	203	133	419	4.34
Rb	216	148	403	4.18
Cs	235	169	376	3.09
Be	89	31	899	9.32
Mg	136	65	738	7.64
Ca	174	99	590	6.11
Sr	192	113	549	5.69
Ba	198	135	503	5.21

*M^+ for alkali metals; M^{2+} for alkaline earth elements

The structures of these ionic compounds are related to the packing of ions as discussed in Chapter 9. These compounds have comparatively high melting points, and in the molten state they are good conductors of electricity; even in the gas phase 'ion pairs' can be detected. Such substances are usually soluble in polar solvents such as water, in which they are almost completely dissociated in dilute solution. Typical salts with such properties are sodium chloride and potassium nitrate.

One particularly interesting series of salts is the hydrides, which have the general formula MH, and the structure M^+H^-, with a hydride ion. Lithium hydride, for instance, has the rock salt structure (Chapter 9, p.108), and when it is electrolysed at a temperature just below the melting point liberates lithium at the cathode and hydrogen at the anode.

The lithium ion is much smaller than ions of the other alkali metals, and as a result it is the most strongly solvated since it interacts strongly with the lone pair electrons of the solvent molecules. Similarly, lithium salts have more 'covalent character' than the analogous salts of the other alkali metals. With the iodides, for instance, the small lithium ion polarizes the charge cloud of the iodide ion, so that there is some measure of electron sharing; the other alkali metal ions have a much smaller effect.

A similar effect is observed with the metal alkyls, which show an interesting variation in properties. Thus the alkyls of sodium and heavier alkali metals are colourless, involatile solids that are insoluble in benzene and react violently with air; these properties indicate an ionic structure M^+R^-, the intense reactivity arising from the localization of the negative charge on a single carbon atom. The lithium alkyls are much less reactive, however, and (except for the methyl) dissolve in hydrocarbon solvents to give tetrameric or hexameric species. X-ray studies have shown that the alkyl groups bridge two or more lithium atoms, so that the bonding is neither ionic nor simple covalent. The best bonding description involves delocalized σ molecular orbitals of the types found in diborance, B_2H_6, and aluminium trimethyl, $Al_2(CH_3)_6$, which will be discussed in detail in Chapter 13.

10.5 Beryllium and the Alkaline Earth Metals

Because of the increased effective nuclear charge, which 'pulls in' the outer electron charge cloud, the beryllium atom is much smaller than the lithium atom and its first ionization energy is greater. As we go down the group from beryllium to barium there is a considerable increase in both atomic and ionic (M^{2+}) size and a marked decrease in ionization energy (see *Table 10.4*), and as a result beryllium occupies a unique position in the group in forming 'covalent' compounds whereas the compounds formed by the remaining Group IIA elements are almost always ionic.

Beryllium ions, Be^{2+}, are present of course in the aqueous solutions of salts such as the nitrate and sulphate, but the ions are strongly solvated – probably by four water molecules – and may be extensively hydrolysed.

Beryllium forms few truly ionic compounds, however, and the oxide for instance is a non-volatile (m.p. 2570 °C) giant molecule (Wurtzite structure), in which the Be–O bonds have around 40% covalent character according to Pauling. Even the fluoride, BeF_2, does not appear to contain discrete Be^{2+} and F^- ions, since it is only poorly conducting in the fused state; the Be–F bonds, however, will be very highly polar.

In Chapter 8 we discussed the shapes of covalent beryllium compounds, and saw that monomeric BeX_2 molecules should have a linear arrangement of Be–X bonds on the grounds of simple repulsion theory. The bonding may be described on the basis of delocalized three-centre bonding σ molecular orbitals (derived from the beryllium $2s$ and $2p_x$ a.o.'s and the appropriate X a.o.'s), or in terms of localized two-electron bonds formed by the overlap of beryllium *sp* hybridized a.o.'s with the X a.o.'s. There are not many such molecules, but the chloride, $BeCl_2$, evidently adopts this configuration in solution since it then has no dipole moment. The solid melts at 404 °C and sublimes readily, and it has been shown that at 745 °C the chloride vapour is monomeric, although vapour density measurements at lower temperatures (~550 °C) indicate around 20% dimer formation. The bromide and iodide have similar linear structures.

In Chapter 8 we pointed out that such molecules would act as Lewis acids and accept electron pairs from Lewis bases and in so doing achieve a higher degree of symmetry. The two examples we gave were $[BeCl_4]^{2-}$ and $BeCl_2,2O(C_2H_5)_2$, and in each case the beryllium atom has four electron pairs with a tetrahedral distribution:

$$\text{Cl–Be–Cl} \xrightarrow{+2L} \text{Cl}_2\text{Be} \leftarrow \text{L}_2$$

In the above representation we have written L to represent the Lewis base, or 'ligand' [either Cl^- or $O(C_2H_5)_2$], which donates an electron pair to beryllium.

Since the four bonds in $[BeCl_4]^{2-}$ are tetrahedrally disposed about the beryllium atom, we may adopt the valence-bond approach and describe the bonding on the basis of beryllium using sp^3 hybridized orbitals. There are several formal ways of looking at this problem. The simplest is to consider the ion as being made up of a Be^{2+} ion to which four Cl^- ions each donate a pair of electrons:

$4Cl^- \rightarrow Be^{2+}$ giving $[BeCl_4]^{2-}$

The Be^{2+} orbitals may be denoted:

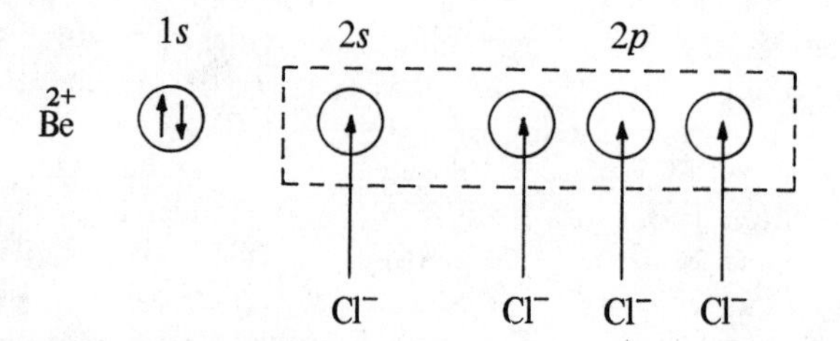

(We shall use this pictorial representation in future with a circle to represent the appropriate orbital capable of containing up to two electrons. An electron in an orbital is shown by an arrow ↑ or ↓; the dotted rectangle encloses the orbitals considered to be involved in the hybridization; the donated electron pairs are represented by the arrow into the circle from the donor.)

In this description we mix (or hybridize) the *s* and *p* atomic orbitals to produce four equivalent sp^3 hybrid orbitals that point towards the corners of a tetrahedron. Each hybrid orbital, which is vacant, is then overlapped by a doubly filled 3*p* atomic orbital from a Cl^- ion, thus giving

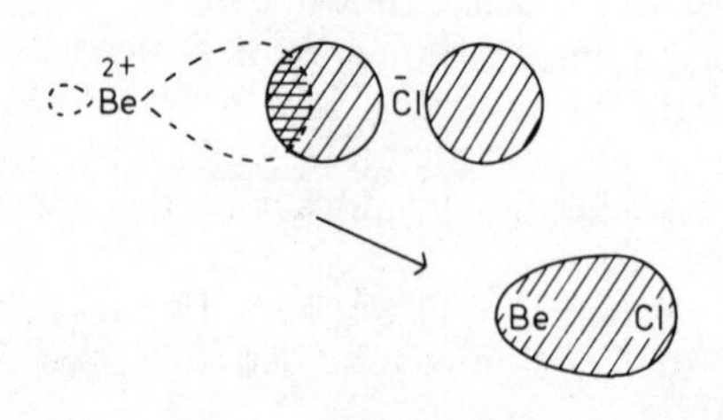

Figure 10.1 Formation of a Be–Cl bond by the overlap of a vacant sp^3 *orbital* of* Be^{2+} *with a doubly filled 3p orbital of* Cl^-

*In this and subsequent orbital diagrams a vacant orbital will be represented by a dotted outline, a singly filled orbital by a full line, and a doubly filled orbital by a full line plus slanting shading. The overlap is shown by horizontal shading.

a two-electron two-centre bond embracing the beryllium and one chlorine. There are four such bonds in all. *Figure 10.1* shows the formation of just one bond.

The other formal way of achieving a valence-bond description based on beryllium sp^3 hybridized orbitals is to place the double negative charge on the central beryllium atom, so that it becomes isoelectronic with carbon, with one electron in each of the 2*s* and 2*p* atomic orbitals. Hybridization gives four equivalent sp^3 orbitals, each containing one electron, and each orbital can then overlap with a singly filled 3*p* atomic orbital of chlorine. Thus:

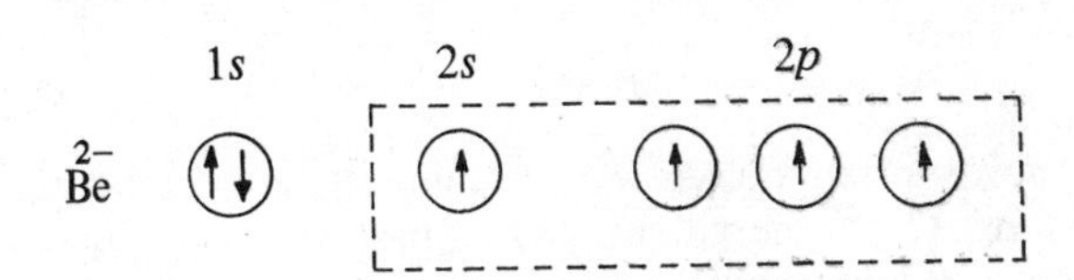

These two methods of formally representing the bonding are both artificial in that they are not intended to indicate the process by which the complex anion is actually formed, although in practice readers may find the first method more helpful, especially when applying the valence-bond approach to complex ions of the transition metals (see Chapter 11).

If readers wish to picture the hypothetical formation of $[BeCl_4]^{2-}$ they may imagine a Be^{2+} ion in the gas phase which is approached by four Cl^- ions. The latter will stay as far apart from one another as possible so that we shall end up with them being disposed tetrahedrally about Be^{2+}. Now we revert to our bonding description on the basis of the electron pairs from the Cl^- ions being donated into the vacant sp^3 hybrid orbitals of beryllium. It must be emphasized, however, that the Be^{2+} ion does not sit around with empty sp^3 hybrid orbitals waiting like buckets for the electron pairs to flow in. Thus we must not say that the beryllium atom is sp^3 hybridized and therefore forms tetrahedral bonds, but rather that beryllium forms tetrahedral bonds that can be described on the basis of sp^3 hybridization. It is *we* who choose to mix, or hybridize, the *s* and *p* orbitals in order to provide a convenient and coherent description of the bonding.

As we saw in Chapter 8, the tetrahedral bonding can also be described by the more elaborate molecular-orbital approach, and this does not hybridize the *s* and *p* orbitals but uses them unhybridized to overlap with all available ligand orbitals of appropriate symmetry, the molecular orbitals then embracing all the nuclei and not just beryllium and one chlorine.

We have emphasized that simple linear beryllium compounds such as the dihalides tend to achieve a higher symmetry (tetrahedral) by accepting electron pairs from such donors as the chloride ion and diethyl ether. If such donors are not available then the urge to acquire electrons is such that in the solid state the chlorine atoms bonded to beryllium donate a lone pair to a neighbouring beryllium atom, so forming a bridge and giving a polymeric continuous chain structure (1) in which each beryllium atom is surrounded by four chlorine atoms. The bond distribution about each beryllium is considerably distorted from the tetrahedral arrangement, of course, because of the constraints imposed by the $BeCl_2Be$ four-membered ring, and the ClBeCl angles in the ring are 98.2°.

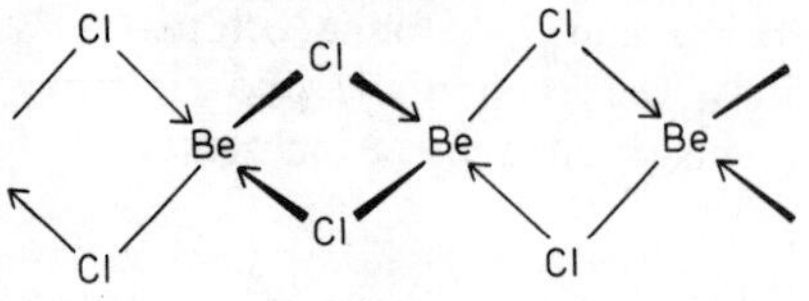

(1) 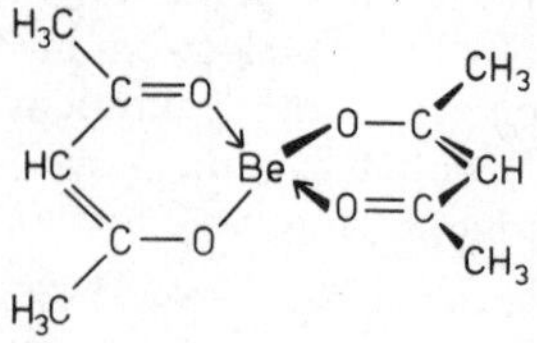 (2)

A distorted tetrahedral structure is also found in the acetylacetone compound (2) of beryllium, where the two acetylacetone groups are at right angles to one another.

In all of the beryllium compounds discussed up until now, the beryllium atoms have had either linear or tetrahedral configurations, but there is a recent report (Guemas-Brisseau *et al.*[3]) on the structure of $[Be(NH_2)_3]^-$, which shows beryllium to have a planar distribution (3) of three nitrogen atoms. The beryllium atom can be assigned the formal negative charge, and is then isoelectronic with the neutral boron atom, and accordingly forms trigonal-planar bonds. Another molecule containing trigonal-planar beryllium is $Be(NMe_2)_2$, which has a linear trimeric structure in which the two terminal beryllium atoms have planar environments; the central beryllium atom, on the other hand, has a 'tetrahedral' configuration (4) (Atwood and Stucky[4]).

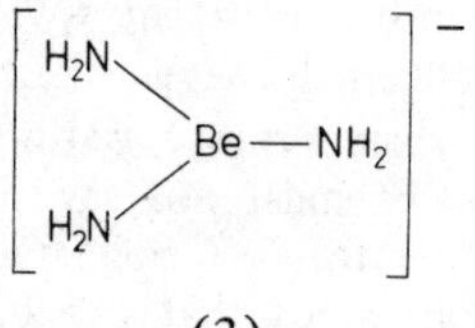

(3) 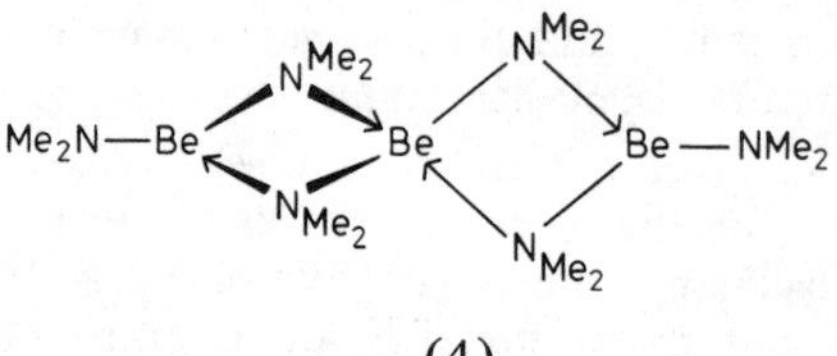 (4)

10.6 Boron and the Group IIIB Elements

Boron forms only covalent bonds, but as we proceed down the group ionic characteristics become more pronounced. Aluminium, for instance, closely resembles beryllium and forms mainly covalent bonds, the ionic character of which is considerably greater than that of the corresponding bonds formed by boron.

We have seen previously (Chapter 8, p.90) that the basic shape to be expected for simple boron compounds of formula BX_3 is trigonal-planar, with XBX bond angles of 120°, although the number of electron pairs about the boron atom can be increased to four – with a tetrahedral configuration – by the formation of donor–acceptor compounds such as BF_3,NH_3 and $[BCl_4]^-$. The stereochemistry of boron compounds is dominated by these two arrangements, although very recently it has been shown (Calvo and Faggiani[5]) that the simple borate anion, BO_2^-, which is found in various minerals, has a linear configuration with a B–O bond

Table 10.5 Stereochemistry of boron compounds

Formal electronic configuration of boron	*Hybridization*	*Shape*		*Examples*
B $1s^2 2s2p^2$	sp^2	Trigonal-planar	>B–	BBr_3, BCl_3. BF_3
B^- $1s^2 2s2p^3$	sp^3	Tetrahedral	B^-	$[BCl_4]^-$, $[BH_4]^-$
	sp $(+p^2)$	Linear	$=\bar{B}=$	BO_2^-

length of 125 pm. This is to be expected since the anion is isoelectronic with CO_2. *Table 10.5* sums up the position.

There are many examples of the trigonal-planar boron compounds, this being the shape of all simple monomeric compounds of general stoichiometry BX_3, where X is Br, Cl, F, CH_3 and OCH_3 for instance. In these compounds the XBX angle is the predicted 120° but, when the three bonded atoms are not identical, small deviations from 120° occur; thus with $C_6H_5BCl_2$ and HBF_2, the ClBCl and FBF angles are both close to 118°. We saw in Chapter 9 that orthoboric acid, $B(OH)_3$, also has a trigonal-planar arrangement of OH groups about the boron atom, with hydrogen bonding between the $B(OH)_3$ units giving a layer structure (5). A more limited degree of polymerization is found in the structures of cyclic boron–oxygen compounds such as methylboronic acid, $(CH_3BO)_3$ (6) and the metaborate anion $B_3O_6^{3-}$ (7). The simple trigonal-planar orthoborate anion BO_3^{2-} is also present in cobalt and magnesium borates.

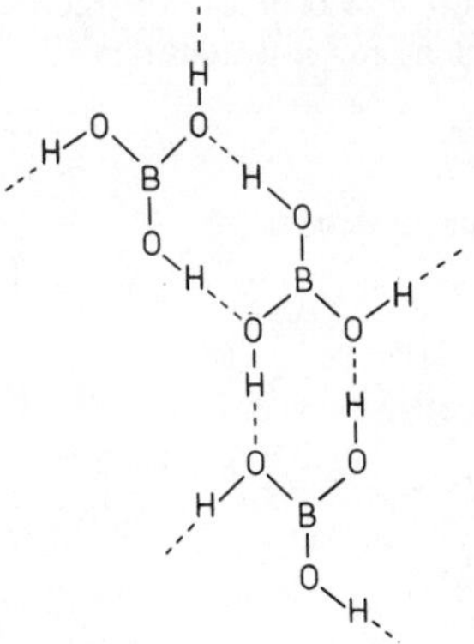

(5) 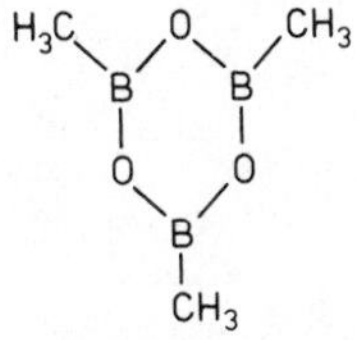 (6) 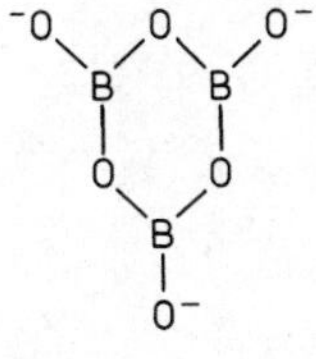 (7)

The boron atom also forms planar bonds in the diboron tetrahalides, B_2F_4 and B_2Cl_4. These halides are especially interesting, because although X-ray experiments show that both have the completely planar structure (8a) in the solid state, spectroscopic investigations show that the chloride takes up a staggered configuration (8b) in the gas phase. A recent electron diffraction study of the chloride shows the B–Cl bond lengths to be 175 pm (Ryan and Hedberg[6]). Measurements over a series of temperatures have shown that the energy barrier to rotation in B_2Cl_4 is only 7.7 kJ

mol^{-1}, so it would seem that although the staggered configuration may be the one of lowest energy, the crystal lattice forces in the solid may be sufficient to impose the planar arrangement.

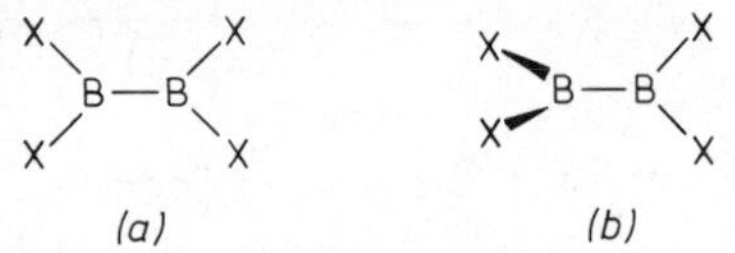

(a) (b)

(8)

Boron forms two structurally interesting compounds with nitrogen, namely borazole (borazine), $B_3N_3H_6$, and the nitride, BN. Borazole, which has a planar cyclic hexagon structure of alternate boron and nitrogen atoms (9), is often referred to as the inorganic benzene. A number of derivatives of borazole have been made in which the hydrogen atoms have been replaced by chlorine atoms or alkyl groups, and all have the same basic structure (9). In this structural formula the formal negative and positive charges have been assigned to the boron and nitrogen atoms respectively, making both atoms isoelectronic with carbon, so that the molecule has the same skeletal configuration as benzene. Indeed we now have the same problem that we had with benzene (pp.78 and 99) about describing the π bonding which is delocalized around the ring in the same sort of way; all B–N distances are 144 pm, which is between the calculated single (154 pm) and double (136 pm) boron–nitrogen bond lengths. The valence-bond approach describes the structure in terms of two canonical forms [one being shown in (9)], whereas the molecular-orbital description involves three bonding π orbitals embracing all six atoms in the hexagon, although these delocalized orbitals differ somewhat from their benzene analogues because the constituent $2p_z$ atomic orbitals of boron and nitrogen are not identical in energy.

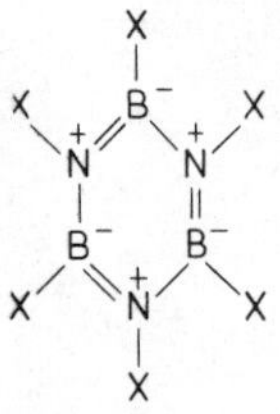

(9) 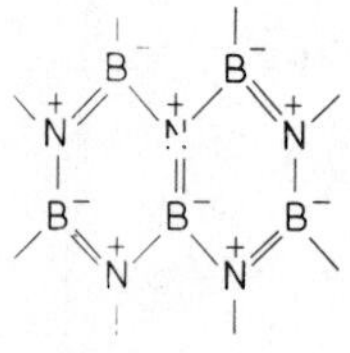 (10)

B_3N_3 hexagons are also present in the structure of boron nitride, the hexagons being interconnected to form an infinite plane (10) as in graphite (see p.144); the B–N distance is 145 pm. The basic skeleton of σ bonds can be accounted for in terms of sp^2 hybrid orbitals of the boron and nitrogen atoms; the remaining electron on each atom is in the $2p_z$ orbital, the mutual overlap of which give π bonding extending over the whole plane The layers are probably arranged so that the B atoms in one layer are immediately above the N atoms in the next layer below.

The tendency of boron to attain a tetrahedral symmetry, which has been stressed previously, is typified by simple anions such as $[BCl_4]^-$, $[BF_4]^-$ and $[BH_4]^-$. It is found also in many simple co-ordination compounds, e.g. boron trifluoride accepts electrons from many donors [NH_3, $N(CH_3)_3$, CH_3CN, H_2O and $(CH_3)_2O$] to give 1 : 1 compounds in which the FBF bond angles range from 107 to 114°. It is interesting to note that although BH_3 does not exist as a monomer (see Chapter 13), it can be stabilized with such donors as $N(CH_3)_3$, $P(NH_2)_3$, $As(CH_3)_3$ (Durig *et al.*[7]) and CO, with the formation of tetrahedral co-ordination compounds.

Also into the category of tetrahedral boron compounds falls the cubic form of boron nitride, which is made by reaction at very high temperatures and pressures. This has a giant molecule diamond-type structure (see p. 144) in which boron uses tetrahedrally arranged orbitals. This compound is especially interesting in view of its extreme hardness and consequent potential uses.

It is perhaps worth returning for the moment to the simple boron compounds, BX_3, as typified by the trihalides, which have monomeric trigonal-planar structures. This monomeric nature contrasts with the polymeric behaviour of beryllium halides which as a consequence of halogen bridging provide the beryllium atoms with a tetrahedral environment of chlorine atoms. In part this difference might be accounted for on steric grounds, since the boron atom is appreciably smaller than the beryllium atom (80 against 89 pm), but this is clearly not the whole story, because the boron halides still act as good electron acceptors and bond with donor ligands – including Cl^- – to give tetrahedral species. It is possible that there may be a certain amount of stabilization of the trigonal-planar arrangement through some delocalization of the doubly filled p_z atomic orbitals of the halogen atoms into the vacant $2p_z$ atomic orbital of boron.

With aluminium, and the heavier elements of Group IIIA, the simple MX_3 compounds are rarely monomeric, except at high temperatures in the gas phase, or with molecules with very bulky groups – as in $Al\{N(SiMe_3)_2\}_3$ (11). Thus aluminium tribomide and tri-iodide are dimeric in all physical states, there being halogen bridges between AlX_2 units (12). The four terminal halogen atoms and the two aluminium atoms are co-planar, with the bridging halogens above and below the plane, giving each aluminium

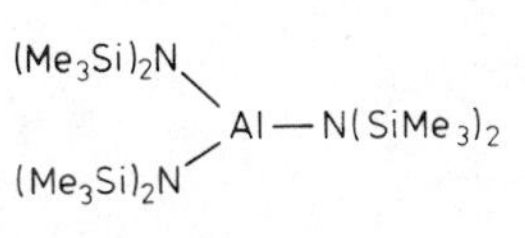

(11)

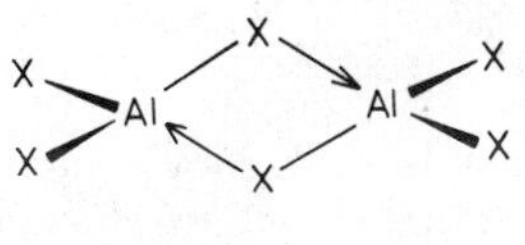

(12)

atom a distorted tetrahedral environment. Aluminium trichloride has an analogous structure except in the solid state when more extensive chlorine bridging gives a layer structure in which each aluminium atom has an octahedral environment (13). The bridging bonding forces in the dimeric halide molecules are not particularly strong, since the enthalpies of dissociation are only in the region of 50–60 kJ mol^{-1}.

Gallium trichloride and tribromide, and indium trichloride, tribromide and tri-iodide are also dimeric with analogous halogen-bridged structures, but it seems that gallium tri-iodide is a monomer.

Similar halogen bridging is known to be present in the dimeric methylaluminium dichloride, $[CH_3AlCl_2]_2$, and dimethylaluminium chloride,

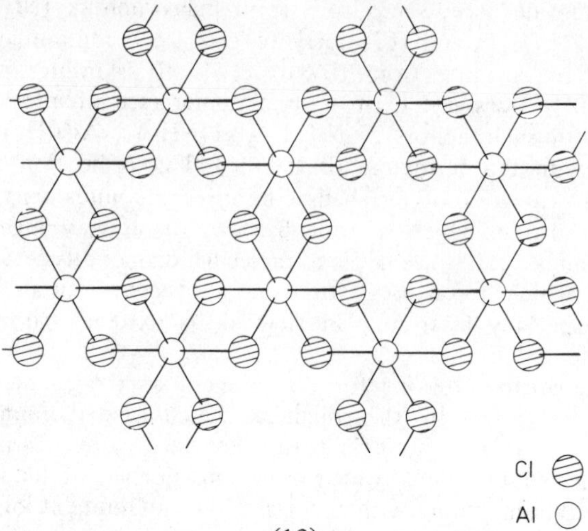

(13)

$[(CH_3)_2AlCl]_2$. Dimeric structures are also found for the trialkyls and triaryls of aluminium, such as the trimethyl (14), triethyl and triphenyl, but since these structures involve bonding of an 'electron-deficient' kind, in which the bridging carbon atom apparently forms more than four bonds, discussion of the bonding will be deferred until Chapter 13.

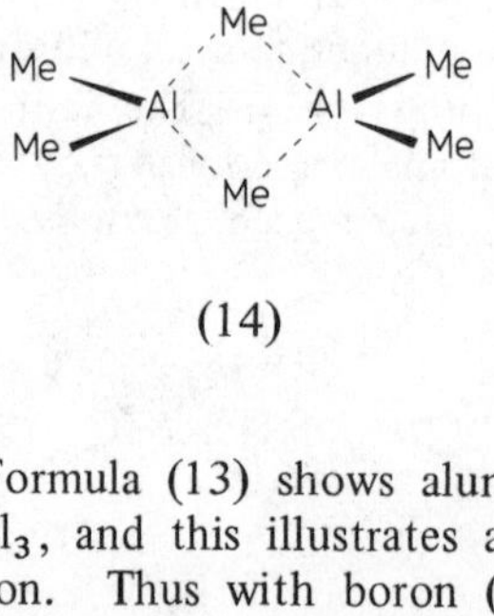

(14) 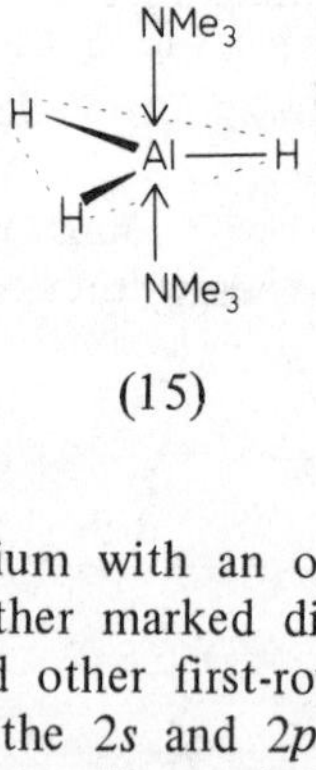 (15) 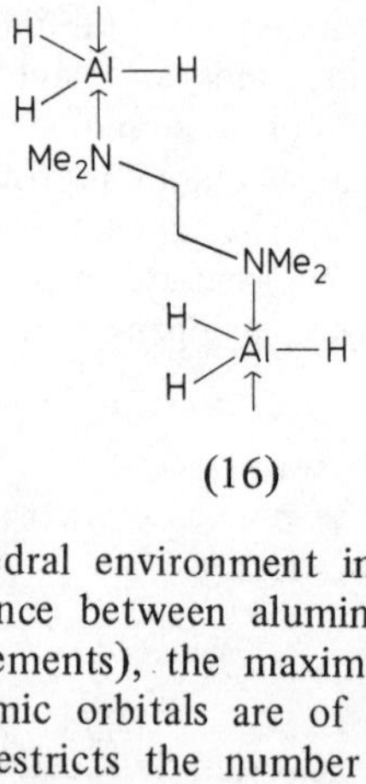 (16)

Formula (13) shows aluminium with an octahedral environment in solid $AlCl_3$, and this illustrates another marked difference between aluminium and boron. Thus with boron (and other first-row elements), the maximum covalency is four, because only the 2*s* and 2*p* atomic orbitals are of appropriate energy to accept electron pairs, and this restricts the number of electron pairs around the boron atom to four (an 'octet' of electrons). Boron has other empty orbitals, of course, but these are very much higher in energy and hence they cannot attract electrons from donor ligands. However, aluminium and the heavier elements of Group IIIA are capable of achieving a higher covalency than four. Thus the 3*d* orbitals are reasonably close in energy to the 3*s* and 3*p* orbitals, which contain the valency electrons, so that electron pairs can be donated into these 3*d* orbitals and the aluminium atom may form more than four bonds. Typical examples are solid $AlCl_3$ and the anion $[AlF_6]^{3-}$.

Table 10.6 summarizes the stereochemistry of the covalent compounds of the tervalent elements. The trigonal-bipyramidal arrangement is found

Table 10.6 Stereochemistry of compounds of tervalent Al, Ga, In and Tl

Shape	*Hybridization scheme*	*Examples*
Trigonal-planar	sp^2	$Al\{N(SiMe_3)_2\}_3$
Tetrahedral	sp^3	$[AlCl_4]^-$ Al_2Br_6
Trigonal-bipyramidal	sp^3d	$AlH_3,2NMe_3$ $InCl_3,2PPh_3$
Octahedral	sp^3d^2	$[AlF_6]^{3-}$

for co-ordination compounds of stoichiometry $MX_3,2L$. A typical example is $AlH_3,2NMe_3$ (15) in which three hydrogen atoms occupy the three equatorial positions (*xy* plane with HAlH angles of 120°) and the two trimethylamine groups occupy the axial positions (*z* axis). A closely related compound is that formed by aluminium hydride (see Chapter 13) with tetramethylethylenediamine (16); the ligand has two donor nitrogen atoms and these co-ordinate to different aluminium atoms, giving rise to a chain structure.

We account for the nitrogen atoms bonding through the axial positions partly because of the bulk of the ligand and partly because nitrogen is more electronegative than hydrogen. Simple repulsion theory requires electronegative and bulky groups to go into the axial positions. $InCl_3,2PPh_3$ also has a trigonal-bipyramidal structure, but the axial positions are occupied by two chlorine atoms; the greater electronegativity of chlorine may be the dominating factor here, but we shall see in the next chapter (p.249) that related trimethylamine adducts of titanium, vanadium and chromium trihalides, $MX_3,2NMe_3$, have *trans*-trigonal-bipyramidal configurations with the bulky trimethylamine groups taking up axial positions.

Comparatively little structural work has been done on compounds of the heavier elements in lower oxidation states. The controversy over the structure of the so-called dihalides of gallium has been resolved by crystal structure and Raman spectra studies on the dibromide and dichloride, which have shown that divalent gallium is not present, but that ions of gallium(I) and gallium(III) are in the structures $Ga^+[GaX_4]^-$; the anion has the expected tetrahedral shape. In the complex $[N(CH_3)_4]_2[Ga_2Cl_6]$, in which the gallium is formally divalent, a recent crystal structure determination (Brown and Hall[8]) has shown the presence of a direct gallium–gallium bond (239 pm) with a tetrahedral distribution of bonds about each gallium. Thus the $[Ga_2Cl_6]^{2-}$ ion has the same shape as C_2Cl_6.

As we go down the group there is a tendency for the univalent state to become more stable, and in particular there are many thallium(I) compounds. The pair of electrons not used in bonding is usually referred to as the 'inert' pair.

10.7 Carbon and the Group IVB Elements

Carbon forms covalent compounds, except for such molecules as the sodium alkyls, which have a Na^+R^- structure, and the carbides formed by metals in periodic groups I, II and III, which contain ionic carbide groups. However, even in the carbides there is appreciable covalent character in the bonds. This tendency for covalent bonding is also found in the quadrivalent compounds of the other Group IV elements, silicon, germanium, tin and lead, although the bonds become increasingly ionic as we go down the group. With the Group III elements we noticed an increasing tendency for the elements to retain an inert pair of electrons, and this is also a feature of Group IV chemistry, and consequently the divalent state becomes increasingly important as we go from carbon to lead.

In Chapter 8 we saw that quadrivalent carbon forms molecules with three basic symmetries, tetrahedral, trigonal-planar and linear, and that in the valence-bond description the carbon atom is assigned sp^3, sp^2 and sp hybridization schemes, respectively. *Table 10.7* summarizes the position.

Table 10.7 Stereochemistry of carbon compounds

Shape	*Hybridization scheme*	*Number of σ bonds*	*Number of lone pairs*	*Number of π bonds*	*Examples*
Tetrahedral	sp^3	4	0	0	CCl_4
Trigonal-planar	sp^2	3	0	1	$COCl_2$
Linear	sp	2	0	2	CO_2; HCN

Carbon itself exists in two main modifications, diamond (17) and graphite (18). In the diamond structure, the carbon atoms form a giant molecule with each carbon linked to four tetrahedrally disposed neighbours, with C–C distances of 154 pm as in ethane; the bonding may be described in terms of sp^3 hybrids. With graphite, however, the carbon atoms are arranged as

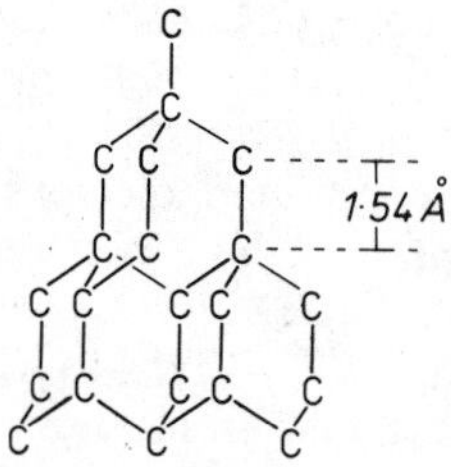

(17)

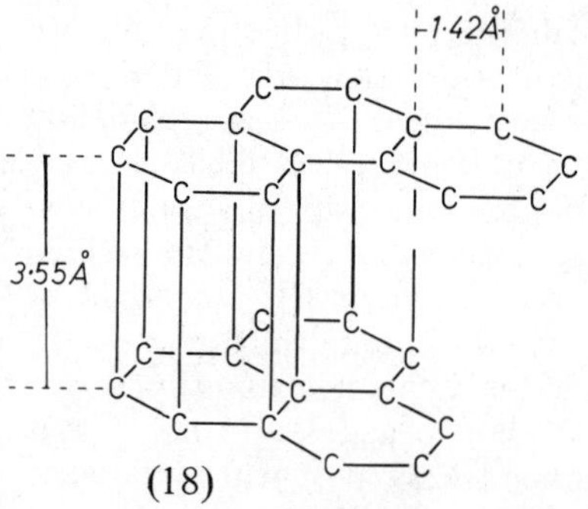

(18)

an infinite series of linked hexagons to give a layer structure like that in boron nitride (p.140). The C–C distances are all the same, being 142 pm. The distance between adjacent planes is 335 pm, so that the layers must be held together only by weak van der Waals' forces. Each layer of carbon atoms resembles a vast collection of fused benzene rings; the carbon atoms can be considered to be sp^2 hybridized, leaving $2p_z$ orbitals (one per carbon atom) sticking out above and below the plane of the carbon atoms. The mutual overlap of these $2p_z$ orbitals will produce delocalized π molecular orbitals in the form of double streamers above and below the plane. The C–C bond length, 142 pm, is much shorter than a single C–C bond, as would be expected.

The diamond-type of structure is also found in elemental silicon and germanium, and with the low temperature modification of tin called 'grey' tin. Lead, however, is more metallic and has a cubic structure.

Carbon forms tetrahedral molecules in the simple CX_4 compounds, such as the tetrahalides, and in all the saturated hydrocarbons, C_nH_{2n+2}. When it forms only three σ bonds, however, as in such compounds as carbonyl chloride, $COCl_2$, carbon adopts a trigonal-planar bonding arrangement. *Table 10.8* lists most of the planar carbon compounds of this type

Table 10.8 Trigonal-planar carbon compounds

Compound (COXY)	*XCY* /degrees	*C=O* /pm	*C–X(Y)* /pm
$COBr_2$	110	113	205
$COCl_2$	111.3	117	175
COF_2	108	117	131
COClF	112	116	175(C–Cl); 130(C–F)
CH_3COCl	112.7	119	179(C–Cl)
CH_3COCN	115	123	–
CH_3COF	110.3	118	135(C–F)
CH_3CHO	117.5	122	–
$HCONH_2$	113.2	119	138(C–N)
HCOF	108	118	134(C–F)
CH_3OCHO	109.3	120	–
$(CH_3)_2CO$	117.2	122	–

for which parameters are known, and it can be seen that although the molecules are planar the bond angles are distorted from the ideal 120°. This distortion can be attributed to the spatial requirements of the C=O bond, the electrons in which take up more room than those in the C–X bonds, thus forcing the C–X bonds closer together. In the simplest

(19)

valence-bond approach we describe the bonding in terms of sp^2 hybridization, although it must be appreciated that the three carbon orbitals used for σ bonding will not be identical, some containing more *s* character than others.

The carbonate ion, CO_3^{2-}, is also planar, and this can be appreciated if we write down just one of the canonical forms (19), which illustrates the similarity to molecules such as $COCl_2$. Of course all three bonds are identical and the π bond is delocalized over all three CO bonds; since the effective π-bond order is only one-third, the C–O bond length (~129 pm) is much greater than in $COCl_2$ (117 pm). A simple description of the carbonate ion involves carbon sp^2 hybrid orbitals for the σ bonding, with the π bonding accounted for by the molecular-orbital scheme shown in *Figure 10.2*. In the π molecular-orbital scheme the combination of the four π $2p_z$ atomic orbitals (one from carbon and one from each oxygen) gives rise to four π molecular orbitals, one bonding (π_1), two non-bonding (π_2 and π_3) and one antibonding (π_4^*). The six π electrons occupy the

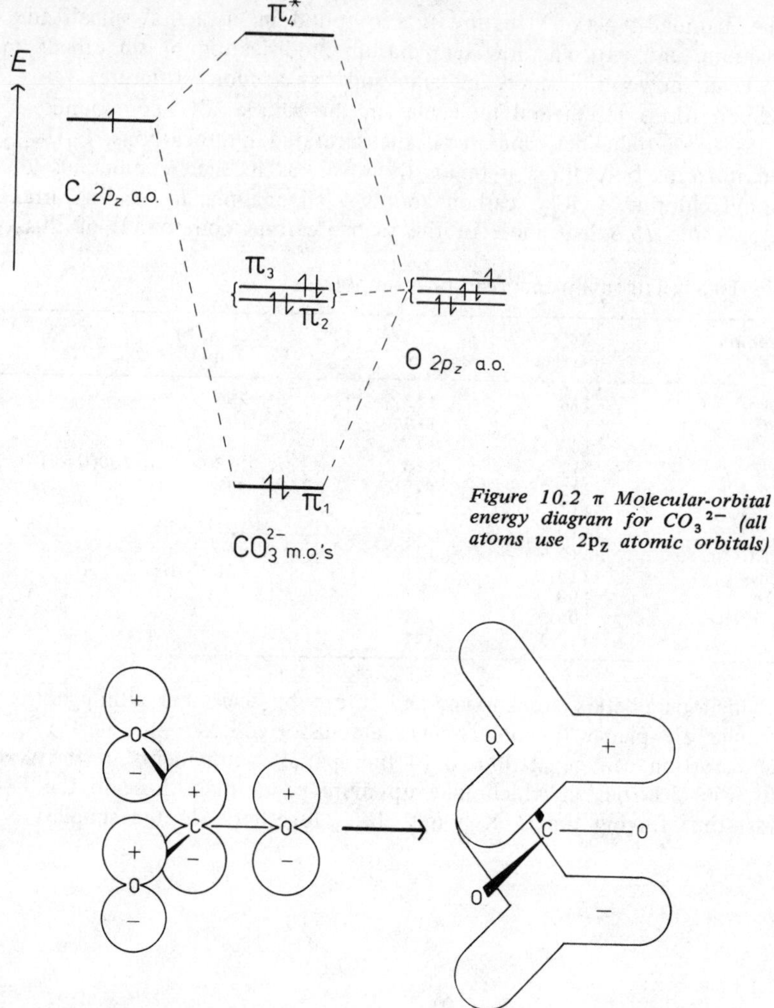

Figure 10.2 π Molecular-orbital energy diagram for CO_3^{2-} (all atoms use $2p_z$ atomic orbitals)

Figure 10.3 Bonding π orbital in CO_3^{2-}

three lowest energy molecular orbitals, π_1, π_2 and π_3, so that the effective π bonding is provided by the two electrons in the π_1 orbital. This π_1 orbital, which is given by the combination

$$\pi_1 = \phi_{C_{2p_z}} + \phi_{O_{2p_z}}$$

is illustrated in *Figure 10.3.*

Some readers may prefer to consider the bonding scheme for the carbonate anion solely on the basis of the molecular orbital approach, in which case it is necessary to combine the π orbital diagram of *Figure 10.2* with a σ orbital diagram similar to that shown in Chapter 8 for boron trichloride. The composite molecular-orbital diagram is shown in *Figure 10.4.*

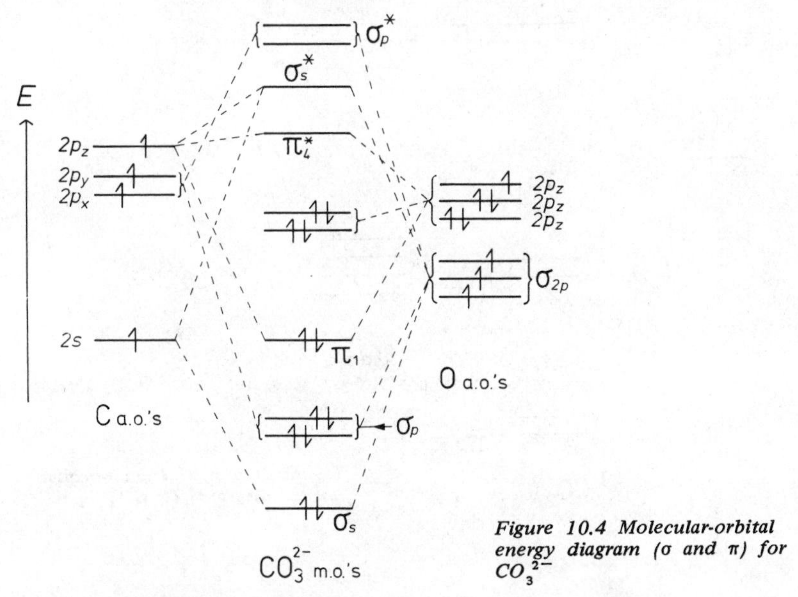

Figure 10.4 Molecular-orbital energy diagram (σ and π) for CO_3^{2-}

In many compounds, carbon forms two σ bonds and two π bonds, and adopts the predicted linear structure. The carbon atom may either form two double bonds (molecules of type *A*) or one single and one triple bond (molecules of type *B*).

Type A: X=C=X

Type B: Y–C≡Z

The structures of quite a number of compounds of type *A* have been determined, e.g. CO_2, COS, COSe, CONH, CS_2, CSSe, CSTe and CSNH, and all have been shown to have linear bonding arrangements about carbon. The linear configuration for CO_2 and CS_2 is confirmed by the zero dipole moments found for each molecule. Throughout this range of compounds the C=O and C=S bond lengths remain constant at 116 and 156 pm, respectively, and these lengths are appreciably shorter than the calculated double bond lengths (122 and 161 pm). The bonding in CO_2 can be described in terms of the three canonical structures:

$$\underset{\text{I}}{\text{O=C=O}} \qquad \underset{\text{II}}{\overset{+}{\text{O}}\text{≡C–}\overset{-}{\text{O}}} \qquad \underset{\text{III}}{\overset{-}{\text{O}}\text{–C≡}\overset{+}{\text{O}}}$$

A slightly more sophisticated description uses carbon *sp* hybrid orbitals to provide the σ bonds, together with delocalized π bonding according to the scheme shown in *Figure 10.5*. The bonding (π_1 and π_2) and non-bonding (π_3 and π_4) molecular orbitals are fully occupied. *Figure 10.6*, which illustrates the formation of one of the bonding π orbitals, is drawn so as to point out that there is a greater electron charge density over the oxygen atoms; this is a reflection of the oxygen 2*p* orbitals being lower in

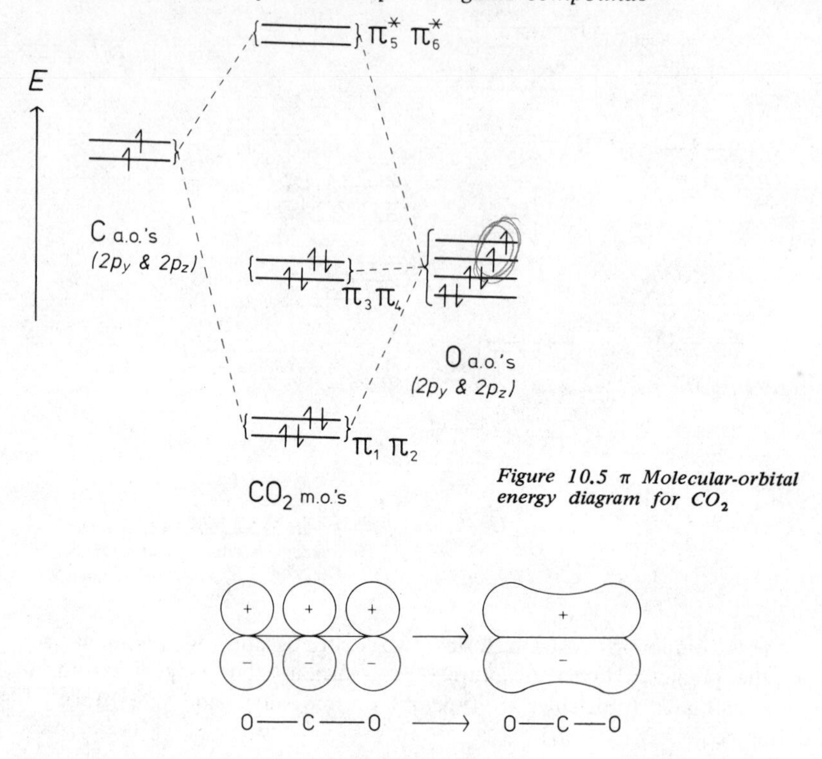

***Figure 10.5** π Molecular-orbital energy diagram for CO_2*

***Figure 10.6** Bonding π molecular orbital for CO_2, showing asymmetry*

energy than their carbon analogues and contributing more to the molecular wave function.

Once again a fuller molecular-orbital energy diagram can be obtained by combining the π scheme in *Figure 10.5* with a σ scheme similar to that proposed for $BeCl_2$ in *Figure 8.7* (p.89). Interested readers may like to set up a composite scheme on this basis.

Linear arrangements are found also in the isocyanate ion, NCO^-, which is isoelectronic with carbon dioxide, with respective C–O and C–N bond lengths of 118 and 120 pm, and in carbon suboxide, C_3O_2, with the skeleton

O–C–C–C–O

and C–O and C–C bond lengths of 116 and 128 pm, respectively.

Linear molecules of type *B*, with a triple bond, are found in the cyanides X–C≡N, where X may be Br, CH_3, Cl, F, H or I, and in every case the carbon–nitrogen bond length remains at 116 pm, a value which would be expected for a triple bond. The phosphorus analogue of HCN, i.e. HCP, has a similar linear structure with a C–P bond distance of 154 pm.

Silicon and the heavier elements of Group IVB differ from carbon in that vacant *d* orbitals are energetically available so that more than four bonds can be established provided there are suitable donor molecules. *Table 10.9* summarizes the more common symmetries that are found.

Table 10.9 Stereochemistry of compounds of Si, Ge, Sn and Pb

Shape	*Hybridization scheme*	*Number of σ bonds*	*Examples*
Tetrahedral	sp^3	4	$SiCl_4$, $SnMe_4$
Trigonal-bipyramidal	sp^3d	5	$PhSi(OCH_2CH_2)_3N$ $[Me_3Sn(OMe)]_n$
Octahedral	sp^3d^2	6	$[SnCl_6]^{2-}$

Unlike carbon, the other quadrivalent elements do not form linear or trigonal-planar molecules, but are invariably bonded to at least four other atoms. The tetrahedral arrangement is always present in simple MX_4 molecules.

The heavier elements do not form multiple bonds to oxygen and sulphur. The dioxide and disulphide of silicon are not monomeric like their carbon analogues but give macromolecular systems with an approximately tetrahedral arrangement of oxygen or sulphur atoms about the silicon atoms. Various structures are known for silica (quartz, cristobalite, tridymite), which differ only in the way in which the tetrahedral SiO_4 units are linked to each other by shared oxygen atoms. The silicate ion, SiO_3^{2-}, in sodium silicate also forms a linked tetrahedral system (20), and a similar Si–O–Si skeleton is found for the dimethylsilicone polymers (21). The many apparently

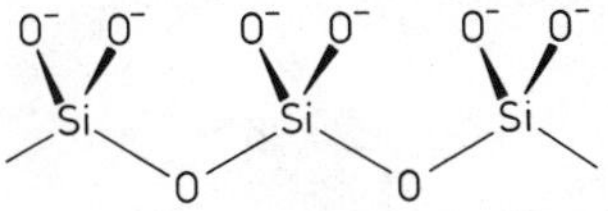

(20)

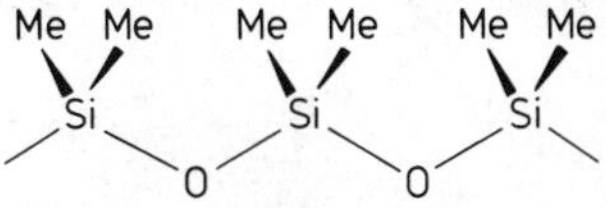

(21)

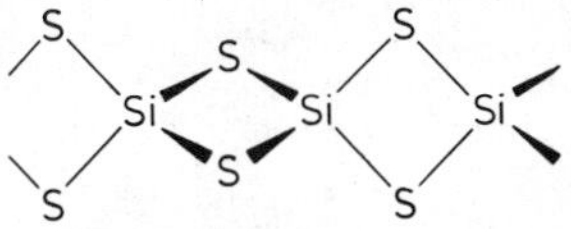

(22)

complex silicate structures can be discussed in terms of chains, sheets and three-dimensional structures of the linked SiO_4 tetrahedral units. As we might expect, the bond angles in such structures are not all the same, being greatest between the external Si–O⁻ bonds (119°) where repulsion is greatest, and least between the linked Si–O bonds (101°). The disulphide, SiS_2, a solid crystallizing in long silky needles, consists of infinite chains of the type shown in (22).

Further discussion is justified in respect of the contrasting behaviour of carbon, which readily forms multiple bonds with itself and with atoms such as oxygen and nitrogen, and the heavier elements that prefer to form single bonds only, if necessary by means of extensive polymerization. We have already seen that while CO_2 and $(CH_3)_2CO$ are monomers with carbon–oxygen double bonds, their silicon analogues are polymeric and contain only σ bonds. The most obvious factors influencing these bonding patterns are the relative sizes and electronegativities of carbon and silicon and their abilities to form π bonds with, say, oxygen. In the first place, the larger size of silicon might give it a preference for an increased number of bonded neighbours, i.e. a tetrahedral environment of four σ-bonded oxygen atoms rather than a linear arrangement with two multiply bonded oxygens. Then since carbon is a small atom the C–O internuclear distance will be short and there can be effective overlap of the carbon and oxygen $2p_\pi$ orbitals and the formation of strong π bonds. Indeed, bond energy data (Johnson[9]) agree with this in that C–O and C=O bond energies are of the order of 350 and 750 kJ mol^{-1}, respectively, showing that the double bond is at least twice as strong as the single bond. With silicon dioxide, on the other hand, the larger silicon atom results in a longer Si–O distance and less extensive π bonding; moreover, the silicon $3p_\pi$ atomic orbitals may be rather too diffuse for good overlap with the $2p_\pi$ orbitals of oxygen. We must not forget that because silicon is much less electronegative than carbon, the Si–O bonds will be appreciably more polar than C–O bonds, and this ionic character should lead to a strengthening of the Si–O σ bond. This enhancement of the strength of the Si–O single bond fits in with the bond energy found for Si–O in SiO_2, which is around 460 kJ mol^{-1}, and appreciably higher than a carbon–oxygen single bond.

A number of five co-ordinate compounds of silicon and tin have been characterized in recent years, and no doubt there are many others awaiting preparation. Two silicon compounds examined recently are $PhSi(OCH_2CH_2)$ (23) and $Ph_2Si(OCH_2CH_2)_2NH$ (24) (Daly and Sanz[10]), both of which have the silicon atoms with a trigonal-bipyramidal environment.

N O O Si O Ph

(23)

N O O Si Ph Ph

(24)

Sn OH Me Me Sn Me HO Sn

(25)

The trigonal-bipyramidal structure is also found for several tin compounds, one typical example being $Me_3Sn(OH)$ (25), which polymerizes through the OH group rather than remain monomeric. Six co-ordinate compounds of these heavier elements are quite common, and in ions such as $[SnCl_6]^{2-}$ they have regular octahedral structures.

The divalent state is important only for tin and lead. The tin compounds $SnCl_2,2H_2O$ and K_2SnCl_4,H_2O have been shown to contain respectively the pyramidal species $SnCl_2,H_2O$ and $[SnCl_3]^-$, and the latter ion is also present in $CsSnCl_3$. It is tempting to look upon these pyramidal species as based on a tetrahedron with the unused electron pair occupying the fourth position, but the bond angles are in fact rather less than 90° and clearly are best accounted for in terms of the tin atom using unhybridized p orbitals for bond formation, with the inert pair in the s orbital. The dihalides form many complexes but there have been few structural studies.

10.8 Nitrogen and the Group VB Elements

10.8.1 NITROGEN

Nitrogen, with the ground state configuration $1s^2 2s^2 2p^3$, has three unpaired electrons, and it is accordingly tervalent in its simple compounds; the next suitable lowest energy orbital, the 3*d*, is too far away to allow the promotion of one of the 2*s* electrons, so the neutral nitrogen atom never exceeds a covalency of three. We have seen that the pyramidal distribution of bonds can be discussed on the basis of a tetrahedral distribution of electron pairs, there being one lone and three bonding pairs. A positively charged nitrogen atom, on the other hand, is isoelectronic with a neutral carbon atom and can have four unpaired electrons and hence four covalent bonds; a negatively charged nitrogen atom is isoelectronic with a neutral oxygen atom, and we get two bonds and two lone pairs of electrons. When some of the electrons are involved in π bonding, trigonal-planar and linear structures can also result. *Table 10.10* summarizes the structural chemistry of nitrogen.

Table 10.10 Stereochemistry of nitrogen compounds

Shape	*Hybridization scheme*	*Number of σ bonds*	*Number of lone pairs*	*Number of π bonds*	*Formal state of N*	*Example*
Tetrahedral	sp^3	4	0	0	N^+	NH_4^+
Pyramidal	sp^3	3	1	0	N	NH_3
Angular	sp^3	2	2	0	N^-	NH_2^-
Trigonal-planar	sp^2	3	0	1	N^+	$(\bar{O})(\bar{O})\overset{+}{N}=O$
Angular	sp^2	2	1	1	N	Cl–N=O
Linear	sp	2	0	2	N^+	$\overset{-}{N}=\overset{+}{N}=O$
		1	2	1	N^-	

10.8.1.1 sp³ *Hybridization*

The tetrahedral distribution of bonds about a positively charged nitrogen atom has been established for the ammonium ion, NH_4^+, and the tetraalkylammonium ions, NR_4^+. Trimethylamine oxide, in which nitrogen is formally positive (i.e. $Me_3N^+–\bar{O}$), is also tetrahedral with N–O = 140 pm.

The neutral nitrogen atom forms quite a number of pyramidal NX_3 molecules (see *Table 10.11*), and in each case the bond angles are less than the tetrahedral angle. The deviation of the bond angles from tetrahedral (109.5°) arises from the presence of the lone pair of electrons, because these electrons occupy more volume than do those in the bonds. Hence the bonding pairs and the lone pair will tend to arrange themselves so as to give maximum mutual separation, thus giving bond

Table 10.11 Bond angles in pyramidal nitrogen molecules

∠*HNH*/degrees			∠*FNF*/degrees		
NH_3	NH_2OH	$NH_2(CH_3)$	NHF_2	$NF_2(CH_3)$	NF_3
107.1	107	105.9	102.9	101	102.4

angles less than the expected tetrahedral value. While the HNH angles are around 107°, the FNF angles are smaller and closer to 102°. This bigger deviation from the tetrahedral angle is to be expected because the fluorine atoms are more electronegative than the hydrogen atoms, so that the electrons in the N–F bonds will be pulled further from the nitrogen atom, and restricted more closely to the line of the nuclei; this gives a 'thinner' molecular orbital, which occupies less volume and leaves more of the space around the nitrogen atom available for the lone pair. The N–F bonds in NF_3 accordingly close up together, compared with the N–H bonds in NH_3. The wave functions for the atomic orbitals used in bonding by nitrogen will not correspond to simple sp^3 orbitals but will have rather less s character. It should be appreciated that the exact description of the orbitals used (i.e. the amount of s and p character) will depend upon the bond angle, which is a consequence of the arrangement of electrons for minimum repulsion.

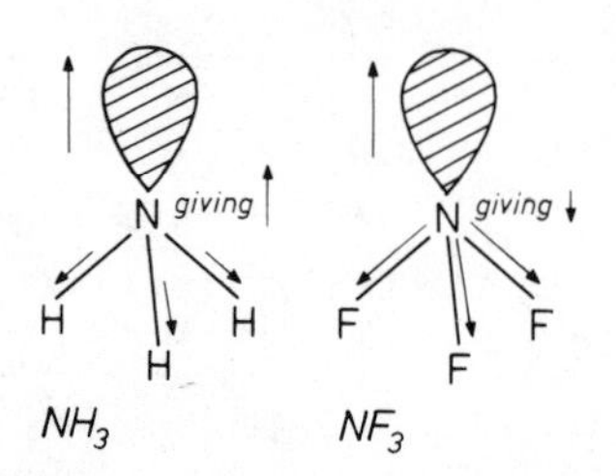

Figure 10.7 Dipole moments for NH_3 and NF_3

The observed dipole moments of NH_3 (1.5 D) and NH_3 (0.2 D) can be accounted for by attributing a dipole to the strongly directional lone pair of electrons. This is illustrated in *Figure 10.7*. Thus the resultant moment of the three N–H or N–F bonds is in the opposite direction to that of the lone pair, but whereas the NF_3 bond dipole just about cancels out that of the lone pair, the NH_3 bond dipole is much less, and a significant overall moment remains. The dipole moment of the N–H bonds is in the direction shown because the two orbitals used (sp^3 for N and s for H) are different in size and the region of maximum overlap is much closer to hydrogen than to nitrogen.

The pyramidal configuration is also found in N_2H_4 and N_2F_4; the molecules adopting a skew configuration (26).

X, X, X, X, N—N

(26)

O^-, $\overset{+}{N}$=O, ^-O

(27)

HO, 141pm, N, 121pm, O, 130·3°, O

(28)

10.8.1.2 sp^2 *Hybridization*

A planar arrangement arises when the nitrogen atom has a formal positive charge, making it isoelectronic with carbon, and uses one electron for π bonding. The simplest example is probably the nitrate anion, NO_3^- (27), which is isoelectronic with the carbonate ion, CO_3^-. All the N–O bond lengths are identical, of course, since the π bonding is delocalized over all three oxygen atoms, and the N–O distances of 126 pm are between the single and double bond distances, as would be expected. The molecular-orbital approach used for the carbonate ion (*Figure 10.5*) can be applied directly to the nitrate ion, and there will be one bonding π molecular orbital (containing two electrons) involving all four nuclei, and two non-bonding π molecular orbitals (each with two electrons) which concern only the three oxygen atoms.

The parent acid, HNO_3, is also planar, and microwave experiments give the parameters shown in (28). The N–O bond leading to the OH group is now much longer and the other N–O distances shorter than in the nitrate ion, since we now have one single bond, with the π bonding concentrated only on two N–O bonds. The bond angles are no longer identical, the angle between the multiple bonds now increasing to 130.3°.

There are several related molecules of general formula NO_2X, where X = Cl, F, Me or NH_2, and all have a planar structure with the ONO bond angle close to 130°. The known parameters are summarized in *Table 10.12.*

Table 10.12 Parameters for molecules of the type NO_2X

Compound	*∠ONO*/degrees	*N–O*/pm	*N–X*/pm
NO_2Cl	129.5	121	183
NO_2F	130	123	135
CH_3NO_2	130	120	–
$NO_2(NH_2)$	130.1	121	143

Neutral nitrogen forms quite a number of angular structures, of the general types XNO, XNCO and XNCS, in which one of the trigonal bonds is replaced by a lone pair of electrons; formulae (29) and (30) show this point clearly. The known parameters are listed in *Table 10.13*. The

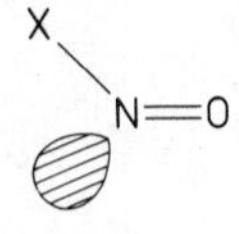

X–N=O (29)   X–N=C=O (30)   F–N=N–F (31)

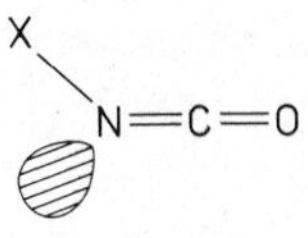

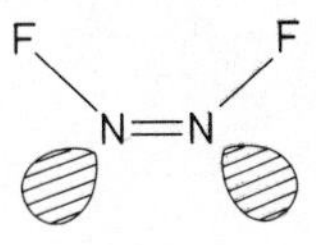

XNO bond angles are well below 120°, so evidently the lone pair of electrons is exerting a considerable repulsion effect. The bond lengths are curious in that the X–N distances are appreciably longer than would be expected for single bonds; the N–O distances are shorter than in the XNO_2 molecules, as we would predict, since the π bonding is now restricted to

Table 10.13 Parameters for XNO and XNCO molecules

Compound	*∠XNO or XNC/ degrees*	*X–N/pm*	*N–O or N–C/pm*
BrNO	114	214	115
ClNO	113.3	198	114
FNO	110	152	113
HONO	110.6	143	118
CH_3NO	112.1	149	–
PhNO	116	–	–
HNCO	128.1	99	121
CH_3NCO	140	144	–
HNCS	135	99	122
CH_3NCS	147.8	145	–

one N–O bond, but the bond shortening is rather more than anticipated and the bond is now much the same as in nitric oxide (p.64) which has a bond order of 2.5. Thus although we rationalize the molecular shape in terms of X–N single and N–O double bonds, it looks as though some of the electron density of the X–N bond has become associated with the N–O bond. By contrast, the N–F bonds in the *cis*-N_2F_2 (31) molecule are of normal single bond length and the N–N distance is reasonable for a double bond.

These XNO molecules clearly warrant fuller theoretical study. Just as nitrous acid has an angular structure so the nitrite ion, NO_2^-, is also angular (ONO = 114.9°, N–O = 124 pm).

Nitrosomethane, CH_3NO, is listed in *Table 10.13* as an angular molecule, but in the solid it is dimeric (32) with a nitrogen–nitrogen bond; both the N–N and N–O bonds have some multiple-bond character.

Nitrogen dioxide, NO_2, which has an angular structure (33), is an especially interesting molecule because it contains an unpaired electron which has been shown by electron-spin resonance experiments to be in a σ rather than a π orbital. The molecule can be considered formally as an $-NO_2$ grouping but with the odd electron in a hybrid orbital that is approximately sp^2. The ONO angle is 134.1° and the NO bond length 119 pm. In addition to the σ bonding, based on sp^2 hybrid orbitals, there will be four electrons in the π system, two bonding and two non-bonding.

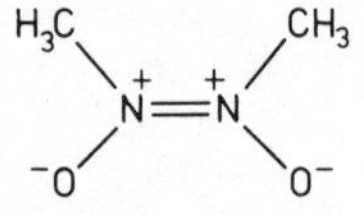

(32) 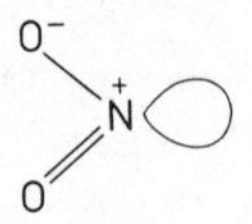 (33) 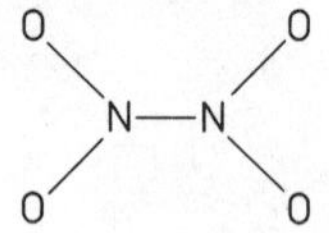 (34)

The pairing up of two NO_2 molecules leads naturally to dinitrogen tetroxide, which has a planar skeleton (34). We can formally regard the molecule as consisting of two NO_2 groups held together by a single σ bond, the π bonding being restricted to the individual ONO systems. This is undoubtedly oversimplified because it does not explain why the molecule is planar rather than eclipsed, and why there is a large energy

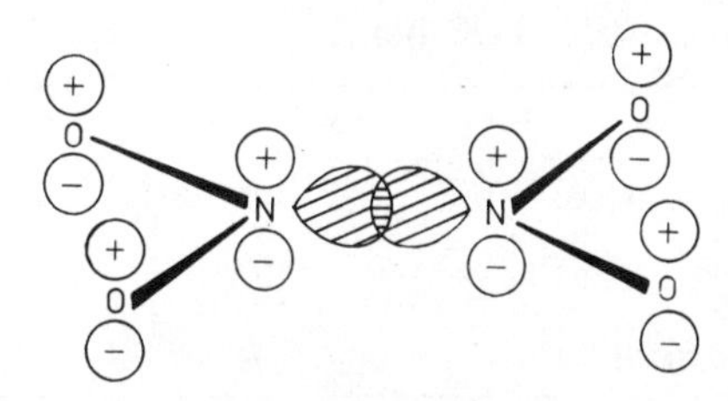

Figure 10.8 Atomic orbitals available for bonding in N_2O_4

barrier to rotation about the N–N bond. We could account for these points by supposing that the π bonding is delocalized over the entire skeleton, but then we have to explain why the N–N bond is so long when it contains considerable π character. Some lengthening would be expected, of course, because the two nitrogen atoms have formal positive charges. One possibility is that the NO_2 groups may be linked by a π bond only, with no σ bonding. Thus we could picture the scheme shown in *Figure 10.8.* In this rather naive representation each nitrogen is sp^2 hybridized, and uses two of these hybrids to σ bond to the two oxygen atoms; this leaves a lone pair of electrons on each nitrogen pointing along the line of the nitrogen nuclei. There will be a single electron in the $2p_z$ atomic orbital of each nitrogen and oxygen atom, so that by simple molecular-orbital theory we shall have three delocalized bonding π molecular orbitals each containing two electrons. However, this scheme is pure speculation and no universally accepted bonding arranged has yet emerged. Interested readers should consult the paper by Bent[12] for a review of the arguments.

Dinitrogen trioxide, N_2O_3, also has a long nitrogen-nitrogen bond (185 pm) and a planar structure (35), with one planar nitrogen and one angular nitrogen. Another set of planar nitrogen compounds is that in which nitrogen is bonded to three silicon atoms as in the molecule $N(SiH_3)_3$ (36), and the planar configuration is retained even when only two SiH_3 groups are linked to nitrogen. *Table 10.14* lists the known parameters. The planarity of the molecules is accounted for by the tendency for nitrogen to donate its $2p_z$ electron lone pair into vacant $3d$ orbitals on the silicon atoms, giving partial double bonding. As would be expected, the N–Si bond distances of 173 pm are less than the calculated single bond distance of 187 pm. Further evidence for the involvement of the nitrogen lone pair in π bonding is the very weak donor ability of the nitrogen atom.

(35) (36) (37)

While the bond angles in $N(SiH_3)_3$ are very close to 120°, the SiNSi angle increases when one group is replaced by H, Me or BF_2, presumably because there is now more room for the residual bulky SiH_3 groups. When two SiH_3 groups are replaced, as in $NMe_2(SiH_3)$, the nitrogen atom no

Table 10.14 Parameters for molecules with N–Si bonds*

Molecule	*∠SiNSi*/degrees	*Si–N*/pm
$N(SiH_3)_3$	119.7	173
$NH(SiH_3)_2$	127.7	173
$NMe(SiH_3)_2$	125.4	173
$N(BF_2)(SiH_3)_2$	123.9	174
$NMe_2(SiH_3)$	120 (SiNC)	175

*From Simpson[13]

longer retains a planar environment, the N–Si bond now being at an angle of 28° to the CNC plane.

10.8.1.3 sp *Hybridization*

We now discuss linear structures, in which the nitrogen atom forms two σ and two π bonds, as in the azide ion, N_3^-, the nitronium ion, NO_2^+, and nitrous oxide, N_2O. It can be seen that the first two are isoelectronic with carbon dioxide:

$$O{=}C{=}O \qquad \overset{-}{N}{=}\overset{+}{N}{=}\overset{-}{N} \qquad O{=}\overset{+}{N}{=}O$$

Hence the bonding can be accounted for in exactly the same way, with the nitrogen and oxygen atoms using *sp* hybrid orbitals to form σ bonds; the terminal atoms have a lone pair in one such hybrid orbital (see *Figure 10.9*). The π molecular-orbital energy diagram is the same as that given in *Figure 10.6* for CO_2 (p.148).

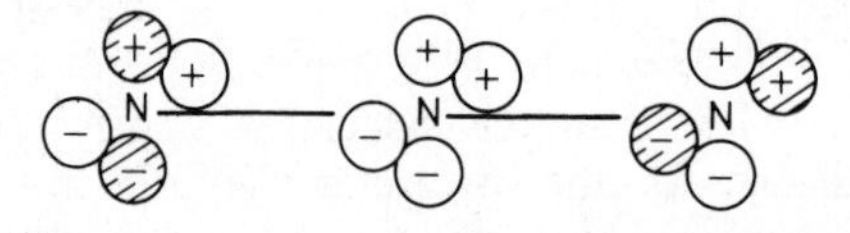

(a) σ system

(b) π system

Figure 10.9 σ and π orbitals used to describe bonding in N_3^-

The parent acid, hydrazoic acid, HN_3, and methyl azide, CH_3N_3, are covalent, and both molecules have a linear N–N–N skeleton (37), the parameters being given in *Table 10.15*. The data show that in each instance the terminal (N_{II}–N_{III}) bond is much the shorter of the two and so has a higher bond order. In valence-bond terms we have two principal contributing structures:

$$X\diagdown N{=}\overset{+}{N}{=}\overset{-}{N} \qquad X\diagdown \overset{-}{N}{-}\overset{+}{N}{\equiv}N$$

Table 10.15 Parameters for XN_3 molecules

Molecule	*XNN*/degrees	N_I–N_{II}/pm	N_{II}–N_{III}/pm
HN_3	114.1	124	113
MeN_3	117	124	112

The σ bonding is much the same as in the azide ion, except that atom N_I is taken as sp^2 hybridized. The π bonding consists of a localized π bond (π_2) between atoms II and III, a π bond (π_1) linking all three nitrogen atoms, and a non-bonding pair of π electrons (π_3) associated with atoms I and III. The π orbitals are illustrated in *Figure 10.10* and the π orbital energy diagram in *Figure 10.11*.

Nitrous oxide has been shown by molecular spectroscopy to have the nuclei arranged N–N–O, rather than N–O–N, the bond lengths being

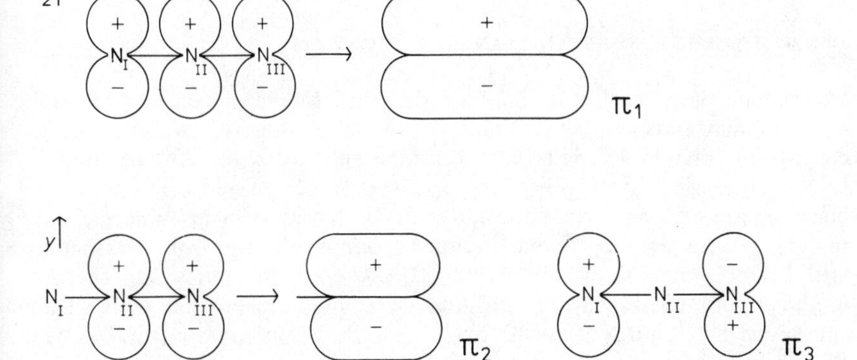

Figure 10.10 π Orbitals for HN_3

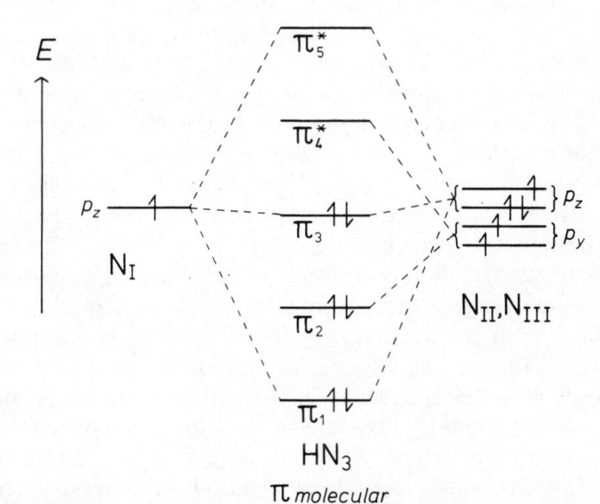

Figure 10.11 π Molecular orbital energy diagram for HN_3

N–N = 113 pm and N–O = 119 pm. The valence-bond description is in terms of two canonical forms:

$$\overset{-}{\text{N}}{=}\overset{+}{\text{N}}{=}\text{O} \qquad\qquad \text{N}{\equiv}\overset{+}{\text{N}}{-}\overset{-}{\text{O}}$$
$$\text{I} \qquad\qquad\qquad\qquad \text{II}$$

We can assume that the nitrogen and oxygen atoms use *sp* hybrid orbitals, and that an orbital interpretation can be made by superposing the orbital forms corresponding to structures I and II. There will be one localized π orbital between the two nitrogen atoms, and a delocalized π orbital extending over all three atoms. The two remaining electrons will occupy a non-bonding π orbital, similar to that described for HN_3.

Solid dinitrogen pentoxide has been found to be ionic, with NO_2^+ and NO_3^- ions. The nitronium ion is linear, $\text{O}{=}\overset{+}{\text{N}}{=}\text{O}$, as would be expected, with N–O bond lengths of 115 pm.

10.8.2 PHOSPHORUS, ARSENIC AND ANTIMONY

Whereas nitrogen itself is a diatomic gas, the remaining elements are solids at room temperature. The common form of phosphorus, white phosphorus, consists of discrete P_4 units held together only by weak van der Waals' forces, although more highly polymeric forms such as red phosphorus and black phosphorus are also known, the latter having a giant molecule structure. P_4 units are also found in the vapour phase, although dissociation into P_2 molecules occurs above 800 °C. Arsenic also forms As_4 species in the vapour phase, but the position is rather more obscure with antimony and bismuth, although it seems likely that their vapours contain Sb_4 and Bi_2 units. Electron diffraction experiments show that the phosphorus atom in the P_4 molecule are at the corners of a regular tetrahedron, the P–P bond distances being 221 pm and the PPP angles being 60°.

As with the Group IV elements, we notice a marked difference between the first element of the group and the remainder. Once again there is a marked reluctance for the heavier elements to form multiple MM bonds (M = P, As, Sb), so that phosphorus prefers to form three σ bonds rather than one σ bond and two π bonds (as with the N_2 molecule). Furthermore, phosphorus, arsenic, antimony and bismuth differ from nitrogen in that the *d* orbitals become energetically available and may be used to increase the covalency of these elements to five or six. The full incorporation of 3*d* orbitals into the σ bonding scheme of, say, PCl_5, is an oversimplification, of course, but is justified at the level of sophistication used in this book. Interested readers who would like to take this matter further should consult a recent review by Mitchell[14], which assesses the role of *d* orbitals in bonding by second and third row elements, and two papers (Rauk *et al.*[15]; Hoffmann *et al.*[16]) that produce molecular-orbital schemes for the hypothetical PH_5 molecule, with a discussion on the extent to which the 3*d* orbital will contribute to the bonding orbitals. We shall see that in molecules such as $POCl_3$ it is convenient to discuss the σ bonding (tetrahedral) on the basis of sp^3 hybrid orbitals together with a π bond between P and O. As far as π bonding is concerned, a

Table 10.16 Stereochemistry of the Group VB elements

Oxidation state	*Shape*	*Hybridization scheme*	*Number of*			*Examples*
			σ bonds	*lone pairs*	*π bonds*	
V	Tetrahedral	sp^3	4	0	0	$[PH_4]^+$, $[AsPh_4]^+$
III	Pyramidal	sp^3	3	1	0	PF_3, AsF_3, $SbCl_3$, $BiBr_3$
V	'Tetrahedral'	sp^3	4	0	1	$POCl_3$
V	Trigonal-bipyramidal	sp^3d	5	0	0	PF_5, AsF_5, $SbCl_5$
III	Distorted tetrahedral	sp^3d	4	1	0	$AsCl_3,NMe_3$; $SbCl_3,NH_2Ph$
V	Octahedral	sp^3d^2	6	0	0	$[PF_6]^-$, $[AsF_6]^-$, $[SbCl_6]^-$, $[BiCl_6]^-$
III	Square-pyramidal	sp^3d^2	5	1	0	$[SbCl_5]^{2-}$

general rule of thumb seems to be that when phosphorus is attached to three, five or six atoms the links are essentially single bonds, whereas some double-bond character is likely to be present when phosphorus is linked to four other atoms.

The essential structural chemistry of the Group VB elements (other than nitrogen) is summarized in *Table 10.16.*

10.8.2.1 sp³ *Hybridization*

Positive charged phosphorus (which is isoelectronic with silicon) and arsenic form simple tetrahedral structures in such ions as $[PH_4]^+$, $[PBr_4]^+$, $[PCl_4]^+$ and $[AsPh_4]^+$.

With the neutral tervalent elements we have a situation similar to that found for nitrogen in which one tetrahedral position is taken up by a lone pair of electrons, so that the three M–X bonds adopt a pyramidal shape. As the values quoted in *Table 10.17* show, the XMX bond angles

Table 10.17 XMX bond angles in MX_3 compounds/degrees

Compound	*XPX*	*XAsX*	*XSbX*
MBr_3	101	97.7	94.5
MCl_3	100.1	98.6	99.5
MF_3	97.8	96	–
MH_3	107.1	93.3	91.8
MI_3	102	100.2	95.8
MMe_3	99.1	96	–

in the MX_3 compounds are appreciably less than the tetrahedral value, and in the hydrides the angles are little more than 90°.

The much smaller bond angle found in the hydrides (compared with NH_3) may possibly be attributed to the increased size of the phosphorus, arsenic and antimony atoms, which allows the bonding pairs to get closer to one another before mutual repulsion becomes important, with a corresponding expansion of the lone-pair electrons. With the halides, repulsion between lone pairs on neighbouring halogen atoms will tend to keep bond angles in the region of 100°. Similar but less symmetrical pyramidal structures are observed for compounds such as Me_2PH, PHF_2 and Me_2AsI.

Quinquevalent phosphorus forms a number of 'tetrahedral' compounds of the type $X_3P{=}Y$ (where X = Br, Cl, F, alkyl or aryl, and Y = O or S), in which four electrons are used for σ bonding and one electron for π bonding. Whereas the nitrogen–oxygen bond in R_3NO molecules has to be written N→O or $\overset{+}{N}{-}\overset{-}{O}$, because nitrogen cannot form more than four two-electron covalent bonds, the PO or PS bonds can be considered as essentially double, because it is possible for the 3*d* orbitals to be used in d_π–p_π bonding; this is illustrated in *Figure 10.12.* It would be an over-simplification to look upon such bonds as full double bonds, because the phosphorus 3*d* orbitals are a little high in energy and in some cases will be rather diffuse for really effective overlap. However, in the molecules under discussion the phosphorus atom is likely to be somewhat positive and this will result in a contraction of the 3*d* orbitals; the π charge

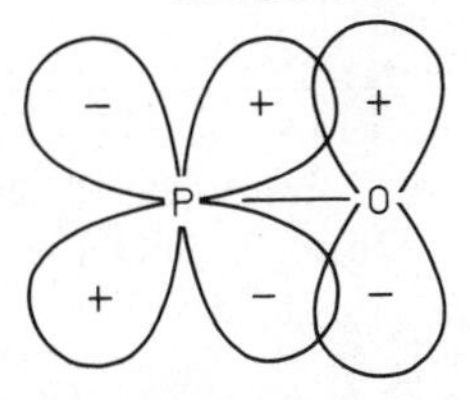

Figure 10.12 d_π–p_π *Bonding*

cloud is expected to be distorted in favour of the oxygen nucleus. We would expect the P=O bonds to be much shorter than the calculated single-bond distance (176 pm) and in practice this is the case, the PO bonds in $POBr_3$, $POCl_3$, POF_3 and $POPh_3$ being around 145 pm. In the analogous PSX_3 compounds the P–S distance averages 186 pm. In all these molecules the XPX angles are less than the tetrahedral value, ranging from 100 to 104°, but this is to be expected because the double bond electrons will take up more room than the electrons in the PX bonds.

Hence with quinquevalent phosphorus we get three single bonds and one double bond whereas with the tervalent element we have a lone pair of electrons in place of the double bond.

The 'tetrahedral' arrangement is also found in a range of oxo compounds of phosphorus, such as P_4O_6 (38) and P_4O_{10} (39), with bridging OPO bond angles of 99 and 101.5°, respectively. In P_4O_{10} the terminal and skeleton P–O bonds are 140 and 160 pm, respectively, in accordance with the double- and single-bond formulation.

(38) (39) (40)

The structure of phosphorous acid (40) is interesting in that one PO bond (147 pm) is appreciably shorter than the other two bonds (154 pm), agreeing with the view that the molecule has one P=O bond with a hydrogen atom attached to each of the other two oxygens. The individual $HPO(OH)_2$ molecules are linked together by hydrogen bonds. The angle between the P–OH bonds is 102°, and those between the P=O and P–OH bonds are 113 and 116°, values consistent with a distorted tetrahedral arrangement in which the electrons in the double bond exert the greatest repulsion effect. Simple orthophosphate ions, such as $H_2PO_4^-$ and HPO_4^{2-}, also have a tetrahedral configuration, as do more condensed phosphate ions such as $P_2O_7^{4-}$, $P_3O_{10}^{5-}$ and $P_4O_{12}^{4-}$. The first two condensed ions form chain structures (41) and the latter has a chair-shaped cyclic arrangement (42). Thus we have represented the triphosphate structure by just one canonical form, although it must be appreciated that all the terminal PO bonds are equivalent (150 pm) with some π character; the bridging PO bonds forming the skeleton are much longer (161 and 168 pm). Similar bond lengths (148 and 161 pm) are found for the terminal and bridging PO bonds in the cyclic $P_4O_{12}^{4-}$ ion (Koster and Wagner[17]).

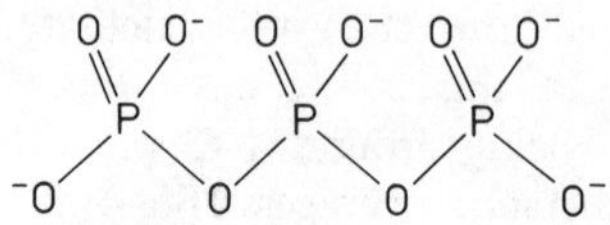

(41)

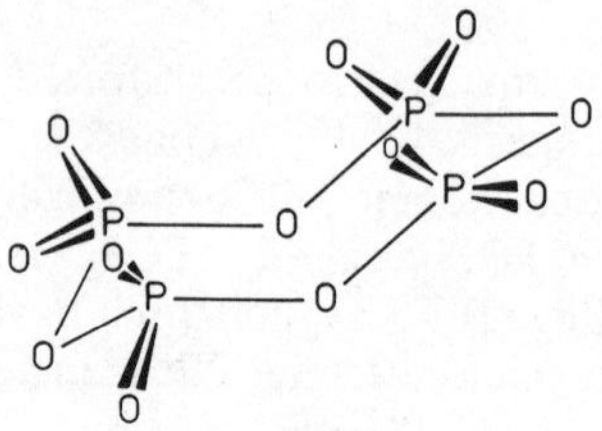

(42)

Phosphorus also forms a number of sulphides which contain four phosphorus atoms in the basic unit. Thus P_4S_{10} has the same structure as P_4O_{10}. P_4S_6, the analogue of P_4O_6, does not appear to have been characterized, but P_4S_3, P_4S_5 and P_4S_7 have been prepared and their structures determined. These are shown in the formulae (43), (44) and (45). An interesting feature of these structures is the tendency of phosphorus to form

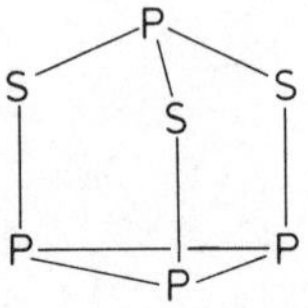

(43)

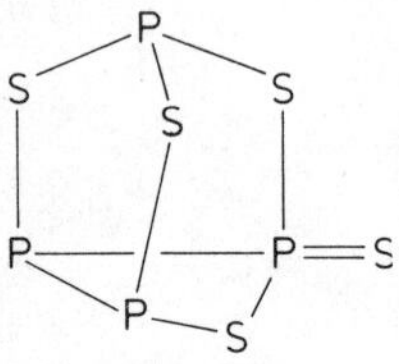

(44)

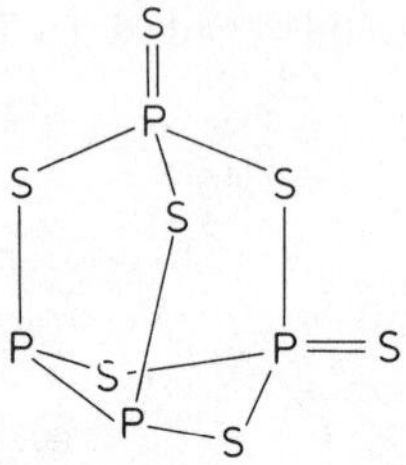

(45)

some terminal P=S bonds while still retaining P–P links. The bridging and terminal PS bonds have values close to 209 and 195 pm, respectively, in all these compounds, in agreement with our assignment of some π character to the terminal bonds. In these sulphides the SPS angles (internal) range from the tetrahedral value in P_4S_{10} down to 99° in P_4S_3. The two selenides, P_4Se_3 and P_4Se_5, have also been examined and found to have exactly the same types of structure as their sulphide analogues. In P_4Se_5 (Penney and Sheldrick[18]) the bridging P–Se distances range from 221 to 234 pm, while the terminal PSe bond is appreciably shorter at 212 pm; by comparison, the PSe bond in Et_3PSe is even shorter at 196 pm. The arsenic sulphides, As_4S_3 and As_4S_5, have also been examined recently (Whitfield[19]). Whereas As_4S_3 in both its α and β forms has exactly the same structure as P_4S_3 and P_4Se_3, the As_4S_5 molecule (46) differs from P_4S_5 in that all the As–S bonds are bridging and there is no terminal As–S bond. Thus arsenic evidently prefers to form single bridging bonds rather than terminal double bonds; in these arsenic sulphides the As–S distances range from 222 to 227 pm. The arsenic analogue of the 'missing' P_4S_6, is known: it is not a discrete molecule but has a layer structure in which each arsenic is linked to three bridging sulphur atoms; As_2Se_3 has the same layer structure (Renniger and Averbach[20]).

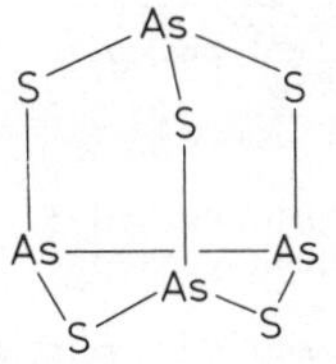

(46) 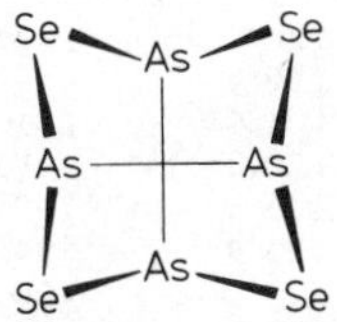 (47) 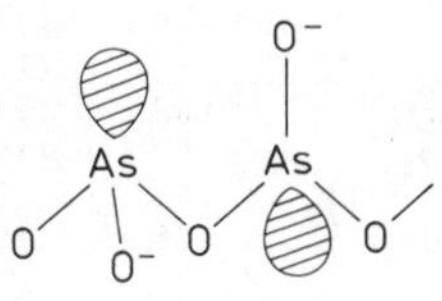 (48)

Arsenic also forms sulphides and selenides of formulae As_4S_4 and As_4Se_4 (47), and in these molecules it has been shown (Whitfield; Goldstreak and Paton[21]) that the arsenic atoms are in pairs linked by the selenium or sulphur atoms; the cage structure can be looked upon as a square of selenium atoms bisecting a distorted tetrahedron of arsenic atoms.

Arsenic also achieves a tetrahedral configuration in the simple arsenate ion AsO_4^{3-}, while one modification of arsenic(III) oxide, As_4O_6, has a structure similar to that of P_4O_6. An essentially tetrahedral configuration is found in linked polyarsenites such as $(NaAsO_2)_x$ (48), in which a lone pair of electrons takes up one bonding position. A recent study (Svennson[22]) of Sb_2O_3 has shown it to have a polymeric structure with each antimony linked to three bridging oxygen atoms. The OSbO bond angles are 79.8, 91.9 and 98.1°, emphasizing the distortion, and the smallness of the angles suggests that the lone pair electrons do not play a very significant role in the structure, being largely *s* in character.

One final observation on the decreasing tendency for π bond-formation with the heavier Group V elements is provided by the structure of $As(SiH_3)_3$ (Beagley *et al.*[23]) which is pyramidal with SiAsSi bond angles of 93°. The AsSi bonding must evidently be described in terms of arsenic using almost pure *p* orbitals, with the lone pair essential in the *s* orbital and unable to contribute to any double bonding by delocalization into the vacant 3*d* orbitals of silicon.

Phosphorus also forms four σ bonds in the phosphazenes (phosphonitrilics), which have the general formula $(PNX_2)_n$. Most of the compounds that have been fully characterized are cyclic compounds (n = 3–8), in which the phosphorus and nitrogen atoms alternate in the ring (49), although some long-chain compounds (50) have been characterized as fibre-like. Despite their polymeric nature these long-chain phosphazenes have no major commercial applications because they are very readily hydrolysed.

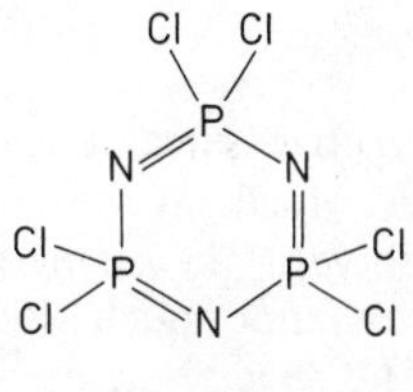

(49) 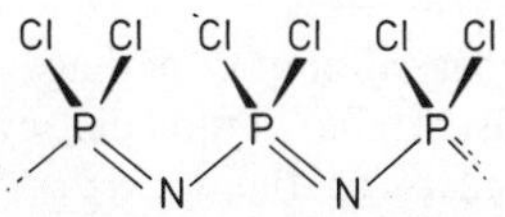 (50)

Table 10.18 Parameters for phosphazenes (phosphonitrilics), PNX_2

Compound	*P–N/* pm	*NPN/* degrees	*PNP/* degrees	*XPX/* degrees	*Comment*
$(PNBr_2)_3$	158	118.3 115.8	126.8 118.4	103.4	Slight puckering
$(PNCl_2)_3$	158	118.4	121.4	101.3	Planar
$(PNF_2)_3$	157	119.4	120.3	99.9	Planar
$[PN(NCS)_2]_3$	158	119	121	100	Slight puckering
$[PN(OPh)_2]_3$	158	117.3	121.9	94.1; 100.1	Slight puckering
$(PNPh_2)_3$	160	117.8	122.1	103.8	Slight puckering
$(PNBr_2)_4$	158	120.1	131.0	103.9	Puckered
$(PNCl_2)_4$	156	120.5	133.6; 137.6	103.1	Puckered
$(PNF_2)_4$	151	122.7	147.4	99.9	Planar
$(PNMe_2)_4$	160	119.8	131.9	104	Puckered
$[PN(NMe_2)_2]_4$	158	120	133	104	Puckered
$[PN(OMe)_2]_4$	157	121	132.2	105	Puckered

Table 10.18 summarizes the structural data that are known for the simpler cyclic compounds. This information has been culled mainly from a recent review by Allcock[24], and readers are referred to it if they are also interested in details of the chemical properties of these compounds.

The trimeric molecules have an essentially planar P_3N_3 hexagon, apart from slight puckering in one or two instances, but nearly all the tetramers and higher cyclic systems have very puckered rings. It is interesting to note that while ring puckering results in wide variations of the PNP angles, the ring bond angles at phosphorus remain surprisingly close to 120°. The length of the P–N bonds is remarkably constant and falls in the range 156–160 pm [except for $(PNF_2)_4$], being very much less than the calculated single-bond value of 180 pm. This bond shortening is attributed to a measure of d_π–p_π bonding arising from the lateral overlap of phosphorus d_π orbitals with the nitrogen $2p_z$ orbitals (see *Figure 10.12*). If we take the plane of the σ bonds as the xy plane, then the phosphorus atom can use either the d_{xz} or d_{yz} orbital for this π bonding (see Chapter 11, p.205). A detailed consideration of orbital overlap suggests that the use of only the d_{xz} orbital leads to the formation of a 'pseudo-aromatic' delocalized π system, but that the use by phosphorus of a linear combination of d_{xz} and d_{yz} orbitals gives a more restricted or 'island' three-centre π orbital (see *Figure 10.13*).

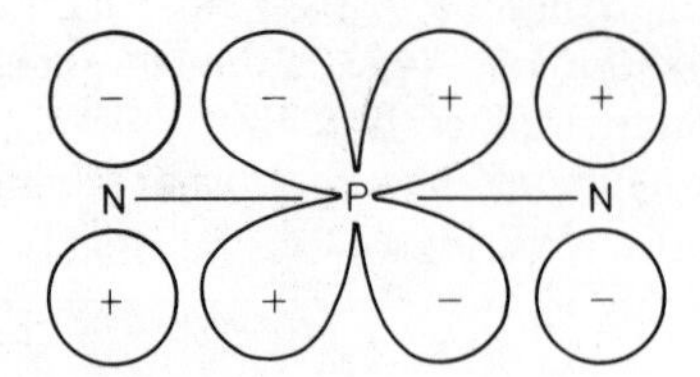

Figure 10.13 'Island' or three-centre π molecular orbitals. P orbital shown as single d, *rather than a hybrid of* d_{xz} *and* d_{yz}

The latter description is perhaps the more attractive since it can be applied equally well to the non-planar cyclic molecules which are present in the higher members of the series, and is consistent with the similarity of spectra found for all the compounds. While there is not much doubt about the existence of π bonding in the phosphazenes, the precise nature of it is by no means settled at this stage.

10.8.2.2 sp^3d *Hybridization*

When the quinquevalent elements form MX_5 compounds, there are two basic configurations that might be expected (*Figure 10.14*), trigonal-bipyramidal and square-pyramidal. The bonding description of such molecules is usually on the basis of sp^3d hybrid orbitals, although the trigonal-bipyramidal

Trigonal-bipyramid Square-pyramid

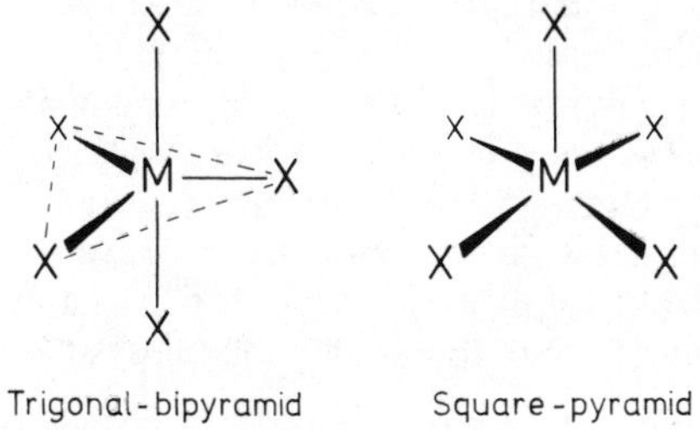

Figure 10.14 Basic configurations for MX_5 compounds

arrangement, for instance, can be explained without involving a d orbital, by using sp^2 hybrid orbitals for the three equatorial bonds and the unhybridized p_z orbital for the two axial bonds – by means of a three-centre delocalized orbital along the z axis. The d_{z^2} orbital can then be incorporated into the axial bonds to any required extent.

In practice the trigonal-bipyramidal arrangement is found for simple MX_5 molecules such as the pentahalides PF_5, PCl_5, PBr_5, AsF_5 and $SbCl_5$ in the gas phase. In such molecules the axial M–X bonds are invariably longer than the equatorial ones, in line with the idea that they are less strong. When trigonal-bipyramidal molecules have two different halogens, as in PCl_3F_2 and PCl_2F_3, the axial positions are occupied by the most electronegative atoms, in this case fluorine. The alkyl derivatives of the antimony halides (R_3SbX_2; X = Cl, Br or I) also have trigonal-bipyramidal structures, with the alkyl groups in the equatorial positions.

The penta-alkyls and -aryls also assume the trigonal-bipyramidal arrangement in most cases, as in Ph_5P and Ph_5As, although Ph_5Sb has a square-pyramidal shape, and this is not apparently a result of crystal packing effects since spectroscopic studies (Beattie and Livingston[25]) suggest that the shape is retained when the compound is dissolved in dichloromethane. There is evidently little to choose between these two bonding arrangements because (p-tolyl)$_5$Sb has a trigonal-bipyramidal structure (Brabart *et al.*[26]).

Even with phosphorus compounds the square-pyramidal configuration can be observed, as in the catechol derivative $MeP(O_2C_6H_4)_2$ (51) (Wunderlich[27]), where the methyl group occupies the apical position. With the corresponding fluoro compound, $FP(O_2C_6H_4)_2$, the molecule adopts a structure in between the two idealized configurations.

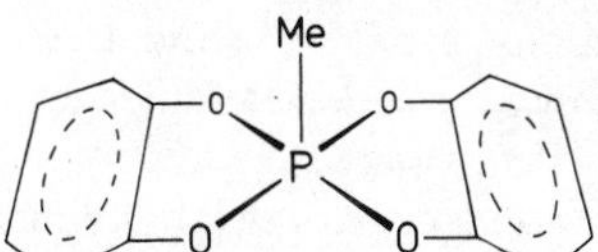

(51) 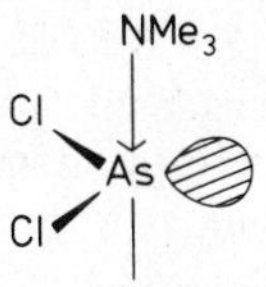 (52) 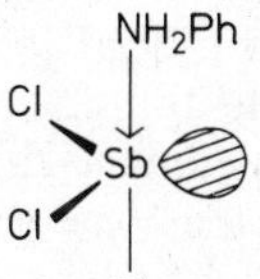 (53)

The pentachloride and pentabromide of phosphorus have ionic structures in the solid state, being $[PCl_4]^+[PCl_6]^-$ and $[PBr_4]^+Br^-$, respectively. The positive ions are tetrahedral, as we have already seen, and the $[PCl_6]^-$ ion is octahedral, so that the trigonal-bipyramidal structure found in the gas phase has become a mixture of the more symmetrical ions in the solid state.

Just as the pyramidal MX_3 compounds can be considered as based on a tetrahedral arrangement of one lone and three bonding pairs of electrons, so there is a category of compounds based on a trigonal-bipyramidal configuration of five electron pairs, one being a lone pair. The lone pair occupies one of the equatorial positions. Two recent examples are $AsCl_3,NMe_3$ (52) (Webster and Keats[28]) and $SbCl_3,NH_2Ph$ (53) (Hulme and Scruton[29]). In the case of the antimony compound the axial bonds are bent away from the lone pair of electrons, the NSbCl bond angle being 166°. Later on we shall see that this type of structure is also found in molecules such as $SeCl_4$.

10.8.2.3 sp^3d^2 *Hybridization*

All the quinquevalent Group VB elements form hexahalogeno ions $[MX_6]^-$ in which there are six σ bonds octahedrally disposed about the central ion atom. There are also several cases in which the pentahalide accepts a pair of electrons from a donor ligand to give octahedral complexes; examples are PF_5,NC_5H_5 and $SbCl_5,OPCl_3$.

In the tervalent state antimony also forms a six-co-ordinate complex $[SbCl_6]^{3-}$. This is a particularly interesting compound because the antimony atom has a lone pair of electrons which could influence the shape of the ion. However, a crystal structure determination of $[Co(NH_3)_6]^{3+}[SbCl_6]^{3-}$ (Schroeder and Jacobson[30]) shows that the anion has a regular octahedral configuration, so the lone pair must be in the *s* orbital. Into a similar

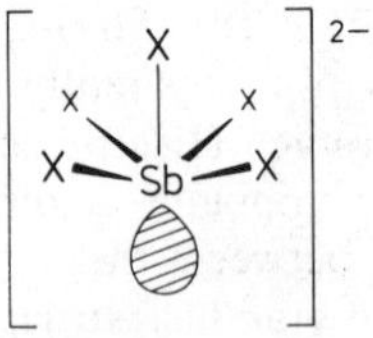

$$\left[\begin{array}{c} \text{X} \\ \text{X} \quad \text{X} \\ \text{X—Sb—X} \\ \text{(lone pair)} \end{array} \right]^{2-}$$

(54)

category come the $[SbX_5]^{2-}$ ions in which we have five bonding pairs and one lone pair of electrons, and the five halogen atoms provide a square-pyramidal environment (54) for the antimony atom. In the ions so far studied, $[SbF_5]^{2-}$ and $[SbCl_5]^{2-}$, the antimony atom is below the plane of the four halogen atoms, and this arrangement can be rationalized if it is assumed that the lone pair occupies one of the octahedral positions, and that the lone pair–bonding pair repulsion is greater than the bonding pair–bonding pair repulsion.

In the next section we shall see that the isoelectronic ion $[TeCl_5]^-$ has a similar structure.

10.9 Oxygen and the Group VIB Elements

10.9.1 OXYGEN

The ground state of the oxygen atom is $1s^2 2s^2 2p_x{}^2 2p_y{}^1 2p_z{}^1$ and accordingly the neutral atom can use the two unpaired electrons to form two covalent bonds. The covalency cannot be increased because the orbitals of principal quantum number 3 are much higher in energy, so that the energy required to promote one of the paired electrons into a higher orbital – giving two more unpaired electrons – is much greater than that which would be saved by forming two further bonds. Similarly these higher orbitals are too high in energy to accept electrons from donor ligands.

The diatomic oxygen molecule, O_2, which was discussed at some length in Chapter 6, is best described on a molecular-orbital basis, with the molecule having the configuration:

$$O_2\,[KK(\sigma 2s)^2(\sigma^* 2s)^2(\sigma 2p)^2(\pi_y 2p = \pi_z 2p)^4(\pi_y^* 2p = \pi_z^* 2p)^2]$$

or in Mulliken notation:

$$O_2\,[KK(z\sigma)^2(y\sigma)^2(x\sigma)^2(w\pi)^4(v\pi)^2].$$

The stereochemistry of oxygen compounds is limited to two bonding arrangements, pyramidal and angular, as *Table 10.19* shows. The pyramidal arrangement is possible when the oxygen atom acquires a positive charge

Table 10.19 Stereochemistry of oxygen compounds

Formal state of O	*Shape*	*Hybridization scheme*	*Number of* σ *bonds*	*Number of lone pairs*	*Number of* π *bonds*	*Example*
O^+	Pyramidal	sp^3	3	1	0	H_3O^+
O	Angular	sp^3	2	2	0	H_2O
O^+	Angular	sp^2	2	1	1	O_3

(making it isoelectronic with a neutral nitrogen atom), whereas the neutral oxygen atom uses its two unpaired electrons to form two angular bonds. A negatively charged oxygen atom, on the other hand, now contains only one unpaired electron and consequently forms only one bond, e.g. OH^-.

10.9.1.1 sp³ *Hybridization*

The water molecule has an angular structure, with lone pairs of electrons taking up two of the tetrahedral positions, the HOH angle being 104.5°. Thus if we compare the series of hydrides CH_4, NH_3 and H_2O, we note that the replacement of a bond by a pair of electrons results in the remaining bonds getting closer together, so that while the HCH bond angle in CH_4 is 109.5° the HNH bond angle in NH_3 is only 107.1°. With the introduction of a second lone pair of electrons in H_2O, the HOH bond angle drops to 104.5°. The bond orbitals are described as having an increas ing amount of *p* character along the series CH_4, NH_3, H_2O.

In Chapter 9 we saw that in ice, the individual water molecules are linke together by hydrogen bonds, a maximum of four hydrogen bonds per H_2O molecule being found at low temperatures. The average number of hydroge bonds formed by each water molecule decreases as the temperature is raised, but even in the liquid state at 40 °C each molecule forms about half of the theoretically possible number of hydrogen bonds. The tetrahedral disposition of the hydrogen bonds around the oxygen atom in the ice structure is explained if we assume that the strongest bond is formed along the axis of an oxygen sp^3 hybrid orbital containing a lone pair of electrons.

Other OX_2 molecules also adopt the angular structure, and XOX angles of 103.1, 110.9 and 111.7° have been reported for F_2O, Cl_2O and OMe_2, respectively. In the latter two molecules the bond angles have increased to values slightly greater than the tetrahedral value, and this we attribute to repulsion between either the lone pairs on the chlorine atoms or the bonding pairs in the C–H bonds of the methyl groups.

Oxygen forms a second hydride, H_2O_2, hydrogen peroxide, which has the structure shown in *Figure 10.15*. The hydrated molecule, $H_2O_2,2H_2O$ has a similar structure, with the hydrogen-bonded water molecules forming planar

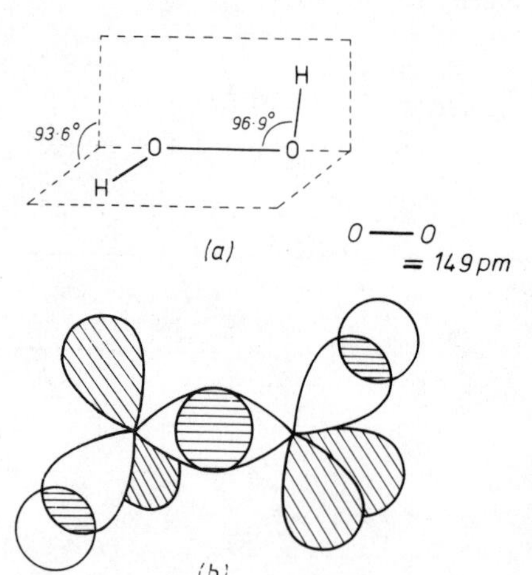

Figure 10.15 Structure of H_2O_2: (a) structural parameters; (b) orbital diagram showing the oxygen atoms using sp^3 hybrid orbitals

chains cross-linked to the hydrogen peroxide molecules by hydrogen bonds. One other molecule with an O–O bond is F_2O_2, and this too has the same general arrangement of the nuclei, although, as the data in *Table 10.20* show, this bond is very much shorter than the one in H_2O_2.

In *Figure 10.15*, the bonding in the H_2O_2 molecule is based on the oxyge atoms being sp^3 hybridized, with two lone pairs of electrons; the O–O bond results from the end-on overlap of an sp^3 orbital from each oxygen atom, and the O–H bonds are formed by the overlap of the remaining sp^3 orbital of oxygen with a hydrogen 1*s* orbital. Repulsion between the two lone pairs on each oxygen atom forces the O–H and O–O bonds together so that the HOO bond angle is well below the tetrahedral value; there is also likely to be some repulsion between the electron pairs on one oxygen atom with those

Table 10.20 Structural parameters for X_2O_2 molecules

Molecule	*X–O*/pm	*O–O*/pm	*XOO*/degrees
H_2O_2	–	149	96.9
H_2O_2,$2H_2O$	–	148	100.4
F_2O_2	158	122	109.5

on the other oxygen atom, and this may account for the O–O bond length being greater than the calculated single bond value.

While the basic bonding in the O_2F_2 molecule can be described in the same way as H_2O_2, account must be taken of the short oxygen–oxygen bond and the relatively long oxygen–fluorine bonds (158 compared with 141 pm in OF_2). The problem is similar to that we came across with XNO molecules (p.153), where we had a short nitrogen–oxygen distance and long X–N bonds, and once again we must assign a significant measure of multiple bonding: in this case to the oxygen–oxygen bond. One interpretation starts with the usual bonding scheme for the oxygen molecule (p.167), and invokes the interaction of the antibonding π^* molecular orbitals (each with a single electron) with the fluorine atomic orbitals, giving two three-centre molecular orbitals, each covering the two oxygen atoms and one fluorine. Alternatively, we can start with the basic atomic orbitals of the oxygen and fluorine atoms, and develop multicentre molecular orbitals to which fluorine and oxygen both contribute.

10.9.1.2 sp² *Hybridization*

When the oxygen atom has a lone pair of electrons and forms one π bond, an angular structure is to be expected. The ozone molecule (55) falls into this category and microwave studies have shown that the OOO bond angle is 116.8° and the oxygen–oxygen distance 128 pm. The bonding is best

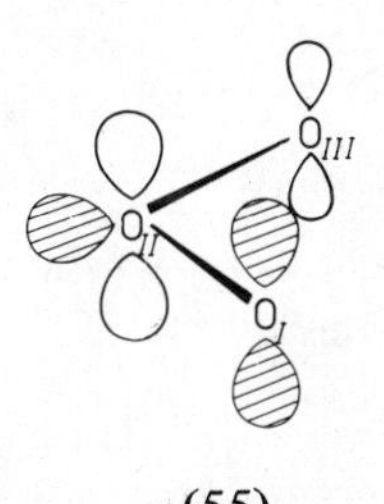

(55)

treated just like that in the nitro group of nitromethane (see Chapter 8, p.101), with the σ bonding of the central oxygen atom described in terms of sp^2 hybrid orbitals; the lone pair of electrons on the oxygen atom takes the place of the σ bond linking nitrogen with carbon in CH_3NO_2. For the π bonding, which embraces all three oxygen atoms, we use a combination of the $2p_z$ orbitals; the central oxygen atom (O_{II}) contributes one π electron, and the outside atoms three π electrons. In formula (55) (one of the canonical forms), O_I is shown as providing two electrons and O_{III} one electron. Hence we have three atomic $2p_z$ orbitals that can be combined to give three π molecular orbitals, one bonding (π_1), one non-bonding (π_2), and one antibonding (π_3):

$$\pi_1 = \phi_{O_I} + \phi_{O_{II}} + \phi_{O_{III}}$$

$$\pi_2 = \phi_{O_I} + \phi_{O_{III}}$$

$$\pi_3 = \phi_{O_I} - \phi_{O_{II}} + \phi_{O_{III}}$$

There will be two electrons in the bonding three-centre delocalized π_1 molecular orbital, and two electrons in the non-bonding π_2 orbital concentrated on the two terminal oxygen atoms.

10.9.2 SULPHUR, SELENIUM AND TELLURIUM

As we go from oxygen to tellurium we find increasing metallic character, as we would expect from the decreasing electronegativity of the elements. There is a marked difference on going from oxygen to sulphur, in that the heavier elements can form up to six bonds, as in SF_6. Once again we shall find it simplest to involve *d* orbitals in our bonding description and this fits in with most theoretical ideas (see Mitchell[14]). For the purposes of our discussions we shall assume that with sulphur, for instance, there is full 3*d* orbital participation in σ bonding, and that d_π–p_π bonding plays a very significant role in bonds such as those between sulphur and oxygen. These heavier elements form molecules in each of the three oxidation states II, IV and VI, so that there are quite a few possible combinations of σ bonds, lone pairs and π bonds. *Table 10.21* summarizes the possible stereochemistries for compounds of these elements, and we will now examine the bonding in each of the main categories, using the various hybridization schemes for the σ bonding as headings for the sections.

10.9.2.1 sp^2 *Hybridization*

This sp^2 σ bonding is found in SO_2, where one orbital is a lone pair, and in SO_3, where there are no lone pairs.

10.9.2.1.1 *Angular molecules:* In SO_2, there are two sulphur electrons available for π bonding, one in the $3p_z$ and the other in a d_π orbital, and these can interact with the oxygen $2p_z$ orbitals (each containing one electron) so as to provide full double bonds. Hence we can represent SO_2 by (56). This fits in with the structure of the molecule, since the experimentally determined parameters are angle OSO = 119.5° and the sulphur–oxygen bond length = 143 pm.

Of course, it is possible to account for the π bonding without using a sulphur *d* orbital, by treating the molecule rather like a nitro group, and taking linear combinations of the sulphur $3p_z$ and the oxygen $2p_z$ atomic orbitals; this would provide a filled delocalized bonding π molecular orbital

Table 10.21 Stereochemistry of S, Se and Te compounds

Hybridization scheme	Shape	Total number of lone pairs and σ bonds	Number of			Oxidation state	Example
			σ bonds	lone pairs	π bonds		
sp^2	Angular	3	2	1	2	IV	SO_2
sp^2	Planar		3	0	3	VI	SO_3
sp^3	Angular	4	2	2	0	II	H_2S
sp^3	Pyramidal		3	1	0	IV	$[SF_3]^+$
sp^3	Pyramidal		3	1	1	IV	$SOCl_2$
sp^3	Tetrahedral		4	0	1	VI	$[SOF_3]^+$
sp^3	Tetrahedral		4	0	2	VI	SO_2Cl_2
sp^3d	Distorted tetrahedral	5	4	1	0	IV	SF_4
sp^3d	Trigonal-bipyramidal		5	0	1	VI	SOF_4
sp^3d^2	Square-pyramidal	6	5	1	0	IV	$[TeF_5]^-$
sp^3d^2	Octahedral		6	0	0	VI	SF_6
?	Octahedral	7	6	1	0	IV	$[TeCl_6]^{2-}$

and a filled non-bonding π orbital concentrated only on the two oxygen atoms. In effect this gives a bond order of 1.5 to the sulphur–oxygen bonds, and in valence-bond terms would be represented in terms of a blend of the canonical structures given in (57). However, it seems likely that the

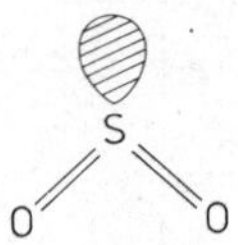

(56) 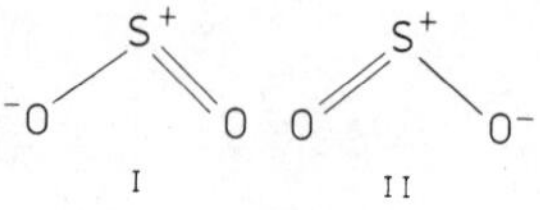 (57) 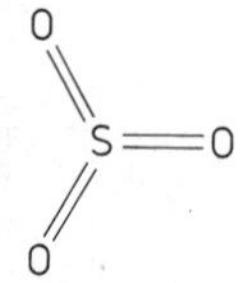 (58)

$3d_\pi$ orbital would be sufficiently low in energy to contribute to the π bon- ing, and for simplicity we take SO_2 to be best represented by the fully π-bonded structure (56).

Selenium dioxide differs from sulphur dioxide in being polymeric in the solid state, there being pyramidal structures with Se–O–Se bridges (see p.17 and some dimeric molecules seem to be present even in the vapour state. However, monomeric SeO_2 species have been studied in the gas phase, and the OSeO angles have been found to be 113.8°. Furthermore, recent devel- opments using the matrix isolation techniques have made it possible to isola and study monomeric SeO_2 and TeO_2 species (Spoliti *et al.*[31]), and both molecules have been found to have the expected angular shapes with bond angles of around 110°.

10.9.2.1.2 *Planar molecules:* In the oxidation state VI, sulphur forms the regular trigonal-planar SO_3 molecule in which the sulphur–oxygen distances are 142 pm (i.e. virtually the same as in SO_2). This molecule is best represented as (58) with full double bonds. Our representation of the σ bonding is based on the sulphur atom using its $3s$, $3p_x$ and $3p_y$ orbitals, either as sp^2 hybrids, giving three two-centre σ bonds (each linking sulphur with one oxygen), or separately, giving three four-centre σ bonds (as in the BCl_3 molecule, see p.91). For the π bonding we take the $2p_z$ atomic orbital of each oxygen atom and the $3p_z$, $3d_{xz}$ and $3d_{yz}$ atomic orbitals of sulphur. Linear combinations of these six atomic orbitals give six molecular orbitals, three bonding and three antibonding; each of the delocalized (four-centre) π orbitals will be doubly filled. However, since the sulphur and oxygen π atomic orbitals do not correspond to exactly the same energy, they will not be mixed equally in the π molecular wave functions, so we should expect the π electrons to be centred more over the oxygen than the sulphur atoms. *Figure 10.16* shows one such π molecular orbital scheme.

10.9.2.2 sp^3 *Hybridization*

Under this heading we include all molecules in which the central atom has a total of four pairs of electrons surrounding it, either as bonding pairs or lone pairs. When there are two lone pairs, the two bonding pairs will show

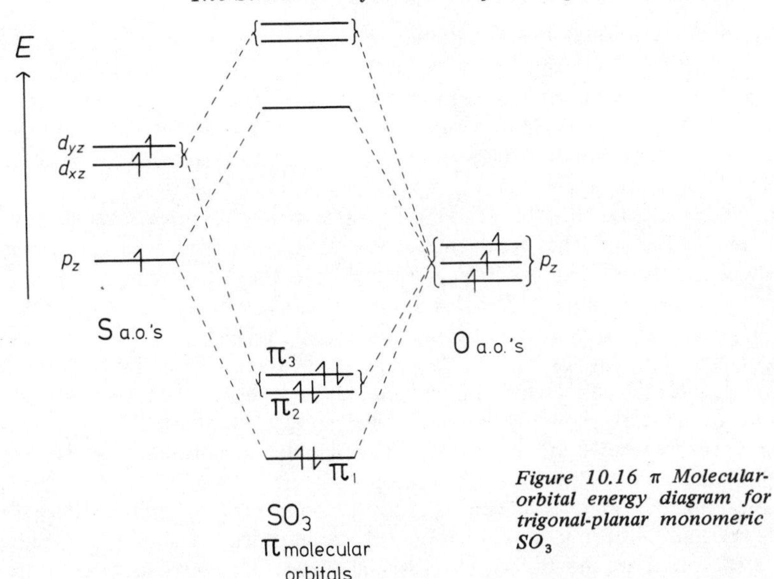

Figure 10.16 π Molecular-orbital energy diagram for trigonal-planar monomeric SO_3

up in an angular structure, while one lone pair and three bonding pairs provide a pyramidal molecule, and four bonding pairs a tetrahedral molecule. We can look at each possibility in turn.

10.9.2.2.1 *Angular molecules:* This arrangement is found for all the simple molecules of the bivalent elements; the known bond angles and bond lengths for these compounds are given in *Table 10.22.*

It is notable that the bond angles are much smaller for the hydrides than for the other compounds. We discussed this problem with the hydrides of the Group VB elements, where there was a similar trend in bond angles, and the explanation was that as the central atom gets bigger so the bonds to

Table 10.22 Structural parameters for X_2M and X_2M_2 molecules (X = S, Se or Te)

Molecule	*Bond angle about S, Se or Te/* degrees	*Bond length*/pm
H_2S	92.1	132 (S–H)
MeSH	96.5	132 (S–H); 182 (S–C)
Me_2S	98.8	180 (S–C)
SCl_2	100.3	–
$S(CN)_2$	98.3	170 (S–C)
MeSCN	99.8	–
SF_2	98.3	159 (S–F)
S_2H_2	91.3	133 (S–H); 206 (S–S)
S_2Me_2	102.8	181 (S–C); 204 (S–S)
S_2F_2	108.3	164 (S–F); 189 (S–S)
H_2Se	90.9	146 (Se–H)
Me_2Se	96.2	194 (Se–C)
$TeBr_2$	98	–

hydrogen get longer and the HMH bond angles can get less before repulsion between the bonding pairs becomes important. The bonding is then described largely in terms of unhybridized *p* orbitals, as the bond angles approach 90°. In the other compounds the bonded atoms (or groups) contain lone pairs or bonding pairs (e.g. NH_3) and the repulsions that arise result in the bond angle opening up to around 100°.

The X_2S_2 compounds have an S–S bond with structures analogous to those of H_2O_2, and the bond angles can be accounted for in the same way However, the bond lengths in F_2S_2 call for further comment, since the sulphur–sulphur distance (189 pm) is much shorter than a single bond (~206 pm) and the sulphur–fluorine bond length (164 pm) is greater than the corresponding bonds in SF_6 (156 pm). We recall the similar situation in the F_2O_2 molecule, when we concluded that there must be a significant amount of multiple bonding between the oxygen atoms at the expense of the oxygen–fluorine bond electrons. The angular arrangement is also found for the puckered chains of sulphur atoms in the ions $S_3{}^{2-}$, $S_4{}^{2-}$ and $S_6{}^{2-}$, and the rings S_6 and S_8. The S_8 ring is the principal component present in the various allotropic modifications of sulphur in the solid state, and it is also found in the liquid and gaseous states. In this puckered ring the sulphur–sulphur bond lengths are 206 pm, and the bond angles 108° – quite close to the tetrahedral angle; a similar puckered Se_8 ring is known. The S_6 molecule has a chair-form hexagon ring with the same bond lengths as those in S_8, but with somewhat smaller bond angles.

10.9.2.2.2 *Pyramidal molecules:* When the quadrivalent elements acquire a positive charge they become isoelectronic with the neighbouring elements of Group VB, and form pyramidal molecules. Three well-characterized examples are the ions $[SF_3]^+$, $[SeF_3]^+$ and $[TeMe_3]^+$, which are found respectively in the compounds $[SF_3]^+[BF_4]^-$ (Gibler *et al.*[32]), $[SeF_3]^+[Nb_2F_{11}]^-$ (Edwards and Jones[33]), and $[Me_3Te]^+[MeTeI_4]^-$ (Einstein *et al.*[34]). The bond angles about S, Se and Te in these pyramidal molecules tie in well with values for the isoelectronic analogues of Group VB, as the data in *Table 10.23* show.

Pyramidal molecules also arise when the quadrivalent elements use one electron for π bonding, and have one lone pair and three bonding pairs of

Table 10.23 Bond angles for pyramidal molecules of S, Se and Te

Molecule	*Angle*/degrees	*Molecule*	*Angle*/degrees
$[SF_3]^+$	97.5	PF_3	97.8
$[SeF_3]^+$	94.2	AsF_3	96.2
$[TeMe_3]^+$	95	–	–

electrons. Good illustrations are provided by the thionyl compounds SOX_2 (X = Br, Cl, F or Me) (59) and the analogous selenium compounds $SeOCl_2$ and $SeOF_2$. The bond angles are rather below the tetrahedral angle, as might be expected, the XSX angles being appreciably smaller than the XSO angles. This can be illustrated by reference to SOF_2 in which the FSF and FSO angles are 92.8 and 106.8°, respectively. Another molecule falling into this category is SSF_2 (an isomer of FSSF which we discussed on p.174), which we can formally represent by (60). The SSF angle is 107.5°, and

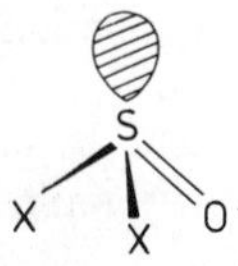

(59) 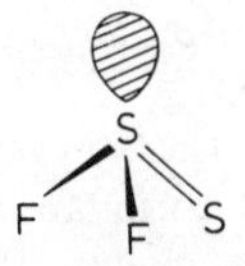 (60)

the sulphur–sulphur and sulphur–fluorine bond lengths are 186 and 160 pm, respectively; these values are fairly close to those (*Table 10.22*) for FSSF, which again emphasizes the multiple bond character of the sulphur–sulphur bond in the latter molecule.

There are a number of oxo species of these elements that have a pyramidal structure. We have already seen that whereas SO_2 is a monomer, SeO_2 is polymeric, and it has a chain structure (61), with a pyramidal environment for the selenium atoms. The isolated sulphite, selenite and tellurite ions, $MO_3{}^{2-}$, are also pyramidal (62) as would be expected.

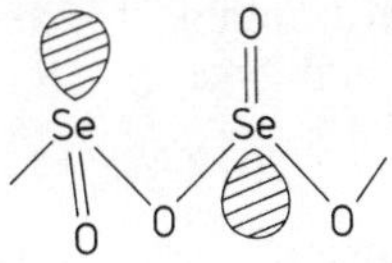

(61) 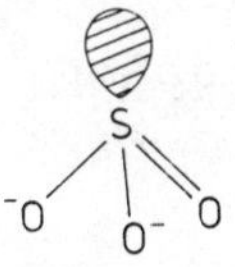 (62)

10.9.2.2.3 *Tetrahedral molecules:* These are formed by the sexivalent elements in such species as the simple oxo anions $MO_4{}^{2-}$ and the sulphuryl halides, SO_2X_2, and in ions such as $[SOF_3]^+$. In the simple sulphate, $SO_4{}^{2-}$, and selenate, $SeO_4{}^{2-}$, anions, all angles are tetrahedral and all bond lengths equal. In the parent acid, H_2SO_4 (63), the sulphur–oxygen distances are not all the same, two being 143 and two 154 pm. These distances seem sensible if we write the molecule as $SO_2(OH)_2$ and associate the two hydrogens with the longer (single) bonds. The OSO angles range from 98 to 117°, the largest angle being between the two shorter (S=O) bonds.

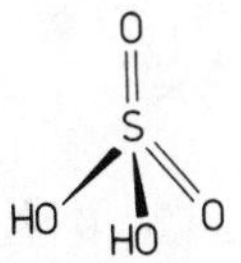

(63) 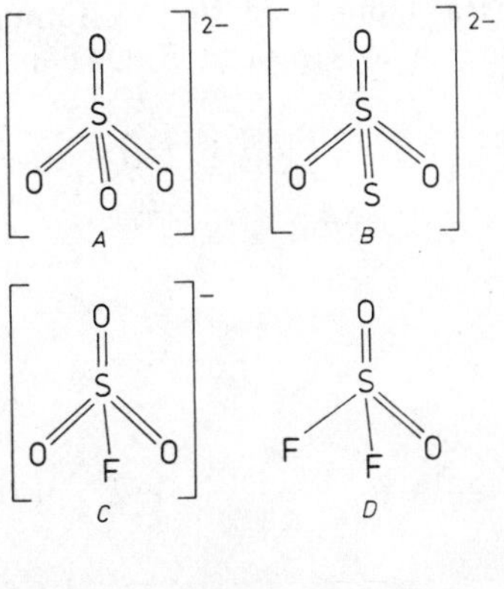 (64)

Other related 'tetrahedral' anions are known in which one or more of the oxygens of the sulphate ion have been replaced. Thus replacement of O^- by F gives the isoelectronic ion $[SO_3F]^-$ (64C), in which the sulphur–oxygen distance is 143 pm, as in other 'double-bonded' species, and the bond angles are OSO = 112.9° and OSF = 105.8°. In the thiosulphate anion (64B) one of the oxygen atoms is replaced by a sulphur atom.

The polythionates form an interesting range of molecules in which SO_3^- groups are linked by chains of sulphur atoms, i.e. $[O_3S–(S)_n–SO_3]^{2-}$. In the dithionates the two end groups link directly, but n can be as large as four, in the hexathionates. Foss[35] has written an excellent account of the structures of these molecules and readers are referred to his article for details of bond lengths and bond angles. The general position may be summarized briefly by emphasizing that there is no branching in the sulphur chain, contrary to earlier speculations. All the sulphur atoms in the chain are bivalent and have an angular structure, the two lone pairs of electrons taking up the two remaining tetrahedral positions; the end sulphur atoms are sexivalent as in the sulphate anion. The S–S distances vary slightly, being 210 pm in the chain itself ($S_{II}–S_{III}$, $S_{III}–S_{IV}$, $S_{IV}–S_V$) of the hexathionate anion (65), but rather shorter (204 pm) between the sulphur atoms of the chain and the sulphur atoms of the end group ($S_I–S_{II}$ and $S_V–S_{VI}$). This rather small difference is probably attributable to the different σ orbitals used by the sexivalent (S_I and S_{VI}) and bivalent (S_{II}, S_{III}, S_{IV} and S_V) sulphur atoms.

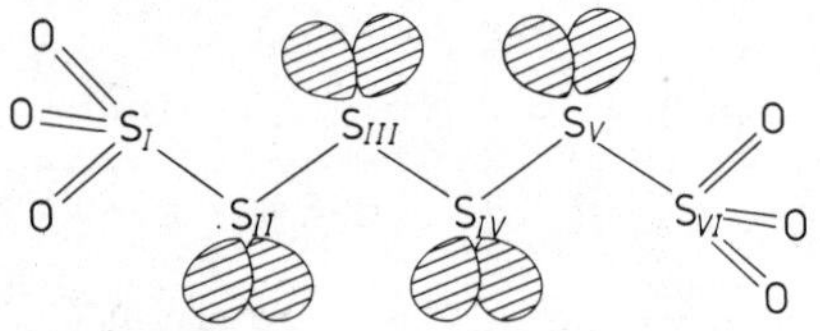

(65)

The sulphuryl halides, SO_2X_2, are other examples of sexivalent 'tetrahedral' molecules, with two π bonds. In sulphuryl fluoride (64D), for instance, the OSO and FSF bond angles are 124 and 96.1°, respectively, and the sulphur–oxygen distances (143 pm) again correspond to double bonds.

Although sulphur trioxide is monomeric in the gas phase, it polymerizes in the solid state to give either linear chains (α and β forms; 66) or a cyclic trimer with a puckered ring (γ form; 67). In both types of structure SO_2 groups are linked through S–O–S bonds, and these linking sulphur–oxygen

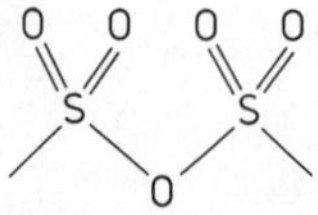

(66) 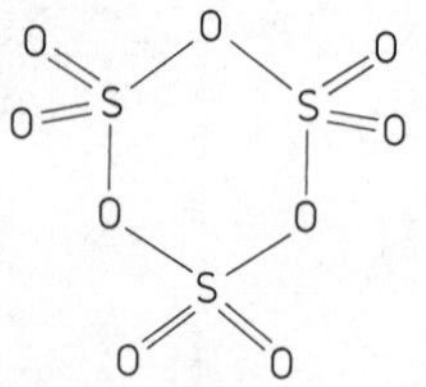 (67) 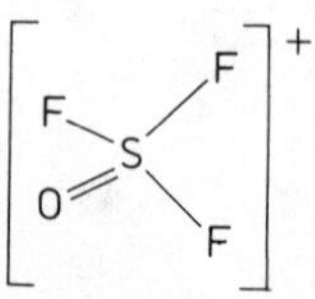 (68)

bonds (161 and 160 pm for chain and cyclic forms, respectively) are much longer than the terminal bonds (141 and 140 pm, respectively). These bond distances fit in with the view that the linking bonds are single and the terminal bonds double. The configurations are not perfectly tetrahedral, by any means, because of constraints imposed by the ring and because mutual repulsions between terminal double bonds will be much greater than those between linking single bonds; in practice the 'outer' and 'inner' OSO angles in the cyclic trimer are 122 and 100°, respectively. If the sexivalent elements acquire a positive charge, then we get a tetrahedral arrangement of bonding pairs with only one π bond. Typical examples are the ions $[SOF_3]^+$ (68) and $[SOMe_3]^+$, the parameters for which are given in *Table 10.24.* These ions are structurally related to the $[SX_3]^+$ ions formed by quadrivalent sulphur, with the sulphur–oxygen double bond replacing the lone pair.

Table 10.24 Parameters for $[SOX_3]^+$ molecules

Ion	*Angles*/degrees	*Bond lengths*/pm
$[SOF_3]^+$	102 (FSF) 116 (OSF)	135 (SO) 144 (SF)
$[SOClF_2]^+$	106.2–111.6	–
$[SOMe_3]^+$	112.1 (OSC)	145 (SO)

It is perhaps worth drawing attention to some trends in bond length changes for a set of 'tetrahedral' sulphur molecules. It can be seen from these data (*Table 10.25*) that an increase in the oxidation state from IV to VI brings about a reduction in the length of both SF and SO bonds, and that as fluorine atoms replace O^- along the sequence SO_4^{2-}, SO_3F^-, SO_2F_2, SOF_3^+ so the bonds again become shorter. This can be rationalized by assuming that both types of change result in a larger $\delta+$ charge on the sulphur atom with a corresponding contraction of the atomic orbitals. The π bonding will be influenced as well as the σ bonding because the sulphur d_π contraction is believed to lead to more effective overlap.

Table 10.25 Bond lengths (pm) in some sulphur compounds

Oxidation state	*Molecule*	*S–O*	*S–F*
IV	SF_3^+	–	150
IV	SO_3^{2-}	152	–
VI	SOF_3^+	135	144
VI	SO_2F_2	141	153
VI	SO_3F^-	142	157
VI	SO_4^{2-}	144	–

10.9.2.3 sp^3d *Hybridization*

When there are five electron pairs around the central atom, the resulting stereochemistry is related to the trigonal-bipyramidal disposition of these electron pairs.

10.9.2.3.1 *Distorted tetrahedral molecules:* We noticed this arrangement for the arsenic atom in the co-ordination compound $AsCl_3,NMe_3$ (p.166), and here it emerges in the simple halides SF_4, SeF_4 and $TeCl_4$ in the gas phase. In SF_4 (69), for instance, the two axial sulphur–fluorine bonds are bent slightly towards the two equatorial bonds, showing the repulsion effect of the lone pair. The axial bonds (161 pm) are appreciably longer than the equatorial ones (155 pm), and this is usually attributed to a difference in the σ orbitals used by sulphur. This is the same basic problem that we discussed with the pentahalides of the Group VB elements (p.165). The structure of the tetrahalides in the solid is still a little obscure, although $TeCl_4$ appears to have a tetrameric structure (Buss and Krebs[36]) which is best described in terms of Cl^- ions bridging pyramidal $[TeCl_3]^+$ ions.

Quadrivalent selenium and tellurium also yield trigonal-bipyramidal molecules in their diphenyl and dimethyl dihalides (R_2MX_2), with phenyl or methyl groups occupying the equatorial positions together with the lone pair of electrons.

If the two electrons in the lone pair are used to form a σ and a π bond to a fifth atom, as in SOF_4, then we shall expect the same basic structure with the S=O in an equatorial position (70).

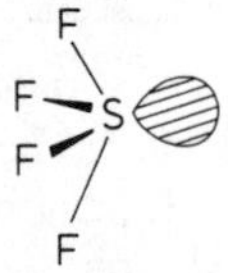

F F S F F (69)

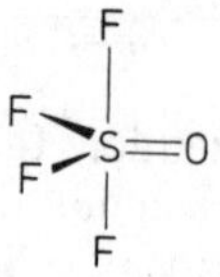

F F F S=O F (70)

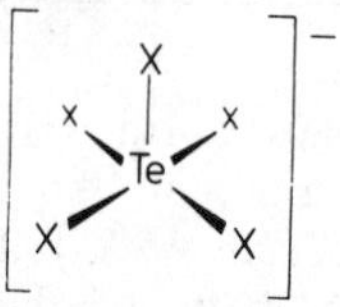

$[TeX_5]^-$ (71)

10.9.2.4 sp^3d^2 *Hybridization*

There are two simple shapes to be expected, octahedral, when the central atom uses all six valency electrons for σ bonding (as in SF_6), and square pyramidal, when there are six electron pairs, one of which is a lone pair.

10.9.2.4.1 *Square pyramidal molecules:* This disposition of bonds is to be found in the tellurium ions $[TeF_5]^-$, $[TeCl_5]^-$ and $[MeTeI_4]^-$, where we can think of the ions as having the lone pair of electrons occupying the sixth position to give an octahedral disposition of the six electron pairs (71). These ions have the same structures as the isoelectronic ions of tervalent antimony, e.g. $[SbF_5]^{2-}$.

10.9.2.4.2 *Octahedral molecules:* When all of the valency shell electrons are used in σ bonding the result is a regular octahedral arrangement as in SF_6, SeF_6 or TeF_6, and substituted molecules such as SF_5Cl and SF_5Br. S_2F_{10} and $S_2O_2F_{10}$ also have octahedral structures in which SF_5 groups are joined by S–S and S–O–O–S linkages, respectively.

The other species that is found to be octahedral is the $[TeCl_6]^{2-}$ anion. Here again we have the problem experienced with the complex anions of antimony(III) (p.166), namely that in addition to the six bonding pairs of electrons we have one lone pair, so that if this pair was stereochemically active we would expect a distorted octahedral structure. The fact that these ions are quite regular is one of the many warnings to us never to believe that theories and rationalizations apply in every case. As with antimony we conclude that the lone pair does not influence the stereochemistry and must therefore be in an s orbital. Readers might now ask why this is not also the case with the $[TeX_5]^-$ ions, where clearly the lone pair is helping to determine the stereochemistry. The simplest explanation is that the central atoms like to be as symmetrical as possible in respect of their electron environment, so the octahedral arrangement dominates, but we shall see in the next section (XeF_6) that this is not necessarily true in every instance.

10.10 Fluorine and the Group VIIB Elements

All the halogens have an unpaired electron in a p orbital which they can use for σ bonding, and accordingly form the homonuclear diatomic molecules F_2, Cl_2, Br_2 and I_2 and a number of heteronuclear diatomic interhalogen compounds such as ICl and IBr; with the heavier halogens this σ bonding may be supplemented by a small amount of π bonding.

The hydrides, HF, HCl, HBr and HI, are also covalent molecules, formed by the overlap of the halogen p_x orbital with the $1s$ orbital of hydrogen (see Chapter 7, p.62); as the halogen atom becomes heavier, so the bond dissociation energy decreases steadily: HF = 574, HCl = 428, HBr = 363, HI = 295 kJ mol^{-1}. The electronegativity of the halogen atom decreases from fluorine to iodine, and this is reflected in the decrease in ionic character of the HX bond along the series HF, HCl, HBr, HI. In orbital terms the decrease in ionic character means a more symmetrical distribution of the bonding electrons between the hydrogen and halogen atoms. When these hydrogen halides dissolve in a polar solvent such as water, they ionize to give the solvated proton and the halide ion:

$$HX + H_2O \rightarrow H_3O^+ + X^-$$

In the case of hydrogen chloride, a monohydrate (HCl,H_2O) may be isolated, and this has been shown to consist of a Cl^- anion and the pyramidal H_3O^+ cation.

This ionization and solvation process can be easily interpreted on an orbital basis, as shown diagrammatically in *Figure 10.17*. Thus as the water molecule approaches HX, the lone pair of electrons on oxygen will point towards the hydrogen atom (which is the 'positive' end of the diatomic molecule), and as the water molecule gets ever nearer to HX so the electron pair in the HX bond will become progressively distorted in favour of X. Ultimately the charge cloud will be concentrated almost entirely around the halogen atom, with hydrogen no longer having any appreciable share, and then the proton becomes associated with the oxygen lone pair to give H_3O^+.

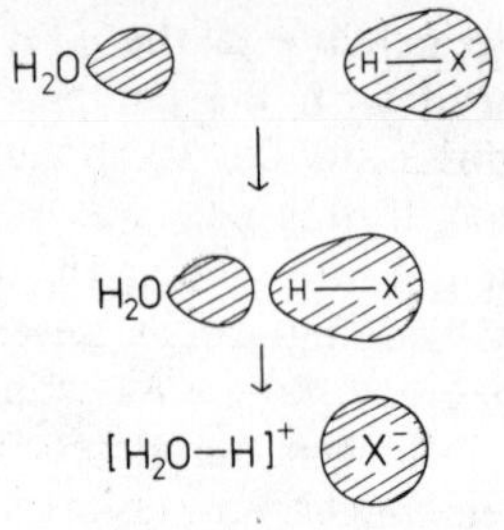

H_2O ... H—X
↓
H_2O ... H—X
↓
$[H_2O—H]^+$ X^-

Figure 10.17 Orbital interpretation of the hydration and hydrolysis of HX molecules

The anhydrous hydrogen halides are hydrogen-bonded in the solid state, but this bonding is appreciably stronger in hydrogen fluoride, because of the greater electronegativity of fluorine. It is for this reason that hydrogen fluoride has a much higher boiling point (19.4 °C) than, say, hydrogen chloride (−85 °C). In the solid state, hydrogen fluoride has a linear zig-zag chain structure (72), with strong bonds along the chains but only weak interchain

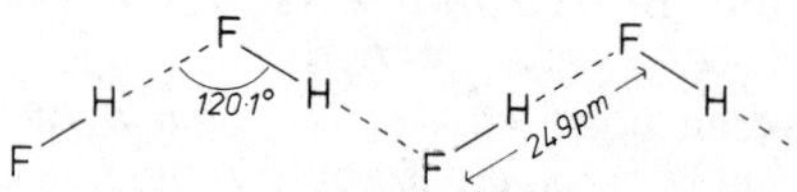

F—H···F(120.1°)—H···F—H···F(249pm)—H···

(72)

forces. The other hydrogen halides probably form similar zig-zag chains in the solid state, although the only structural data are a neutron-diffraction study (Sandor and Farrow[37]) of DCl, which confirms the zig-zag arrangement but with a much smaller ClClCl angle (93.5°) and a Cl—Cl distance of 369 pm.

In the anhydrous liquid state, hydrogen fluoride is obviously associated, probably in the form of long linear zig-zag polymers, but there is no three-dimensional network of hydrogen bonds such as that in water, since the viscosity is much lower. Hydrogen fluoride is also associated to some extent in the vapour phase, although there is some argument about the nature of the association, some workers postulating chain polymers and others an equilibrium between monomers and hexamers; the infrared spectra and electron diffraction data favour the equilibrium picture.

The hydrogen bonding can be interpreted on the basis of interaction between the hydrogen of one molecule and the lone pair electrons of the next-door fluorine. Hence the zig-zag nature of the chains can be explained if we consider the fluorine atom to have sp^3 (or sp^2) hybrid orbitals, one with an electron for bonding and the others doubly filled. An angular chain is also found for the $[H_2F_3]^-$ ion (73).

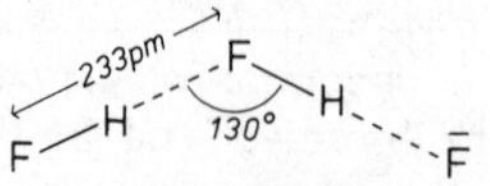

F—H···F(130°, 233pm)—H···F

(73)

$[X\cdots H\cdots X]^-$

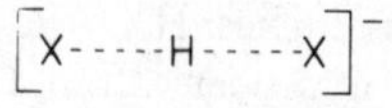

(74)

The simple anions HF_2^-, HCl_2^-, HBr_2^- and HI_2^- are well characterized, and have linear structures (74). The hydrogen bonding in these ions is normally symmetrical, but in the *p*-toluidinium salt of $[HF_2]^-$ neutron diffraction studies have shown the bonding to be asymmetric with H–F distances of 103 and 123 pm; this is attributed to a very asymmetric nearest-neighbour environment. The bonding is much stronger with the fluoride anion than the others, and calculations suggest values of $[HF_2]^-$ = 243, $[HCl_2]^-$ = 75, $[HBr_2]^-$ = 54 and $[HI_2]^-$ = 52 kJ mol^{-1}.

Iodine monochloride, ICl, also forms zig-zag chains in both its α and β forms (75), the forms differing only in whether the Cl side branches are *cis* (α) or *trans* (β).

(75)

Whereas fluorine is limited to a covalency of one, the other halogens can form more than one bond with electronegative elements such as fluorine and oxygen, and in the case of iodine the covalency can be as high as seven. As with the earlier groups of elements, we attribute this higher covalency of the heavier elements to their ability to use *d* orbitals that are energetically available. In sophisticated molecular-orbital treatments these *d* orbitals are not usually considered to participate as fully as our simpler treatment implies.

Because there are several oxidation states to consider, the stereochemical possibilities are quite considerable, as the summary in *Table 10.26* shows.

10.10.1 sp^3 HYBRIDIZATION

When the sum of the bonding and lone pairs of electrons is four, we would expect the distribution of these electron pairs about the central atom to be essentially tetrahedral. Hence molecules with two bonding pairs will be angular, molecules with three bonding pairs will be pyramidal, and molecules with four bonding pairs will be tetrahedral. π-Electrons may be present as well but these will not upset the basic symmetry of the σ-bond structure.

10.10.1.1 *Angular Molecules*

The tervalent elements form two types of angular molecules as typified by $[ICl_2]^+$ and $[ClO_2]^-$; these are shown in formulae (76) and (77), respectively. The first type includes the species $[ClF_2]^+$ and $[BrF_2]^+$, as well as

(76) (77)

Table 10.26 Stereochemistry of compounds formed by the halogens

σ bond hybridization scheme	*Total number of σ bonds and lone pairs*	*Shape*	*Number of*			*Oxidation state*	*Example*
			σ bonds	*lone pairs*	*π bonds*		
sp^3	4	Angular	2	2	0	III	$[ICl_2]^+$
		Angular	2	2	1	III	$[ClO_2]^-$
		Pyramidal	3	1	2	V	$[ClO_3]^-$
		Tetrahedral	4	0	3	VII	$[ClO_4]^-$
sp^3d	5	Linear	2	3	0	I	$[I_3]^-$
		'T-shaped'	3	2	0	III	ClF_3
		Distorted tetrahedral	4	1	0	V	$[IF_4]^+$
		Distorted tetrahedral	4	1	1	V	$[IO_2F_2]^-$
sp^3d^2	6	Planar	4	2	0	III	$[ICl_4]^-$
		Square-pyramidal	5	1	0	V	BrF_5
		Square-pyramidal	5	0	1	V	$[ClOF_4]^-$
		Octahedral	6	0	1	VII	IOF_5
sp^3d^3	7	Pentagonal-bipyramidal	7	0	0	VII	IF_7

Table 10.27 Parameters for $[ClF_2]^+$, $[BrF_2]^+$ and $[ICl_2]^+$

	$[ClF_2]^+$ *in*		$[BrF_2]^+$ *in*	$[ICl_2]^+$ *in*	
	$[ClF_2]^+[AsF_6]^-$	$[ClF_2]^+[SbF_6]^-$	$[BrF_2]^+[SbF_6]^-$	$[ICl_2]^+[SbCl_4]^-$	$[ICl_2]^+[AlCl_4]^-$
Bond angle/ degrees	103.2	95.9	93.5	92.5	96.7
Bond length/ pm	154	158	169	231	228

$[ICl_2]^+$. These positively charged species, which are isoelectronic with SF_2 SeF_2 and $TeCl_2$, have been identified as components of compounds such as BrF_3,SbF_5 that were once believed to be simple adducts but now known to be ionic. The known bond angles and bond lengths are listed in *Table 10.27*, from which it can be seen that the bond angles depend to some extent upon the anion; indeed, a detailed examination of the crystal structures show that there are long-range interactions between the cation and anion through halogen bridges.

In the $[ClO_2]^-$ anion the OClO angle is 110.5° and the chlorine–oxygen bond length 156 pm. An interesting molecule with a similar angular structure is chlorine dioxide, ClO_2, in which the OClO angle is 117.4° and the chlorine–oxygen distance 147 pm. This molecule is unusual because it is paramagnetic, and yet it shows no tendency to dimerize like NO_2 (p.154). Since the Cl–O bonds are much shorter than in the chlorite ion, the bond order must be greater. The simplest way of describing the bonding is on the basis of that in sulphur dioxide, with the extra electron in an antibonding orbital. If we take the plane of ClO_2 as the xy plane, then for σ bonding the chlorine atom will be able to use its $3s$, $3p_x$ and $3p_y$ orbitals. For convenience we can use the valence-bond approach for the σ bonding, so that chlorine will have three sp^2 hybrid orbitals, one containing a lone pair of electrons and the other two overlapping with the oxygen $2p(\sigma)$ orbitals. Of course, a fuller molecular-orbital treatment of the σ bonding is possible along the lines of the scheme laid down for BCl_3 (p.91).

For the π bonding, chlorine can contribute three electrons and the oxygens one electron each. The simplest π molecular-orbital scheme will take linear combinations of the chlorine $3p_z$ and $3d_{xz}$ and the oxygen $2p_z$ atomic orbitals, the actual combinations taking the form:

$$\pi_1 = \phi_{Cl(3p_z)} + \phi_{O_I(2p_z)} + \phi_{O_{II}(2p_z)}$$

$$\pi_2 = \phi_{Cl(3d_{xz})} - \phi_{O_I(2p_z)} + \phi_{O_{II}(2p_z)}$$

$$\pi_3{}^* = \phi_{Cl(3p_z)} - \phi_{O_I(2p_z)} - \phi_{O_{II}(2p_z)}$$

$$\pi_4{}^* = \phi_{Cl(3d_{xz})} + \phi_{O_I(2p_z)} - \phi_{O_{II}(2p_z)}$$

This scheme is oversimplified, of course, since it excludes mixing coefficients and ignores the equal possibility of the $3d_{yz}$ orbital of chlorine contributing to the π bonding, i.e. a combination of $3d_{xz}$ and $3d_{yz}$ should be used. *Figure 10.18* shows a pictorial representation of π_1 and π_2, and *Figure 10.1* gives a simple molecular-orbital energy diagram. The π_1 and π_2 orbitals will each contain two electrons, giving in effect double bonds (as in SO_2), and the odd electron will go into the antibonding $\pi_3{}^*$ orbital, thus reducing the overall bond order.

10.10.1.2 *Pyramidal Molecules*

When the halogen is linked to three other atoms with σ bonds and has one lone pair, the resultant molecule is pyramidal. These molecules will

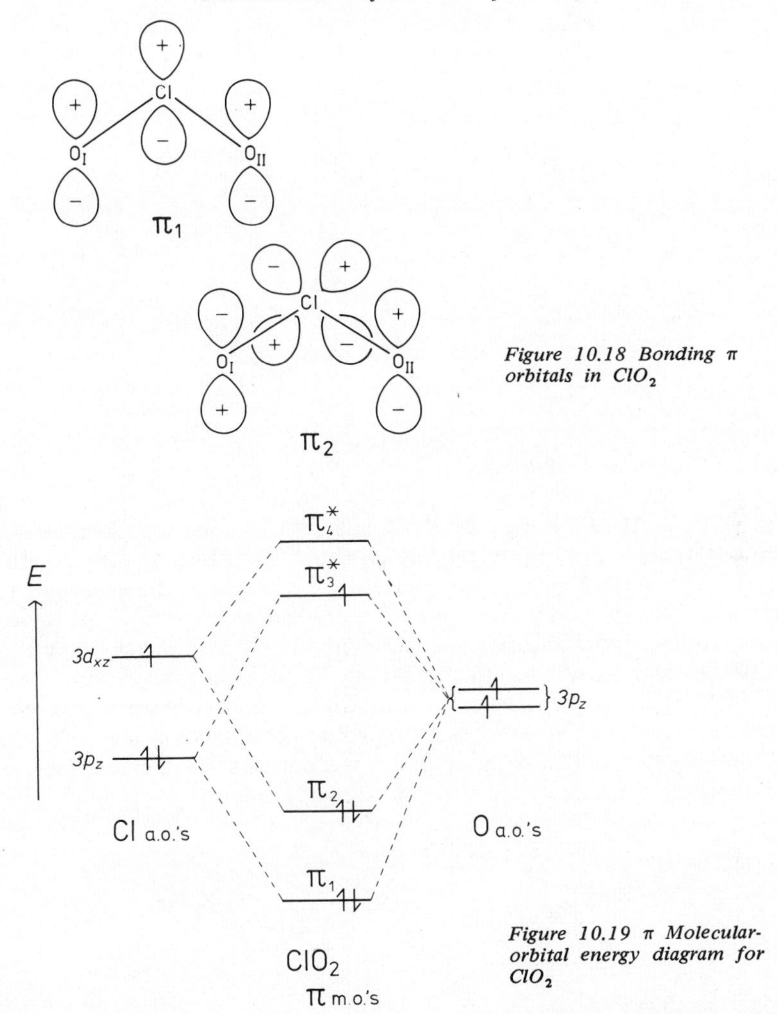

Figure 10.18 Bonding π orbitals in ClO_2

Figure 10.19 π Molecular-orbital energy diagram for ClO_2

arise with the quinquevalent elements in such molecules as chloryl fluoride, $ClFO_2$ (78) (Parent and Gerry[38]), for which microwave studies have shown that the bond angles are 113.5° (OClO) and 103.2° (FClO), with Cl–F and Cl–O bond lengths of 167 and 148 pm, respectively. The pyramidal bonding arrangement is also found for the $[ClO_3]^-$, $[BrO_3]^-$ and $[IO_3]^-$ anions (79). I_2O_5 also has a pyramidal arrangement of oxygens about each iodine (80), one oxygen being bridging; HI_3O_8, which may be obtained by the action of acid on I_2O_5, contains HIO_3 hydrogen-bonded to an I_2O_5 molecule (81). The bond lengths are given in *Table 10.28.*

Table 10.28 Bond lengths (pm) in I_2O_5 and HIO_3

Molecule	*I–O (terminal)*	*I–O (bridge)*
I_2O_5	179 (average)	194 (average)
HIO_3	180 (average) 190 (to OH)	–

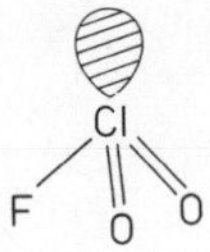

(78) 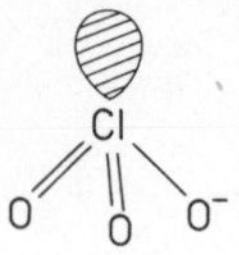 (79)

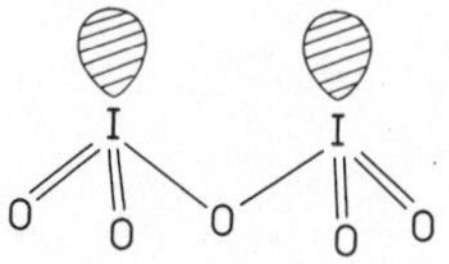

(80) 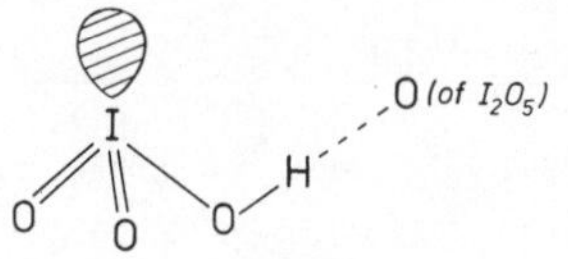 (81)

10.10.1.3 *Tetrahedral Molecules*

The $[ClO_4]^-$, $[BrO_4]^-$ (Siegel *et al.*[39]) and $[IO_4]^-$ ions have symmetrical tetrahedral arrangements of oxygen atoms about the halogen, with bond lengths Cl–O = 146, Br–O = 161 and I–O = 178 pm. The perbromate ion had not been prepared until recent years, and it is perhaps worth recalling that various orbital theories had been invoked to explain the 'non-existence' of the ion, so not for the first time the experimental chemist has caused some embarrassment to the theoretician. These perhalate ions can be directly related to the analogous oxo anions of oxidation states V and III, the decrease in oxidation state being accompanied by the loss of an

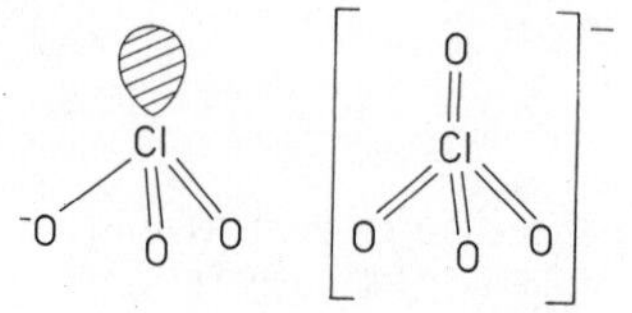

(82) 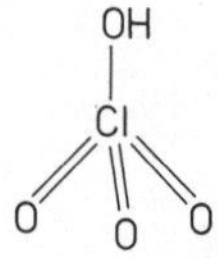 (83) 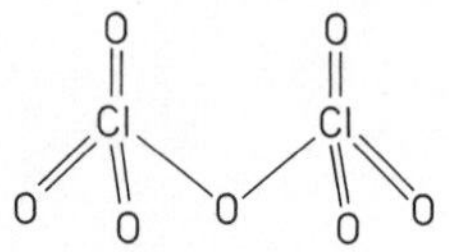 (84)

oxygen (doubly bonded) and its replacement by a lone pair (82). The general trend towards longer bond lengths along the series of ions $[ClO_4]^-$, $[ClO_3]^-$ and $[ClO_2]^-$ fits with the idea of decreasing π-bond character. In $HClO_4$ itself (83) the bond angles are 112.8° (OClO) and 105.8° [OCl(OH)], and the bond lengths 141 pm (three bonds) and 164 pm (one bond).

The tetrahedral arrangement is also found for Cl_2O_7, in which two ClO_3 units are bridged by an oxygen (84). The chlorine–oxygen distances are 141 pm and 171 pm for the terminal and bridging bonds, respectively; the bond angles at chlorine to the terminal oxygens are 115.2°.

10.10.2 sp^3d HYBRIDIZATION

Under this heading we are considering molecules in which the central halogen atom has an environment of five electron pairs, which will provide a trigonal-bipyramidal arrangement. We shall see that the shapes of the molecules can be rationalized if we suppose that the lone pairs of electrons occupy one or more of the three equatorial positions. Hence with three lone pairs the molecule will be linear, with two lone pairs we shall have a

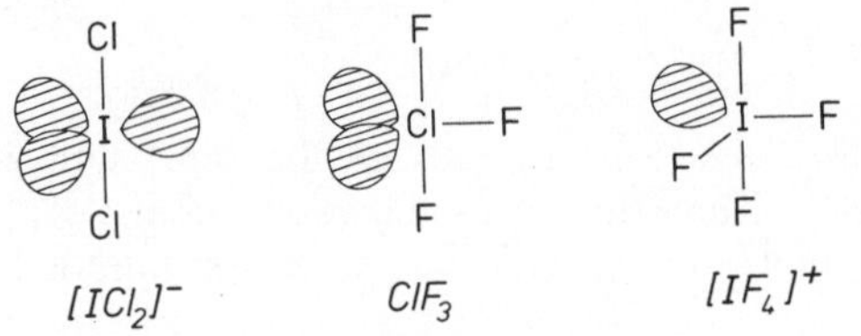

$[ICl_2]^-$ ClF_3 $[IF_4]^+$

(85)

T-shaped molecule, and with one lone pair the molecule should have a distorted tetrahedral configuration. These three shapes may be illustrated by the molecules $[ICl_2]^-$, ClF_3 and $[IF_4]^+$ (85). We will now discuss each of these possibilities in rather more detail.

10.10.2.1 *Linear Molecules*

Iodine gives a series of linear anions $[ICl_2]^-$, $[IBrCl]^-$, $[IBr_2]^-$ and $[I_3]^-$. The $[ICl_2]^-$ and $[IBr_2]^-$ anions are symmetrical, with equal I–X bond lengths, in their potassium salts, but in $Cs[IBr_2]$ the two I–Br bonds are not identical. In the salts of the triodide anion, $[I_3]^-$, the bond lengths are not necessarily the same, as the values in *Table 10.29* show. The tetraethylammonium salt is especially interesting since it exists in two forms, one

Table 10.29 I–I bond lengths (pm) in the $[I_3]^-$ ion

Compound	I_I–I_{II}	I_{II}–I_{III}
NH_4I_3	282	310
CsI_3	283	304
Ph_4AsI_3	290	290
Et_4NI_3 (I)	293	294
(II)	292	296
	289	298

with a symmetrical anion, and the other with two different types of anion with different degrees of asymmetry (Migchelsen and Vos[40]). It is clear that the ion can be appreciably asymmetric, particularly where there are considerable cation–anion interactions or where the crystal structure provides the two ends of the ion with a different environment of nearest neighbours.

It seems reasonable to assume that the $[I_3]^-$ anion would be symmetrical if it were not influenced by neighbouring ions, and on this assumption the shape can be accounted for if we remember that the central iodine atom has three lone pairs of electrons in addition to forming two σ bonds. The five pairs of electrons should give us a trigonal-bipyramidal arrangement (86), with the three lone pairs occupying equatorial positions. The bonding can be described in simple molecular-orbital terms, if we take linear

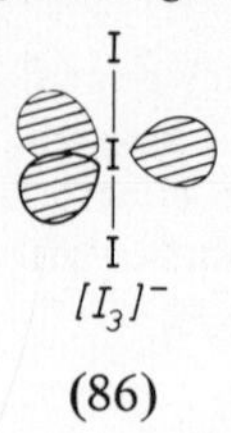

$[I_3]^-$

(86)

combinations of the 5*p* atomic orbitals of the three iodine atoms (written as σ_I, σ_{II} and σ_{III}). Three linear combinations result: ψ_1, which is bonding; ψ_2, which is non-bonding; and ψ_3, which is antibonding.

$$\psi_1 = \sigma_I - \sigma_{II} - \sigma_{III}$$

$$\psi_2 = \sigma_I + \sigma_{II}$$

$$\psi_3 = \sigma_I + \sigma_{II} - \sigma_{III}$$

Figure 10.20 shows how the combinations arise, and *Figure 10.21* gives a simplified molecular-orbital energy diagram. It should be noted that for simplicity all the mixing coefficients are taken as unity.

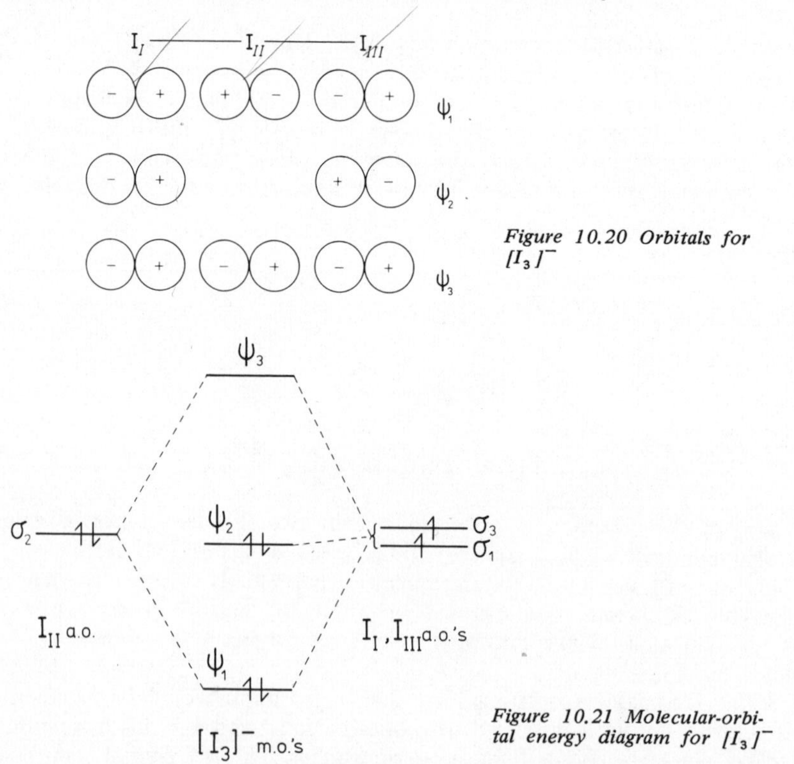

Figure 10.20 Orbitals for $[I_3]^-$

Figure 10.21 Molecular-orbital energy diagram for $[I_3]^-$

Many readers will have noticed that the molecular orbitals are essentially the same as those produced in Chapter 9 (p.126) for a simple hydrogen bond. Thus we have two electrons in a bonding three-centre molecular orbital (ψ_1) embracing all three nuclei, and two electrons in a non-bonding molecular orbital (ψ_2) which involves only the two terminal atoms. Readers interested in a more sophisticated molecular-orbital treatment should read the recent paper by Gabes and Nijman-Meester[41].

The $[I_3]^-$ ion can also be linked into zig-zag chains by I_2 molecules, for instance in the Cs_2I_8 compound, which contains an $[I_8]^{2-}$ anion (87) with

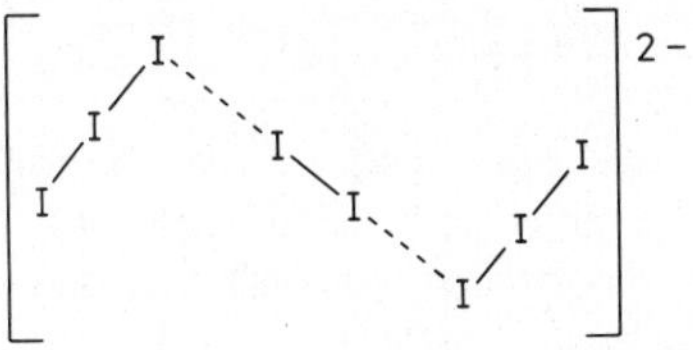

(87)

unequal I–I distances (284 and 300 pm) in the I_3^- units that are linked together. The starch–iodine complex which gives the familiar blue colour in volumetric analysis contains long-chain $[I_3^- {-} I_2]_n$ anions surrounded by a starch amylose helix.

10.10.2.2 *T-shaped Molecules*

Detailed parameters are known for both ClF_3 and BrF_3, and in each molecule the axial bonds (Cl–F = 170, Br–F = 181 pm) are appreciably longer than the equatorial bonds (Cl–F = 160, Br–F = 172 pm). The bond angles are slightly less than a right angle (FClF = 87.5°, FBrF = 86.2°), which is to be expected if the lone pairs exert a repulsive force on the axial bonding pairs. Although detailed parameters are not yet available, it is known that IF_3 is a similar T-shaped molecule.

Although the T-shape fits neatly into the logical sequence shown in formulae (85), some readers might still wonder why the molecules do not adopt the more symmetrical trigonal-planar bonding arrangement with axial lone pairs. When simple, though crude, calculations are made of mutual repulsions between bonding pairs and lone pairs for the alternative structure, the T-shape is favoured, especially if we remember the tendency for electronegative groups to occupy the axial positions. Even so, there is perhaps just a touch of hindsight in this rationalization.

10.10.2.3 *Distorted Tetrahedral Molecules*

The compound IF_5,SbF_5 has been shown to have an ionic structure $[IF_4]^+[SbF_6]^-$, and the structure of the $[IF_4]^+$ cation is directly analogous to that of $TeCl_4$, with the lone pair of electrons occupying one of the equatorial positions. In $[IO_2F_2]^-$ (88), two of the equatorial fluorines are replaced by double-bonded oxygens.

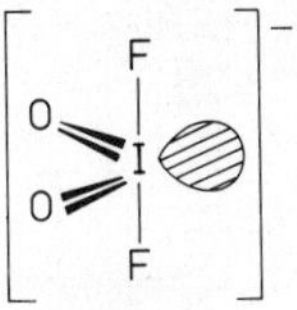

(88) 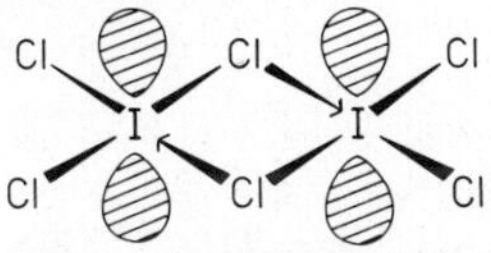 (89) 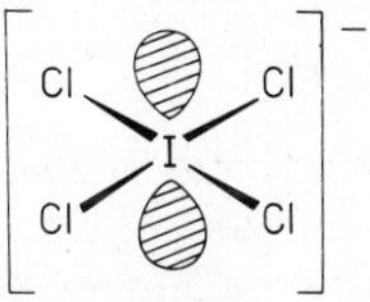 (90)

10.10.3 sp^3d^2 HYBRIDIZATION

Six electron pairs provide an octahedral environment. When one pair is a lone pair we are left with a square-pyramidal arrangement of the five bonds, and this becomes square-planar when a second lone pair is present.

10.10.3.1 *Square-planar Molecules*

Whereas the trifluorides are monomeric T-shaped molecules, ICl_3 is dimeric in the solid state, with a chlorine-bridged structure (89). Each iodine atom is now bonded to four chlorine atoms and has two lone pairs of electrons, so we would expect the planar structure. ICl_3 can accept an electron pair from a chloride ion to give the $[ICl_4]^-$ ion, and this is also planar (90) with the two lone pairs in the axial positions; $[BrF_4]^-$ and $[IF_4]^-$ have analogous structures.

10.10.3.2 *Square-pyramidal Molecules*

Chlorine, bromine and iodine in their quinquevalent states form pentafluorides, and each molecule has the expected square-pyramidal structure (91) with a lone pair of electrons occupying the position that completes an octahedral arrangement of electron pairs. The IF_5 molecule is isoelectronic with $[TeF_5]^-$, which we have seen has the same structure. A recent single crystal structure study of IF_5 by Burbank and Jones[42] has shown that the iodine atom is slightly below the plane of the four basal fluorines, the angle $F_{apical}IF_{basal}$ being 81.9°; the apical I–F bond (173 pm) is appreciably shorter than the basal bonds (187 pm). Spectroscopic studies of the $[ClOF_4]^-$ ion (Christe and Curtis[43]) have shown that this has the expected square-pyramidal structure, with the Cl=O in the apical position (92).

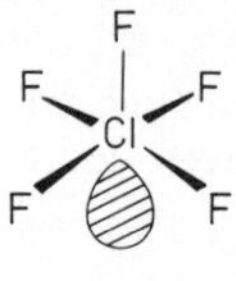

(91)

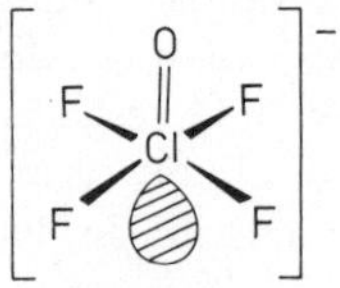

(92)

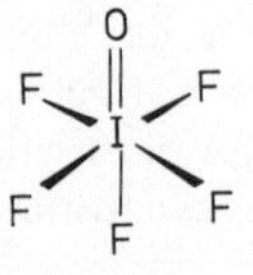

(93)

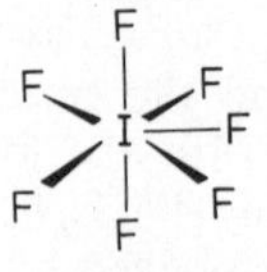

(94)

10.10.3.3 *Octahedral Molecules*

With six σ-bonding electron pairs the $[IF_6]^+$ should have a regular octahedral arrangement of bonds, and this is also the case with IOF_5 (93).

Another molecule in which iodine has a regular octahedral environment is the paraiodate ion $[IO_6]^{5-}$.

10.10.4 sp^3d^3 HYBRIDIZATION

The final bonding arrangement to be discussed for the halogen compounds is the one found in IF_7, where we have to surround the iodine atom with seven electron pairs. The simplest symmetrical arrangement that could arise is the pentagonal-bipyramid (94) in which we have two axial I–F bonds with the remaining five bonds uniformly distributed in the *xy* plane. The available evidence suggests that the molecule has a structure quite close to this ideal one, but with some puckering of the planar IF_5 unit.

10.11 Compounds of the Noble Gases

Despite various early attempts, no chemical compounds of the noble gases were prepared until 1962, and indeed most chemists felt that such compounds would not exist. However, Bartlett and co-workers isolated and characterized the compound $[O_2]^+[PtF_6]^-$, and they reasoned that it ought to be possible to make the xenon analogue $[Xe]^+[PtF_6]^-$ since O_2 and Xe have almost the same ionization energies. This compound was prepared, and before long XeF_4 was also characterized. The breakthrough into noble gas chemistry (the previous name for the elements – inert gases – no longer seemed appropriate) led to an intense effort by a considerable number of research teams, and quite a few oxo and fluoro compounds of xenon were prepared and studied. Readers interested in the study of the chemistry of these compounds over the ten-year period up until 1972 will find a valuable account in Sladky's review[44].

Table 10.30 summarizes the stereochemistry of the compounds that have been characterized, and it can be seen that a considerable number of configurations have now been established, most of which correspond to those listed for halogen compounds in *Table 10.26* (p.182). Apart possibly from XeF_6, the shapes of all the listed molecules correspond to prediction based on the simple ideas of electron-pair repulsion. We will consider briefly each of the configurations and then comment on the bonding in xenon compounds.

The first group of structures is that in which the xenon atom is surrounded with a total of four electron pairs (i.e. σ bonds plus lone pairs), where we should expect a tetrahedral configuration. With XeO_4 (95), which is isoelectronic with the periodate ion, IO_4^-, the oxygen atoms are symmetrically arranged about xenon, with Xe–O distances of 174 pm. XeO_3 (96) has three σ bonds and one lone pair, being isoelectronic with IO_3^-, and the structure is pyramidal with Xe–O = 176 pm. It is interesting to note that the Xe–O distances are very similar to the I–O distances of 178 pm found in the IO_4^- ion, and accordingly we formally draw the xenon–oxygen bonds as double bonds. *Table 10.31* lists the Xe–O and Xe–F bond distances in xenon compounds.

Table 10.30 Stereochemistry of noble gas compounds

Shape	*Total number of σ bonds and lone pairs*	*Number of*			*Oxidation state*	*Example*	*Iodine analogue*
		σ bonds	*lone pairs*	*π bonds*			
Pyramidal	4	3	1	3	VI	XeO_3	$[IO_3]^-$
Tetrahedral	4	4	0	4	VIII	XeO_4	$[IO_4]^-$
Linear	5	2	3	0	II	KrF_2; XeF_2	$[ICl_2]^-$
T-shaped	5	3	2	0	IV	$[XeF_3]^+$	IF_3
Distorted tetrahedral	5	4	1	1	VI	$[XeOF_3]^+$	—
	5	4	1	2	VI	XeO_2F_2	$[IO_2F_2]^-$
Square-planar	6	4	2	0	IV	XeF_4	$[ICl_4]^-$
Square-pyramidal	6	5	1	0	VI	$[XeF_5]^+$	IF_5
	6	5	1	1	VI	$XeOF_4$	—
Octahedral	6	6	0	2	VIII	$[XeO_6]^{4-}$	$[IO_6]^{5-}$
?	7	6	1	0	VI	XeF_6	$[IF_6]^-$

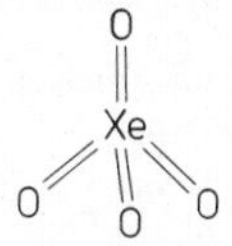

(95) 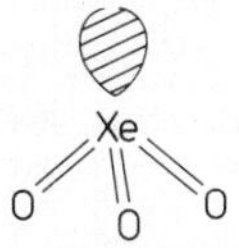 (96)

The second group of structures is based on a trigonal-bipyramidal arrangement of five electron pairs. The linear structure (97) found for the difluorides, KrF_2 and XeF_2, may be rationalized on the basis of two bonds in the axial positions, just as with the iodine compounds $[ICl_2]^-$ and $[I_3]^-$. The same linear arrangement about xenon is found in the molecule $FXeOSO_2F$. The XeF_2 molecule forms a 2 : 1 adduct with arsenic(V) fluoride, $2XeF_2,AsF_5$, but X-ray studies have shown it to be $[Xe_2F_3]^+[AsF_6]^-$, the $[Xe_2F_3]^+$ ion having the linear structure (98) with unequal Xe–F distances. In the adduct $XeF_2,2SbF_5$ the structure is $[Xe–F]^+[Sb_2F_{11}]^-$.

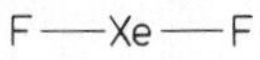

F—Xe—F

(97)

$[F—Xe—F \rightarrow Xe—F]^+$

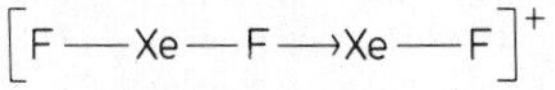

(98)

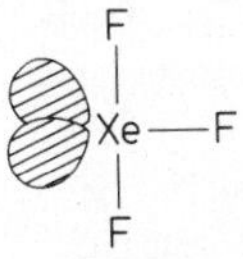

(99) 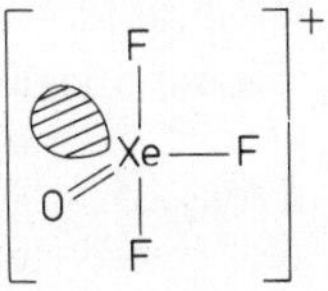 (100) 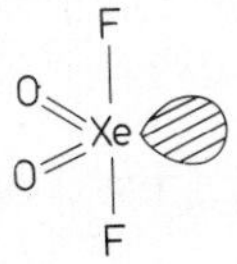 (101)

The adduct formed between XeF_4 and SbF_5 appears to contain the $[XeF_3]^+$ cation, which is isoelectronic with IF_3, and ^{19}F n.m.r. studies (Gillespie and Schrobilgen[45]) are consistent with the cation having the expected trigonal-bipyramidal configuration (99), in which the two lone pairs take up equatorial positions. The same investigations suggest that the $[XeOF_3]^+$ cation has the related structure (100) with the xenon–oxygen bond taking the place of one of the lone pairs. Structurally related to these molecules is XeO_2F_2 (101) for which vibrational spectral studies suggest that the equatorial positions are occupied by a lone pair and two oxygens.

Table 10.31 Xe–F and Xe–O bond lengths (pm)

Oxidation state	*Molecule*	*Xe–F*	*Xe–O*
II	$[Xe–F]^+$	184	–
II	XeF_2	200	–
II	$[Xe_2F_3]^+$	190 (terminal) 214 (bridge)	–
IV	XeF_4	195	–
VI	$[XeF_5]^+$	177 (apical) 190 (basal)	–
VI	$[Xe_2F_{11}]^+$	184 (terminal) 221 (bridge)	–
VI	XeF_6	189 (average)	–
II	$FXeOSO_2F$	194	216
VI	$XeOF_4$	190	170
VI	$[XeO_3F]_n^-$	236; 248	177 (average)
VI	XeO_3	–	176
VIII	XeO_4	–	174

The third series of structures is based on an octahedral disposition of six electron pairs. With XeF_4, which is isoelectronic with $[ICl_4]^-$, the xenon atom has four bonding pairs to fluorine atoms and two lone pairs, giving a square-planar arrangement of Xe–F bonds (102). The $[XeF_5]^+$ cation, which is found in such adducts as XeF_6,PtF_5 and XeF_6,AsF_5, is square-pyramidal (103), as would be expected by analogy with the iso-electronic IF_5 molecule, and in much the same way the apical Xe–F bond

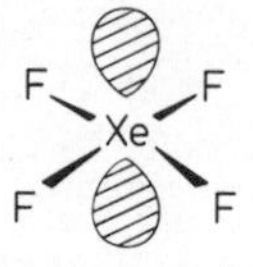

(102) 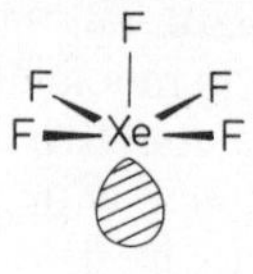 (103) 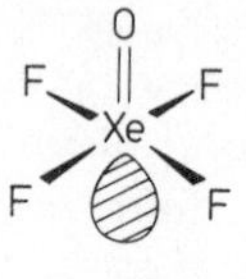 (104)

is appreciably shorter than the basal bonds, and the xenon atom is below the plane of the four fluorines ($F_{apical}XeF_{basal}$ = 83°; 79° in the $[RuF_6]^-$ salt). Closely related is $XeOF_4$ (104), in which the Xe–O bond takes up the apical position, with OXeF bond angles of 91.8°. The xenon atom has a regular octahedral configuration in the perxenate anion, $[XeO_6]^{4-}$, which is isoelectronic with $[IO_6]^{5-}$; all the bond angles are close to 90° and the Xe–O distances are between 184 and 186 pm.

The final configuration to be discussed is that found in XeF_6. In principle this molecule should have a distorted octahedral arrangement of bonds if the lone pair of electrons plays a part in the stereochemistry, and yet we have already seen that in ions such as $[SbCl_6]^{3-}$ and $[TeCl_6]^{2-}$ the structure is perfectly regular. Unfortunately X-ray studies do not help because solid XeF_6, which exists in several modifications, contains $[XeF_5]^+$ cations bridged by long bonds through F^- ions to give tetramers or hexamers; in the latter case the Xe–F bonds and bond angles are very similar to those found in $[XeF_5]^+[PtF_6]^-$, and the bridging Xe–F distances are 256 pm. Electron diffraction data on XeF_6 vapour seems to favour a non-octahedral flexible molecule.

10.12 Bonding in Xenon Compounds

This has been the subject of intense discussion, and the matter is by no means settled. Both the valence-bond and molecular-orbital methods have been used and we can illustrate their application to the simplest molecule, XeF_2. In the valence-bond approach, the xenon atom is considered to lose an electron to give Xe^+, the unpaired electron remaining in the $5p_x$ orbital then being used to bond with one or other of the fluorine atoms. In valence-bond language we refer to the structure as being made up of the two canonical forms shown in (105).

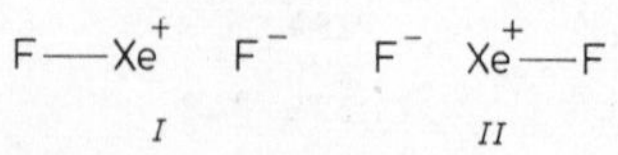

$$\underset{I}{F{-}Xe^{+}\ \ F^{-}} \qquad \underset{II}{F^{-}\ \ Xe^{+}{-}F}$$

(105)

The molecular-orbital method uses the same atomic orbitals ($2p_x$ for F and $5p_x$ for Xe) and takes linear combinations of them to form three molecular orbitals, one bonding, one non-bonding and one antibonding. This is exactly the same type of situation that we had with the $[I_3]^-$ ion (see p.188); for further clarity the combinations used for XeF_2 are given now, but without mixing coefficients.

$$\sigma_1 = \phi_{F_a} - \phi_{Xe} - \phi_{F_b}$$

$$\sigma_2 = \phi_{F_a} + \phi_{F_b}$$

$$\sigma_3 = \phi_{F_a} + \phi_{Xe} - \phi_{F_b}$$

The pictorial representation of these combinations is shown in *Figure 10.22* and the resultant energy diagram in *Figure 10.23.* The bonding between xenon and the two fluorine atoms is provided by the two electrons in the

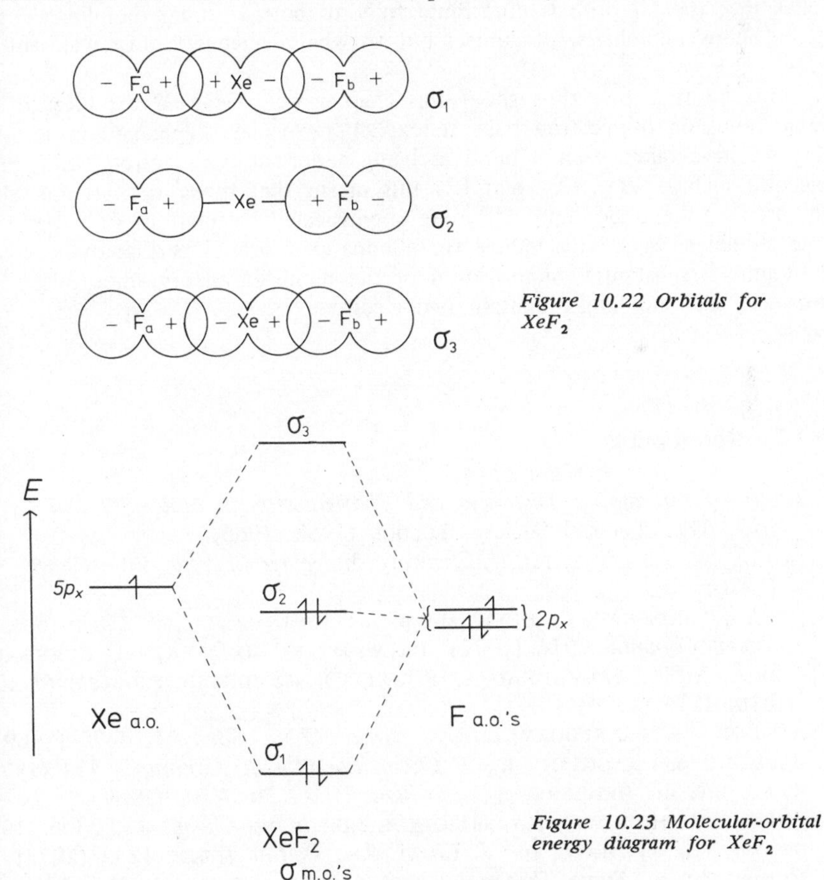

Figure 10.22 Orbitals for XeF_2

Figure 10.23 Molecular-orbital energy diagram for XeF_2

delocalized σ orbital σ_1. It must be made clear that here we have only two electrons linking the three atoms, so that the bonding force between Xe and F is on average one electron.

Logical extensions of this molecular-orbital approach use the other 5*p* orbitals to account for the bonding in XeF_4 ($5p_x$ and $5p_y$ orbitals used) and XeF_6 (all three 5*p* orbitals used). Of course, bonding for XeF_6 based solely on 5*p* Xe orbitals would imply an octahedral shape.

Although the molecular-orbital models using only 5*p* xenon orbitals are attractive in their simplicity, they again raise the problem of the extent to which the *d* orbitals may contribute to the bonding. We have already com across this point in discussing the bonding in the heavier elements of Group VB, VIB and VIIB, as for instance in the fluorides PF_5, SF_6 and IF_7, and here it seemed reasonable to incorporate *d* orbitals. Similarly we felt that at the level of sophistication of our discussions it was sensible to consider π bonding in P–O, S–O and I–O bonds as involving the appropriate *d* orbitals. There seems to be no reason for treating xenon differently from these other elements, and accordingly for consistency the bonding in xenon compounds (as represented by the structural formulae) has been shown simply as single bonds to fluorine and double bonds to terminal oxygens. This implies that the 5*d* orbitals contribute fully to both σ and π bonding; this will be an over-emphasis, of course, but to what degree we must wait and see.

A final point is that throughout this chapter we have used the ideas of mutual repulsion of electron pairs to explain molecular shape, and in so doing we have taken each σ bond as being a normal two-electron bond; in molecules such as PF_5, SF_6 and IF_7 this means that the *d* orbitals are fully used in the σ bonding. If more sophisticated calculations or experiments should subsequently reduce (or eliminate) *d* orbital participation, this will require a substantial modification of the repulsion theory since the bonds will then no longer contain two electrons.

10.13 References

1. *Tables of Interatomic Distances and Configuration in Molecules and Ions*, The Chemical Society, London (1958; 1965)
2. GORDY, W. and COOK, R.L., *Microwave Molecular Spectra*, Interscience (1970)
3. GUERNAS-BRISSEAU, L., DREW, M.G.B. and GOULTER, J.E., *J. Chem. Soc. Chem. Commun.*, 916 (1972); DREW, M.G.B., GOULTER, J.E., GUERNAS-BRISSEAU, L., PALVADEAU, P., ROUXEL, J. and HERPIN, P., *Acta Cryst.*, **B30**, 2579 (1974)
4. ATWOOD, J.L. and STUCKY, G.D., *J. Amer. Chem. Soc.*, **91**, 4426 (1969)
5. CALVO, C. and FAGGIANI, R., *J. Chem. Soc. Chem. Commun.*, 714 (1974
6. RYAN, R.P. and HEDBERG, K., *J. Chem. Phys.*, **50**, 4986 (1969)
7. DURIG, J.R., HUDGENS, B.A. and ODOM, J.D., *Inorg. Chem.*, **13**, 2306 (197
8. BROWN, K.L. and HALL, D., *J. Chem. Soc. Dalton Trans.*, 1843 (1973)
9. JOHNSON, D.A., *Some Thermodynamic Aspects of Inorganic Chemistry*, Cambridge University Press (1968)
10. DALY, J.D. and SANZ, F., *J. Chem. Soc. Dalton Trans.*, 2051 (1974)
11. DALY, J.D. and SANZ, F., *Acta Cryst.*, **B30**, 2768 (1974)
12. BENT, H.A., *Inorg. Chem.*, **2**, 747 (1963)

13. SIMPSON, J., *Main Group Elements, MTP International Review of Science, Inorganic Chemistry Series One,* Vol.1, 231, Butterworths, London (1972)
14. MITCHELL, K.A.R., *Chem. Rev.,* **69**, 157 (1969)
15. RAUK, A., ALLEN, L.C. and MISLOW, K., *J. Amer. Chem. Soc.,* **94**, 3035 (1972)
16. HOFFMANN, R., HOWELL, J.M. and MUETTERTIES, E.L., *J. Amer. Chem. Soc.,* **94**, 3047 (1972)
17. KOSTER, D.A. and WAGNER, A.J., *J. Chem. Soc. A,* 435 (1970)
18. PENNEY, G.J. and SHELDRICK, G.M., *J. Chem. Soc. A,* 245 (1971)
19. WHITFIELD, H.J., *J. Chem. Soc. Dalton Trans.,* 1737, 1740 (1973)
20. RENNINGER, A.L. and AVERBACH, B.L., *Acta Cryst.,* **B29**, 1583 (1973)
21. WHITEFIELD, H.J., *J. Chem. Soc. Dalton Trans.,* 1739 (1973); GOLDSTEIN, P. and PATON, A., *Acta Cryst.,* **B30**, 915 (1974)
22. SVENSSON, C. *Acta Cryst.,* **B30**, 458 (1974)
23. BEAGLEY, B., ROBIETTE, A.G. and SHELDRICK, G.M., *J. Chem. Soc. A,* 3006 (1968)
24. ALLCOCK, H.R., *Chem. Rev.,* **72**, 315 (1972)
25. BEATTIE, I.R. and LIVINGSTON, K.M.S., *J. Chem. Soc. Dalton Trans.,* 784 (1972)
26. BRABANT, C., HUBERT, J. and BEACHAMP, A.L., *Canad. J. Chem.,* 2952 (1973)
27. WUNDERLICH, H. and MOOTZ, D., *Acta Cryst.,* **B30**, 935 (1974); WUNDERLICH, H., *Acta Cryst.,* **B30**, 939 (1974)
28. WEBSTER, M. and KEATS, S. *J. Chem. Soc. A,* 836 (1971)
29. HULME, R. and SCRUTON, J.C., *J. Chem. Soc.,* 2448 (1968)
30. SCHROEDER, D.R. and JACOBSON, R.A., *Inorg. Chem.,* **12**, 210 (1973)
31. SPOLITI, M., GROSSO, V. and CESARO, N.N., *J. Mol. Struct.,* **21**, 7 (1974)
32. GIBLER, D.D., ADAMS, C.J., FISCHER, M., ZALKIN, A. and BARTLETT, N., *Inorg. Chem.,* **11**, 2325 (1972)
33. EDWARDS, A.J. and JONES, G.R., *J. Chem. Soc. A,* 1491 (1970)
34. EINSTEIN, F., TROTTER, J. and WILLISTON, C., *J. Chem. Soc.,* 2018 (1967)
35. FOSS, O., *Adv. Inorg. Chem. Radiochem.,* **2**, 237 (1960)
36. BUSS, B. and KREBS, B. *Angew. Chem.,* **82**, 446 (1970)
37. SANDOR, E. and FARROW, R.F.C., *Discuss. Faraday Soc.,* **48**, 78 (1969)
38. PARENT, C.R. and GERRY, M.C.L., *J. Chem. Soc. Chem. Commun.,* 285 (1972)
39. SIEGEL, S., TANI, B. and APPELMAN, E., *Inorg. Chem.,* **8**, 1190 (1969)
40. MIGCHELSEN, T. and VOS, A., *Acta Cryst.,* **23**, 796 (1967)
41. GABES, W. and NIJMAN-MEESTER, *Inorg. Chem.,* **12**, 589 (1973)
42. BURBANK, R.D. and JONES, G.R., *Inorg. Chem.,* **13**, 1071 (1974)
43. CHRISTE, K.O. and CURTIS, E.C., *Inorg. Chem.,* **11**, 2209 (1972)
44. SLADSKY, F., *MTP International Review of Science, Inorganic Chemistry Series One,* Vol.3, 1, Butterworths, London (1973)
45. GILLESPIE, R. and SCHROBILGEN, G.J., *Inorg. Chem.,* **13**, 2370 (1974)

11

COMPLEX COMPOUNDS

11.1 Introduction and Nomenclature[1,2]

It is difficult to give a formal definition of the term 'complex compound', but in general it applies to all those compounds in which the number of bonds formed by one of the atoms is greater than that expected from the usual valency considerations. Thus, tervalent cobalt forms six bonds with ammonia molecules in the complex cation $[Co(NH_3)_6]^{3+}$, while divalent nickel forms four bonds with chlorine in the complex anion $[NiCl_4]^{2-}$. Of course, there are compounds such as $[C(diars)_2]Br_4$ in which the number of covalent or co-ordinate bonds is the same as the group covalency, but which we instinctively think of as complex. In this unusual cation the four arsenic atoms each donate an electron pair to the formally 4+ carbon, so that four donor–acceptor, or co-ordinate, bonds are formed. Because so many complex molecules contain this type of bond, they are commonly referred to as 'co-ordination compounds'. We use the more general term 'complex compound' since it covers a much wider range of compounds and does not necessarily imply a particular type of bond.

Complex compounds may contain complex cations, e.g. $[Cu(NH_3)_4]^{2+}$, or anions, e.g. $[Fe(CN)_6]^{3-}$, or they may be simple neutral adducts such as $TiCl_4$,2py or $Fe(acac)_3$, where py is an abbreviation for pyridine and acac represents the acetylacetone molecule less a hydrogen, i.e. $CH_3COCHCOCH_3$. We have now introduced three abbreviations for groups attached to the central atom, and these and others used in this chapter are listed in *Table 11.1*. We have named the compounds using the procedures adopted by the Chemical Society and the American Chemical Society, which are based on the IUPAC recommendations. However, these conventions have changed from time to time, and readers who consult the original literature – particularly the older papers – may find some confusion over naming and abbreviations. Thus while acetylacetone is more correctly named pentane-2,4-dione, the complexes derived from it are usually referred to as acetylacetone complexes and the common abbreviation is Hacac.

The atoms, groups of atoms, or ions which bond to the central atom are referred to as 'ligands', and ligands may be characterized according to how many donor centres they have. Hence, ligands that bond through only one atom (e.g. N in NH_3 and C in CN^-) are said to be 'monodentate' (literally one-toothed). Ethylenediamine ($NH_2CH_2CH_2NH_2$) and diethylenetriamine ($NH_2CH_2CH_2NHCH_2CH_2NH_2$), which have two and three donor nitrogen atoms, respectively, are said to be bidentate and tridentate. When both donor atoms of a bidentate ligand bond to the same metal centre, thus giving a ring system, the compound formed is said to be a

Table 11.1 Abbreviations for ligands mentioned in Chapter 11

Abbreviation	*Formula*	*Name of ligand*
acacH	$CH_3COCH_2COCH_3$	acetylacetone (pentane-2,4-dione)
bdpPme	$(Ph_2PCH_2CH_2)_2NCH_2CH_2OMe$	*N,N*-bis-(2-diphenylphosphino-ethyl)-2-methoxyamine
bipy		2,2′-bipyridyl
bPyaen	$CHNCH_2CH_2NCH$	*N,N′*-bis(2-pyridylmethylene)-ethane-1,2-diamine
bPyDAH	$NHCH_2CH_2CH_2NH$	*N,N′*-bis-(2-pyridyl)propane-1,3-diamine
bzacH	$PhCOCH_2COCH_3$	benzoylacetone
dacoDa	$^-O_2CCH_2N(CH_2CH_2CH_2)_2NCH_2CO_2^-$	1,5-diazacyclo-octane-*N,N′*-diacetate
dapip	$H_2NCH_2CH_2CH_2N(CH_2CH_2)_2NCH_2CH_2CH_2NH_2$	*N,N′*-bis-(3-aminopropyl)-piperazine
diars	$C_6H_4(AsMe_2)_2$	1,2-bis(dimethylarsino)benzene [*o*-phenylenebis(dimethylarsine)]
dimim	HN–$CH{=}CH$–$CMe{=}NMe$ (ring)	1,2-dimethylimidazole
disalen	$CHNCH_2CH_2NCH$, O^-, ^-O	*N,N′*-bis(salicylideneimine)
dmp		2,9-dimethyl-1,10-penanthroline
dmso	Me_2SO	dimethyl sulphoxide
dpe	$Ph_2PCH_2CH_2PPh_2$	1,2-bis(diphenylphosphino)ethane
dppea	$HN(CH_2CH_2PPh_2)_2$	bis-(2-diphenylphosphinoethyl)-amine

Abbreviation	*Formula*	*Name of ligand*
dsp	PhP( )$_2$ MeS	bis-(2-methylthiophenyl)phenylphosphine
en	$H_2NCH_2CH_2NH_2$	1,2-diaminoethane (ethylenediamine)
Et_4dien	$HN(CH_2CH_2NEt_2)_2$	bis-(2-diethylaminoethyl)amine (1,1,7,7-tetraethyldiethylenetriamine)
2-Meim	HN(CH=CH)(CMe=NH)	2-methylimidazole
Me_2dmpa	$HN(CH_2$ N Me $)_2$	bis-(6-methylpyridyl-2-methyl)-amine
Me_5dien	$MeN(CH_2CH_2NMe_2)_2$	bis-(2-dimethylaminoethyl)-methylamine
Me_6tren	$N(CH_2CH_2NMe_2)_3$	tris-(2-dimethylaminoethyl)amine
nnp	$Et_2NCH_2CH_2NHCH_2CH_2PPh_2$	(2-diethylaminoethyl-2-diphenylphosphinoethyl)amine
nnpMe	$Et_2NCH_2CH_2NMeCH_2CH_2PPh_2$	(2-diethylaminoethyl-2-diphenylphosphinoethyl)methylamine
nnpO	$Et_2NCH_2CH_2NHCH_2CH_2P(O)Ph_2$	(2-diethylaminoethyl-2-diphenyloxaphosphinoethyl)amine
n_2p_2	$Et_2NCH_2CH_2N(CH_2CH_2PPh_2)_2$	(2-diethylaminoethyl-bis(diphenylphosphinoethyl))amine
ox	$^-O_2C{-}CO_2^-$	oxalate
paphy	N CH=N–NH N	pyridine-2-aldehyde 2-pyridylhydrazone
phen	N N	1,10-phenanthroline
2-picNO	NO Me	2-methylpyridine *N*-oxide
PNNP	$(Ph_2PCH_2CH_2NMe)CH_2CH_2(NMeCH_2CH_2PPh_2)$	*N,N′*-bis-(2-diphenylphosphinoethyl)ethane-1,2-diamine

Abbreviation	Formula	Name of ligand
py	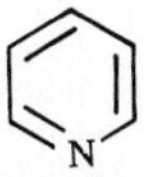 N	pyridine
pyO	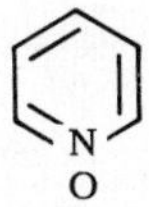 N O	pyridine *N*-oxide
QAS	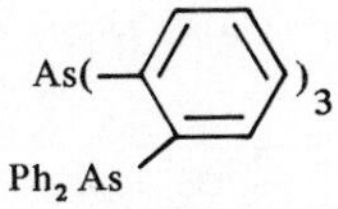 As( )$_3$ Ph_2As	tris-(2-diphenylarsinophenyl)arsine
quin	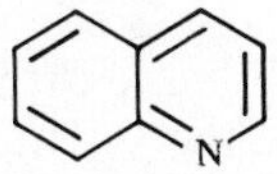 N	quinoline
QP	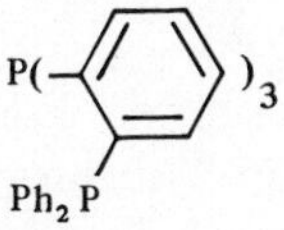 P( )$_3$ Ph_2P	tris-(2-diphenylphosphinophenyl)-phosphine
SAs_3	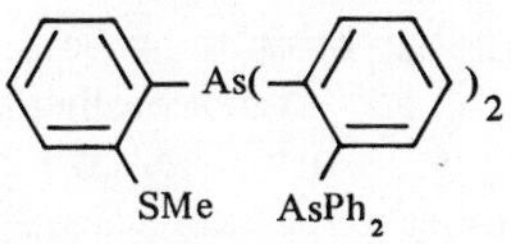 As( )$_2$ SMe $AsPh_2$	bis-(2-diphenylarsinophenyl)-2-methylthiophenylarsine
TAP	$P(CH_2CH_2CH_2AsMe_2)_3$	tris-(3-dimethylarsinopropyl)-phosphine
terpy	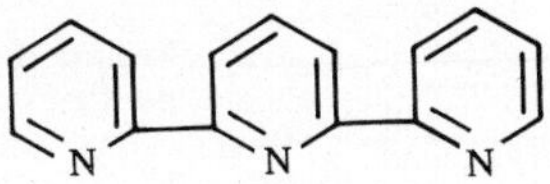 N N N	terpyridyl
thf	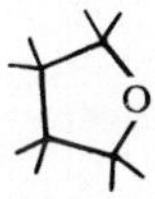 O	tetrahydrofuran
tmpo	Me_3PO	trimethylphosphine oxide
tpas	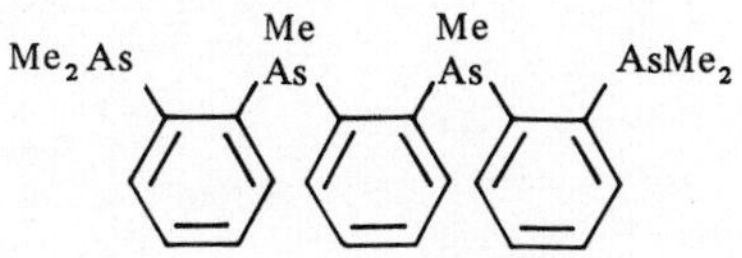 Me_2As Me As Me As $AsMe_2$	1,2-bis-[(2-dimethylarsinophenyl)-methylarsinyl]benzene
tepen	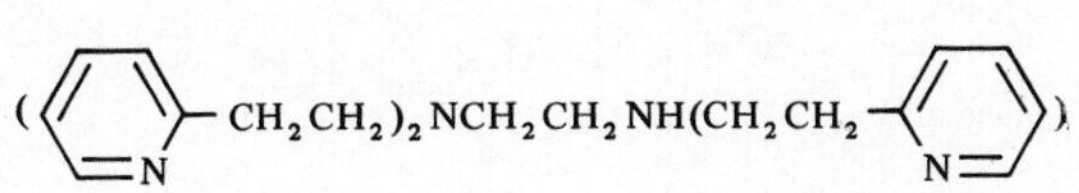 (N $-CH_2CH_2)_2NCH_2CH_2NH(CH_2CH_2-$ N)	*N,N'*tris-[2-(2'-pyridyl)ethyl]-ethane-1,2-diamine

Abbreviation	Formula	Name of ligand
TPN	$N(CH_2CH_2PPh_2)_3$	tris-(2-diphenylphosphinoethyl)-amine
tpn	$N(CH_2CH_2CH_2NH_2)_3$	tris-(3-aminopropyl)amine
triars	$(Me_2AsCH_2CH_2CH_2)_2AsMe$	bis-(3-dimethylarsinopropyl)-methylarsine
TSP	P(2-MeS-C_6H_4)$_3$	tris-(2-methylthiophenyl)phosphine

'chelate' compound. The name comes from the Greek word *chela*, which refers to the claw of a lobster or crab. The word chelate does not apply, however, if the bidentate ligand forms a bridging system with its two donor atoms bonded to separate acceptor centres. We shall see later that while most ligands donate σ lone-pair electrons to the metal ion acceptor, there are some ligands (e.g. C_2H_4 and C_6H_6) that donate π electron pairs; such ligands are referred to as π donors. *Table 11.2* summarizes the position.

In general, a metal ion will tend to achieve as high a co-ordination number as possible, provided that it has suitable orbitals available to receive electrons from the ligands, and subject to any steric problems that may arise. Thus ions of transition metals of the first series are rather small, so that the maximum co-ordination number is usually six; it is frequently less than this, particularly towards the end of the series, and we shall see later that electronic or steric factors account for this. With larger ions of the second and third transition series, the co-ordination number may increase

Table 11.2 Types of ligand

Type	Example
Monodentate σ-bonding	$M^{n+} \leftarrow NH_3$
Bidentate, chelating σ-bonding	M^{n+} ← H_2N-CH_2 / ← H_2N-CH_2 (ring, CH_2-CH_2 bonded)
Bidentate, bridging σ-bonding	$M^{n+} \leftarrow H_2N-CH_2-CH_2-NH_2 \rightarrow M^{n+}$
Tridentate σ-bonding	$CH_2(CH_2NH_2)$–CH_2–NH–CH_2–$CH_2(H_2N)$: $NH_2 \rightarrow M^{n+} \leftarrow H_2N$, NH → M^{n+}
Monodentate π-bonding	$M^{n+} \leftarrow$ ($CH_2=CH_2$)

to eight, so that while titanium normally has a maximum co-ordination number of six, zirconium frequently bonds to as many as eight donor atoms.

It is important that co-ordination numbers be assigned only to complexes that are fully characterized, since the stoichiometry of a compound is not an infallible guide to its structure. Thus the fluorides CuF_2 and MnF_3 do not contain two-co-ordinate copper and three-co-ordinate manganese, respectively, because the fluorides are polymeric in the solid state, and both metal atoms are surrounded by six fluorines in a distorted octahedral arrangement. It is even more necessary for care to be taken before odd co-ordination numbers are assigned, because polymeric or ionic species are commonplace. Thus in Cs_3CoCl_5, cobalt is not five-co-ordinate because the compound contains $Cs^+[CoCl_4]^{2-}$ and Cl^- ions, but no $[CoCl_5]^{3-}$ ions. The solid pentahalides of niobium, tantalum and molybdenum do not contain five-co-ordinate metal atoms, since all the halides are dimeric through bridging halogen atoms (1). In general, therefore, co-ordination numbers should not be

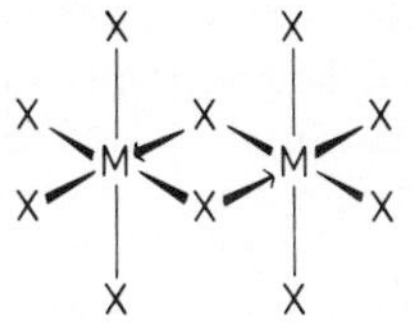

(1)

assigned unless full structural information is available, such as from single-crystal X-ray studies for instance. This point will be discussed again under the appropriate co-ordination numbers.

11.2 Bonding in Complex Compounds

There is a wide range of types of bonding in complex compounds, ranging from a simple trapping of molecules in a host lattice (clathrate compounds) to normal covalent or co-ordinate bonds. In simple hydrates or ammoniates of salts such as $BaCl_2$, the bonding is weak and the water or ammonia molecules can be easily removed by heating. These bonds between the metal ions and the water or ammonia ligands are termed ion–dipole bonds, because they have their origin in the electrostatic attraction of the ion for the polar ligand molecules. Not only do the ammonia and water molecules have permanent dipoles, but additional dipoles are induced by the metal ion, and ion–dipole bonds result from the electrostatic attraction between the metal ion and the polarized molecules. Inasmuch as this attraction is between an ion and a dipole rather than between two ions, the ion–dipole bond is fairly weak, and readily broken, in comparison with an ordinary ionic bond.

The bonding becomes very much stronger if the metal ion has suitable low-energy orbitals which can accept electrons from the ammonia or water molecules. Thus with the ammoniates of $BaCl_2$, we picture the Ba^{2+} ion polarizing the electron charge cloud of the ammonia molecules (in particular the lone pair electrons on the nitrogen atoms), so that in effect the metal ion takes some share of these electrons. The effect is much more profound in, say, the cobaltammine ion $[Co(NH_3)_6]^{3+}$, since the cobalt ion has vacant 3*d*, 4*s* and 4*p* orbitals that can accept the lone pair electrons from the nitrogen atoms; it should, perhaps, be emphasized once again that the co-ordinate bonds so formed are essentially covalent bonds since they involve the sharing of two electrons between two atoms, and that the term 'co-ordinate' merely signifies that the two electrons were originally associated with one atom.

The formation of co-ordinate bonds is often taken to be a characteristic of transition metal compounds, but this type of bonding is by no means limited to these elements, and main group elements such as boron, aluminium, silicon and tin form many such compounds. Indeed, in Chapter 8 we discussed this bonding in the simple co-ordination compound BF_3,NH_3, in which BF_3 was the electron acceptor (or Lewis acid).

This approach, which describes the bonding in terms of the donation of electron pairs from the ligands into vacant orbitals of the acceptor metal, makes use of the valence-bond technique. The acceptor atom is considered to make available vacant hybrid orbitals, which will yield strongly directional bonds and determine the stereochemistry of the complex that is formed. Thus the use of d^2sp^3 hybrid orbitals produces an octahedral complex and sp^3 orbitals a tetrahedral complex. *Table 11.3* summarizes the orbitals involved in bonding in the more likely stereochemistries. Before we get too involved in the complications of molecular shape let us discuss in some detail the bonding in octahedral complexes. The two bonding descriptions already discussed at length in Chapter 8 for covalent polyatomic molecules, namely the valence-bond and molecular-orbital methods, are again the two main methods, but in addition we shall discuss a third approach, the crystal-field approach (and its extension the ligand-field approach), which is especially useful in explaining in simple terms such matters as the electronic spectra and the thermodynamic properties of transition metal complexes. We shall see that all three methods have their limitations.

Table 11.3 Directional properties of orbitals

Co-ordination number	*Orbitals used*	*Spatial arrangement of bonds*
2	sp or dp	linear
3	sp^2 or sd^2	trigonal-planar
4	sp^3 or sd^3 dsp^2	tetrahedral square-planar
5	dsp^3	trigonal-bipyramidal or square-pyramidal
6	d^2sp^3	octahedral

11.2.1 VALENCE-BOND METHOD

For this discussion of bonding let us take a specific element of the first transition series, so that we will be concerned especially with the role played by the $3d$ orbitals. With tervalent chromium, for instance, we assume that the Cr^{3+} ion makes six orbitals available, the $3d_{z^2}$, $3d_{x^2-y^2}$, $4s$, $4p_x$, $4p_y$ and $4p_z$, and that linear combinations of these atomic orbitals provide six equivalent d^2sp^3 hybrid orbitals. The six linear combinations are those given earlier (p.96), namely:

$$\psi_{(i)} = \sqrt{\frac{1}{6}}\phi_s + \sqrt{\frac{1}{2}}\phi_{p_z} + \sqrt{\frac{1}{3}}\phi_{d_{z^2}}$$

$$\psi_{(ii)} = \sqrt{\frac{1}{6}}\phi_s - \sqrt{\frac{1}{2}}\phi_{p_z} + \sqrt{\frac{1}{3}}\phi_{d_{z^2}}$$

$$\psi_{(iii)} = \sqrt{\frac{1}{6}}\phi_s + \sqrt{\frac{1}{12}}\phi_{d_{z^2}} + \frac{1}{2}\phi_{d_{x^2-y^2}} + \sqrt{\frac{1}{2}}\phi_{p_x}$$

$$\psi_{(iv)} = \sqrt{\frac{1}{6}}\phi_s + \sqrt{\frac{1}{12}}\phi_{d_{z^2}} + \frac{1}{2}\phi_{d_{x^2-y^2}} - \sqrt{\frac{1}{2}}\phi_{p_x}$$

$$\psi_{(v)} = \sqrt{\frac{1}{6}}\phi_s + \sqrt{\frac{1}{12}}\phi_{d_{z^2}} - \frac{1}{2}\phi_{d_{x^2-y^2}} + \sqrt{\frac{1}{2}}\phi_{p_y}$$

$$\psi_{(vi)} = \sqrt{\frac{1}{6}}\phi_s + \sqrt{\frac{1}{12}}\phi_{d_{z^2}} - \frac{1}{2}\phi_{d_{x^2-y^2}} - \sqrt{\frac{1}{2}}\phi_{p_y}$$

The σ bonding in a complex cation $[CrL_6]^{3+}$ is provided by the interaction of each of these vacant d^2sp^3 hybrid orbitals with a doubly filled (lone pair) orbital of the appropriate ligand. Thus we have six bonding pairs of electrons, each linking the Cr^{3+} ion with one ligand. This limitation of each bonding pair to the central metal ion and just one ligand is a principal characteristic of the valence-bond approach, as we have seen earlier when discussing more simple molecules such as $BeCl_2$ (see p.87).

We have used only the $3d_{z^2}$ and $3d_{x^2-y^2}$ in our valence-bond description of the σ bonding in the $[CrL_6]^{3+}$ ion because only these two d orbitals point along the x, y and z bonding axes. The other $3d$ orbitals, the $3d_{xy}$, $3d_{xz}$ and $3d_{yz}$, cannot be involved in the σ bonding because their lobes point between the axes. In much the same way in the HF molecule we could not use the $2p_y$ and $2p_z$ atomic orbitals of fluorine to provide σ bonding along the H–F bond axis (taken as the x axis), the overlap integral being zero in each case. The point is illustrated in *Figure 11.1.*

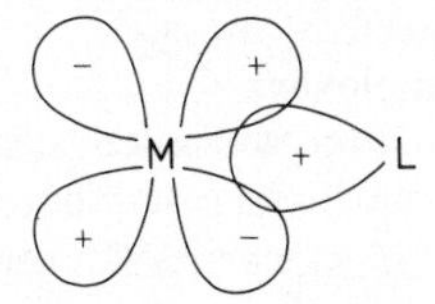

Figure 11.1 d ***Orbital that does not form a σ bond***

However, these $3d$ orbitals that are not used for σ bonding can be used for π bonding since they have the correct symmetry for overlap with π-type orbitals on the ligands. This is illustrated in *Figure 11.2(a)*, in which the σ bond axis is taken as the x axis, so that π bonding results from the overlap of the metal $3d_{xy}$ orbital with a similar d_π on the ligand; the ligand may equally well use a p_π orbital [*Figure 11.2(b)*]. The most common instance of this π bonding occurs with later elements of the transition series which have doubly filled d_π orbitals that overlap with vacant ligand π orbitals [*Figure 11.2(c)*] and so donate electrons to the ligand. This is referred to as 'back donation', and it helps to reduce the build-up of electron density on the metal ion caused by the donation of

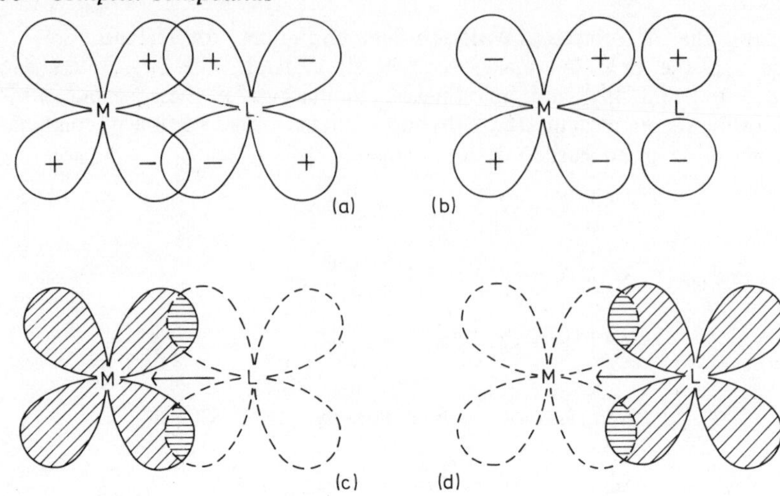

Figure 11.2 Use of d *orbital in* d_π–d_π *bonding*

electron pairs by the ligands in the σ bonding. Even if the metal ion has all three of its d_π orbitals doubly filled, as with Co^{3+}, for instance, it can only provide enough electrons for three double bonds, so that in an octahedral complex in which all six ligands have vacant π-acceptor orbitals the maximum π-bond order is 0.5. With Cr^{3+}, the d_π orbitals are only singly filled, so that at best there can be only partial delocalization of the d electrons towards the ligands, which amounts to quite a small degree of π bonding.

In principle, π bonding could arise through donation of electrons from ligand doubly filled π orbitals to the metal vacant d_π orbitals [see *Figure 11.2(d)*], but in practice this is not likely to be very important since it would lead to an even greater build-up of electron density on the metal. The type of complex ion where this might take place to a very limited extent includes $[TiBr_6]^{2-}$, where the titanium (formally Ti^{4+}) has all its $3d_\pi$ orbitals vacant and available to receive electrons from the doubly filled $4p_\pi$ orbitals of bromide.

We shall discuss the magnetic properties of octahedral complexes later in this chapter, but it may be mentioned at this point that qualitative aspects of the magnetic behaviour of most transition metal complexes can be accounted for on the basis of a simple valence-bond model. We have already mentioned that Cr^{3+} has three $3d$ orbitals not used in σ bonding, and that each of these orbitals contains one electron, and this is confirmed by magnetic measurements. Similarly, most six co-ordinate complexes of tervalent cobalt are found to be diamagnetic, corresponding to each $3d_\pi$ orbital containing two electrons. However, the valence-bond approach, as we describe it, is purely qualitative and does not explain all observations, particularly in respect of the variation of magnetic moment with temperature. The treatment cannot give even relative energies for the various possible configurations, and since it does not provide information about the energy levels between which electrons move it cannot be used to account for the electronic spectra of complexes.

11.2.2 MOLECULAR-ORBITAL METHOD[3-5]

In the valence-bond approach, the metal 3d, 4s and 4p atomic orbitals were hybridized to provide six equivalent d^2sp^3 orbitals, each of which was involved in the formation of a localized two-electron σ bond with just one ligand. With the molecular-orbital treatment the bonding is no longer restricted to the metal and just one ligand, and the approach is on the basis of taking each metal orbital in turn (i.e. the $3d_{z^2}$, $3d_{x^2-y^2}$, $4s$, $4p_x$, $4p_y$ and $4p_z$) and letting it interact with σ orbitals on any of the ligands surrounding the metal. Thus, we use the LCAO technique to compound twelve molecular orbitals (six bonding and six antibonding) from the six atomic orbitals of the metal and the six σ orbitals of the ligands; the twelve electrons from the six ligands are placed in the six bonding molecular orbitals. The linear combinations for the six bonding molecular orbitals are as follows:

$$\phi_1 = ad_{z^2} + \left(\frac{1-a^2}{12}\right)^{1/2} [2\sigma_z + 2\sigma_{\bar{z}} - \sigma_x - \sigma_{\bar{x}} - \sigma_y - \sigma_{\bar{y}}]$$

$$\phi_2 = ad_{x^2-y^2} + \left(\frac{1-a^2}{4}\right)^{1/2} [\sigma_x + \sigma_{\bar{x}} - \sigma_y - \sigma_{\bar{y}}]$$

$$\phi_3 = as + \left(\frac{1-a^2}{6}\right)^{1/2} [\sigma_x + \sigma_{\bar{x}} + \sigma_y + \sigma_{\bar{y}} + \sigma_z + \sigma_{\bar{z}}]$$

$$\phi_4 = ap_x + \left(\frac{1-a^2}{2}\right)^{1/2} [\sigma_x - \sigma_{\bar{x}}]$$

$$\phi_5 = ap_y + \left(\frac{1-a^2}{2}\right)^{1/2} [\sigma_y - \sigma_{\bar{y}}]$$

$$\phi_6 = ap_z + \left(\frac{1-a^2}{2}\right)^{1/2} [\sigma_z - \sigma_{\bar{z}}]$$

To simplify the form of the expressions we have used d_{z^2} to represent the wave function rather than ϕd_{z^2}. σ_x and $\sigma_{\bar{x}}$ refer to the wave functions for the σ orbitals of the ligands on the $+x$ and $-x$ axes respectively. a is the usual mixing coefficient, so that when $a^2 = 0$ there is no contribution to the wave functions from the metal orbitals and the bonding is ionic with the electrons remaining associated only with the ligands; at the other extreme, with $a^2 = 1$, there will be no contribution from the ligand orbitals and the electrons will be entirely in metal orbitals. By appropriate adjustment of this mixing coefficient we can provide molecular wave functions that will allow for such factors as the relative electronegativities of the metal and the ligands.

A study of these linear combinations shows that while the d_{z^2} orbital interacts with all six ligand σ orbitals, the orbitals of the ligands along the z axis contribute more than the orbitals of the other ligands because they will overlap more effectively with the lobes of the d_{z^2} orbital which protrude furthest along the z axis. The $d_{x^2-y^2}$ orbital interacts only with orbitals of ligands along the x and y axes, and the signs of the ligand orbital functions are chosen to ensure a match with the $d_{x^2-y^2}$ orbital. This is illustrated in *Figure 11.3(a)*. Since the s orbital protrudes equally in all directions, and has a positive sign, interaction occurs equally with all the ligand σ orbitals. Each of the metal p orbitals can overlap only with σ orbitals of the two ligands along the particular axis, and the mathematical signs of the functions are chosen to match; this is illustrated in *Figure 11.3(b)*.

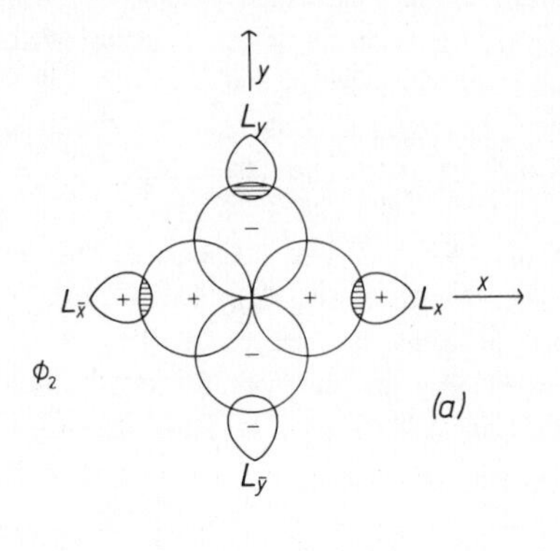

Figure 11.3 (a) Overlap of metal $d_{x^2-y^2}$ orbital with σ orbitals of ligands; (b) overlap of metal p_x orbital with σ ligand orbitals

Figure 11.4 is the molecular-orbital energy diagram that shows the relationship between the energy of the various orbitals of the metal and ligands, and the energy of the resultant molecular orbitals.

As with the valence-bond method we can get a description of π bonding in terms of the interaction of the d_{xy}, d_{xz} and d_{yz} orbitals of the metal with appropriate p_π or d_π orbitals of the ligands, but we do not

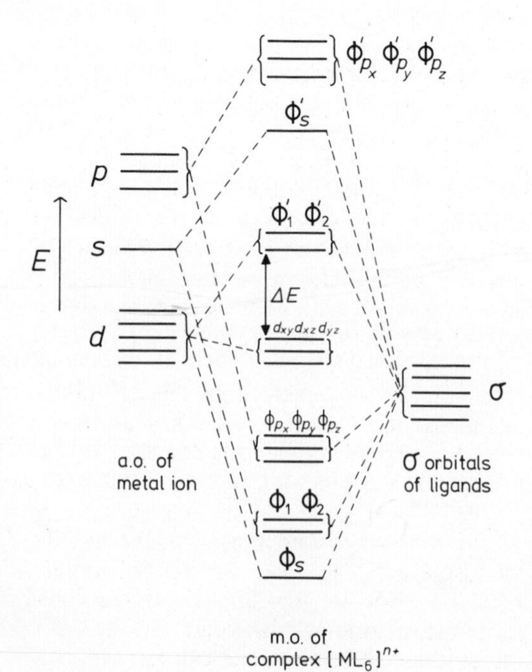

Figure 11.4 Energy levels of molecular orbitals (degenerate levels are slightly separated and bracketed together for clarity)

limit our π bonding to the metal and one ligand. The d_π orbitals of the metal are taken in turn and considered to interact with π orbitals of all suitable ligands, giving delocalized bonding (π) and antibonding (π^*) molecular orbitals. If we consider the d_{xy} orbital, then it can interact with π orbitals of ligands along the x and y axes, as illustrated in

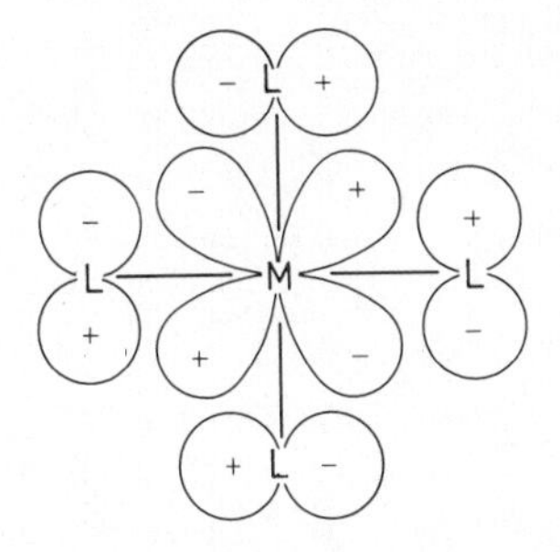

Figure 11.5 Overlap of d_{xy} orbital by p_π orbitals of each of the four ligands in the xy *plane*

Figure 11.5 (the L π orbitals are shown as p_π for simplicity). The function describing this bonding molecular orbital is given by:

$$\phi_{\pi_1} = \beta d_{xy} + \left(\frac{1-\beta^2}{4}\right)^{1/2} [\pi_x - \pi_{\bar{x}} + \pi_y - \pi_{\bar{y}}]$$

β is the mixing coefficient, so that when $\beta^2 = 1$ there is no contribution from the ligand orbitals and the $3d$ orbitals are then described as non-bonding.

Qualitative calculations and predictions about spectra and magnetic properties can be made by means of the molecular-orbital method and we shall discuss these problems in rather more detail later in this chapter and again in Chapter 12.

11.2.3 CRYSTAL-FIELD AND LIGAND-FIELD METHODS[3,6,7]

The earliest electrostatic models attempted to account for the properties of complex compounds on the basis of interactions between the metal ions and ligands considered as point charges or dipoles, and while these successfully explained why tetrahedral and octahedral configurations were to be expected for four-co-ordinate and six-co-ordinate complexes, respectively, they could not account satisfactorily for the existence of square-planar complexes. Difficulties also arise with ligands such as carbon monoxide, which formed stable complexes despite being non-polar. Moreover, the theory predicted that for a given set of ligands the smallest ions should form the strongest bonds, and yet second- and third-row transition elements are known to form more stable complexes than the smaller elements of the first row.

This simple electrostatic theory was subsequently extended to include the effects of the ligand point charges or dipoles on the d electrons of the metal ion. This extension, called the 'crystal-field' theory, was originally applied to the behaviour of metal ions in a crystal lattice, but it may be applied equally well to complexes in which the d electrons of the metal

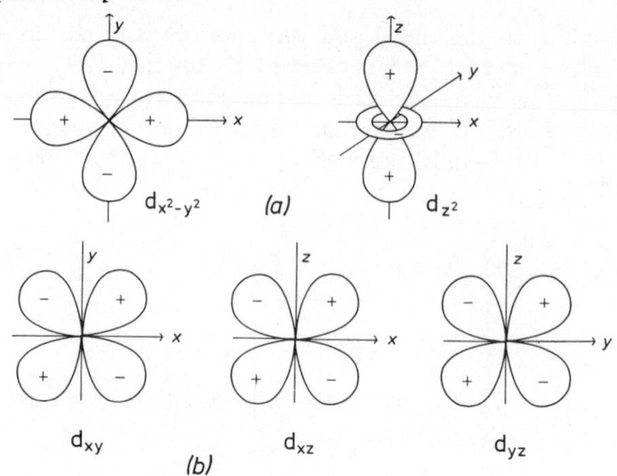

Figure 11.6 d *Orbitals: (a)* $d_{x^2-y^2}$ *and* d_{z^2}*; (b)* d_{xy}, d_{xz} *and* d_{yz}

will be affected by the charge field produced by the six ligands (the ligand field).

If we look again at the shapes of the *d* orbitals (see *Figure 11.6*), we see that they may be divided into two groups:

(a) d_{z^2} and $d_{x^2-y^2}$ which are concentrated along the axes; these orbitals are called the e_g orbitals
(b) d_{xy}, d_{xz} and d_{yz}, which are concentrated between the *x*, *y* and *z* axes; these orbitals are called the t_{2g} orbitals

It is worth noting that the e_g orbitals are those that were involved in σ bonding in both the valence-bond and molecular-orbital treatments, while the t_{2g} orbitals were either non-bonding or concerned with π bonding.

If we now consider a typical octahedral complex, $ML_6{}^{n+}$, then the σ lone pairs of the six ligands that surround the M^{n+} ion will exert an electrostatic field concentrated along the *x*, *y* and *z* axes. We picture the free ion in space gradually experiencing a field as the six ligands begin to approach. At first we consider this field as just a uniform spherical field which will have an influence on all the *d* electrons and gradually raise their energy, but as the ligands get closer so the field becomes more clearly defined as being concentrated along the axes and then some *d* electrons will be influenced more than others. Since the e_g orbitals are concentrated along the axes, the electrons in these orbitals will be repelled more strongly by the field of the ligand lone pairs than will electrons in the t_{2g} orbitals. Because of this field produced by the octahedrally disposed ligands, the five degenerate *d* orbitals will split into two groups, the high-energy e_g doublet and the lower-energy t_{2g} triplet. The energy diagram illustrating the influence of the field is shown in *Figure 11.7*.

The difference in energy between the upper doublet and the lower triplet is usually given the label Δ, although some workers prefer to use the expression $10Dq$. It is convenient to discuss the energies of the *d* orbitals by reference to an energy zero which is taken as the 'weighted' mean energy of the five *d* orbitals. Thus the t_{2g} orbitals have an energy of

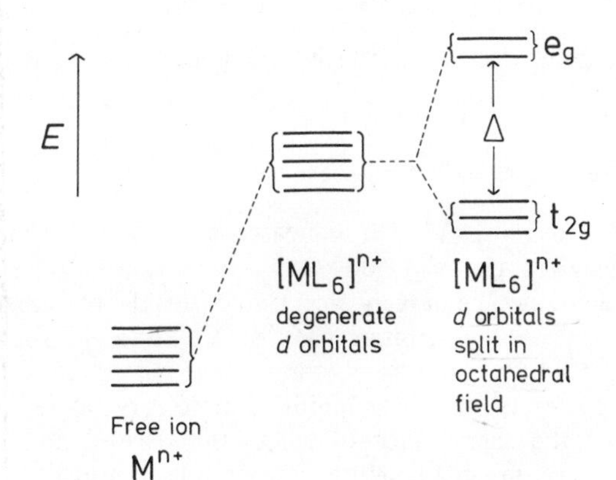

Figure 11.7 Splitting of d *orbitals in an octahedral field*

$-2\Delta/5$ (see *Figure 11.7*) while the energy of the e_g orbitals is $+3\Delta/5$, so that the sum of the energies of the three t_{2g} orbitals ($3 \times -2\Delta/5 = 6\Delta/5$) balances that of the two e_g orbitals ($2 \times 3\Delta/5 = 6\Delta/5$).

The value of the splitting parameter Δ varies from complex to complex and depends upon the nature of the ligands and on the metal ion. Thus, the field produced by the ligands depends on their nature, and especially on the ease with which the ligand electrons are distorted by the ion. In practice the ligands may be placed in an order of decreasing crystal-field effect, an order that is essentially independent of the particular metal ion to which they are attached:

$$CN^- > NH_3 > H_2O > F^- > Cl^- > Br^- > I^-$$

This 'pecking' order, often referred to as the 'spectrochemical series', may seem a little odd at first, since F^- might be expected to produce a bigger field than CN^-, but it must be remembered that the electrons are held more tightly to the nucleus of the fluoride ion, and not distorted by the metal ion as much as they are in the 'softer' cyanide ion.

In the simplest crystal-field theory we would expect the size and charge of the metal ion to be an important factor in determining the value of Δ. Thus the ion with the largest charge should polarize the ligand electrons most and so produce the largest value of Δ, and this is the case with the hexa-aquo ions of vanadium, for instance, where Δ for $[V(H_2O)_6]^{3+}$ is 18 000 cm^{-1} but drops to 11 800 cm^{-1} for $[V(H_2O)_6]^{2+}$. With the first transition metal series, the hexa-aquo ions of the tervalent elements, $[M(H_2O)_6]^{3+}$, generally have Δ values in the 15,000–20,000 cm^{-1} range, whereas analogous complexes of the divalent ions usually have Δ in the region of 10,000 cm^{-1}. We have commented already that the simplest electrostatic theory predicts that, for two ions of the same charge, the smaller one should exert the greater influence on the ligands (cf. the greater polarizing power of Li^+ compared with Na^+), and at first sight this should apply to transition metal compounds as well. However, as we noted earlier, the prediction is not correct if we compare ions of transition metals from different series. Indeed, in general we shall find that for analogous complexes Δ increases by 30–40% on going from an element of the first transition series to the second-row element in the same group, and by a further 30–40% on going to the third-row element. A typical sequence would be:

$$[VL_6]^{4+}, \quad [NbL_6]^{4+} \text{ and } [TaL_6]^{4+} .$$

Thus, on going from $[VL_6]^{4+}$ to $[NbL_6]^{4+}$, Δ increases even though the Nb^{4+} ion is appreciably bigger than the V^{4+} ion.

This apparent anomaly, which may be largely accounted for if we remember that we are dealing with different *d* orbitals (3*d* for V^{4+} and 4*d* for Nb^{4+}) with different radial distribution functions, shows the danger of overestimating the predictive power of the oversimplified basic crystal-field theory. A detailed account of the significance of calculations based on this theory would be out of place in an elementary textbook, however, so more theoretically inclined readers are referred to Freedman and Watson[8] for such a discussion. One basic difficulty is that the simple crystal-field theory considers the forces at work to be purely electrostatic, and there is no reference to electron sharing (covalent bonding) between the metal ion and the ligands, except that which may be implied from the polarization of the lone-pair electrons of the ligands by the metal ion.

Such electron sharing, whether as a result of σ or π bonding, must have an effect on the splitting of the *d* orbitals. Thus, the large ligand-field of the 'soft' cyanide ion, which we attributed to the ease with which the charge clouds could be polarized and pulled towards the metal ion, is perhaps better described on the basis of the cyanide ion being bonded by both σ and π bonds to the metal ion, the latter shortening the expected metal–carbon distance and so increasing the ligand field. When we modify the crystal-field theory to take into account this bonding, the new approach is described as the 'ligand-field' theory; it combines features of the crystal-field and molecular-orbital descriptions. In the ligand-field approach we discuss the value of the splitting Δ, which depends on the ligand field, no matter whether the field arises through electrostatic or bonding forces. In the molecular orbital approach, the equivalent splitting is described as ΔE, which depends on bond strength, since the energy of antibonding orbitals increases as that of the bonding orbitals decreases.

The actual values of Δ or ΔE may be calculated from measurements of electronic (visible or ultraviolet) spectra, and in the next section we shall discuss the influence of Δ upon the magnetic properties and the stability of octahedral complexes.

11.3 Absorption Spectra[9–11]

Compounds of the transition elements with unpaired *d* electrons are characterized by being coloured, and their absorption spectra accordingly show peaks in the visible and near-infrared regions (25 000–8000 cm^{-1}; 4000–12 500 Å). These peaks are of low intensity, with extinction coefficients* usually between 1 and 50, but many complexes show further peaks in the

$$^*\epsilon_{max} = \frac{\log I_0/I \times \text{molecular weight}}{\text{concentration (g l}^{-1}) \times \text{cell length (cm)}}$$

ultraviolet region (50 000–25 000 cm^{-1}) which are much more intense, with extinction coefficients in the 1000–10 000 cm^{-1} range. If we limit our discussion for the moment to complexes formed by the first-row transition elements, then this division into weak visible-region peaks and much more intense ultraviolet-region peaks is surprisingly clear cut.

The weak peaks arise from the transition of electrons from one *d* level to another, and they are accordingly referred to as *d–d* transitions. In theory, such transitions should have zero intensity, because transitions between two levels of the same ℓ quantum number are 'forbidden' since they do not bring about a change of electric dipole. The fact that we do observe such peaks, even if they are weak, is accounted for by describing either the ground or excited states by wave functions that are not pure *d* functions but involve a small amount of *p* character. This becomes possible if the complex lacks a centre of symmetry, either because all the ligands are not the same or because there is some distortion in the bond lengths or bond angles. Tetrahedral complexes, which have no centre of symmetry, show appreciably more intense peaks in their spectra than octahedral complexes. We shall return again to these points in Chapter 12, but readers are referred to references 9, 10 and 11 for a more detailed consideration of the factors that effect the intensity of *d–d* peaks.

We shall see that the qualitative aspects of the *d–d* peaks found for octahedral complexes of d^1 and d^9 ions can be explained fairly simply, since the peaks can be related directly to the promotion of an electron between particular *d* orbitals. In other cases, for complexes in which the number of *d* electrons falls in the range 2–8, we shall be unable to make such simple correlations, because of the complications arising through interaction of the various electrons which results in each d^n configuration producing several 'terms'. A discussion of these matters is deferred until Chapter 12, and for the moment we will limit ourselves to an account of the spectra of octahedral complexes of Ti^{3+} (d^1) and Cu^{2+} (d^9) ions, and simply comment that Δ values can be evaluated from the experimentally observed spectra of any complex.

To start with, let us consider the spectrum of a complex of tervalent titanium, $[TiCl_6]^{3-}$, which for the moment we can consider as based on Ti^{3+} being surrounded by six Cl^- ions. The spectrum (*Figure 11.8*),

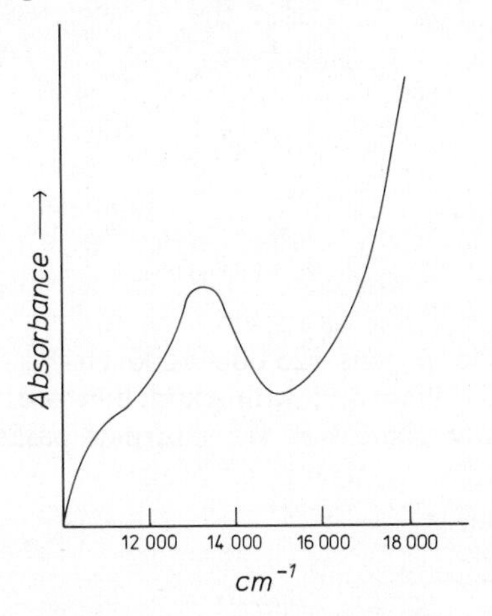

Figure 11.8 Spectrum of the $[TiCl_6]^{3-}$

which shows a broad peak at 12 750 cm^{-1}, is typical of a d–d transition. Now this anion is an octahedral complex, involving the d^1 Ti^{3+} ion, and we can use either the molecular-orbital diagram (*Figure 11.4*) or crystal-field diagram (*Figure 11.7*) to explain the spectrum. The latter is somewhat simpler and using it we say that in the ground state the electron is in a t_{2g} orbital but may be excited to an e_g orbital. The promotion is accordingly referred to as a $e_g \leftarrow t_{2g}$ transition, and 12 750 cm^{-1} represents the splitting Δ. In the molecular-orbital description, in which ΔE is the equivalent to Δ, the peak results from the excitation of the electron from one of the non-bonding d orbitals to one of the lowest-energy anti-bonding orbitals (either $\phi_1{}'$, or $\phi_2{}'$). The peak is not symmetrical, however, but shows a distinct shoulder, which means that there must be two excited levels into which the electron can be promoted. This is explained by the splitting of the e_g orbitals into two separate levels because of distortion of the complex from a perfect octahedron (either because of packing effects or because of the operation of the Jahn–Teller theorem: see later in this chapter).

If we change the environment of the Ti^{3+} ion by surrounding it with six water molecules, with the formation of the complex cation $[Ti(H_2O)_6]^{3+}$ then the spectrum shows a peak at 20 300 cm^{-1}, so that the six water molecules produce a much bigger ligand field than do the six chloride ions. In principle, a whole series of tervalent titanium complexes $[TiL_6]^{3+}$ might be prepared, and Δ values determined from the spectra; in this way we can establish the spectrochemical series referred to on p.211. In practice it is not easy to prepare a wide range of Ti^{3+} complexes in which the ion is surrounded octahedrally with six identical ligands, and most experiments have involved complexes of the type TiX_3,3L (where X = Cl, Br or I). It is then assumed that the ligand field is established by equal contributions from the X and L ligands. The procedure is known as Jorgensen's Rule of Average Environment. In recent years a large number of tervalent titanium complexes have been prepared and ligand-field splitting values established for a range of oxygen and nitrogen donor ligands.

Another configuration that is easy to understand is the d^9 (e.g. Cu^{2+}) in which the ground state is $(t_{2g})^6(e_g)^3$. The electronic spectra octahedral of complexes of divalent copper, such as $[Cu(H_2O)_6]^{2+}$, show a single peak which results from the promotion of one of the t_{2g} electrons into the e_g orbital that is only half filled. It should be noted that because Cu^{2+} carries only a 2+ charge it attracts the surrounding ligands much less strongly than Ti^{3+}, and the ligands accordingly have less effect on the d electron charge clouds, and the splitting of the d levels is less. Thus the peak for the $[Cu(H_2O)_6]^{2+}$ is found at 12 500 cm^{-1}, compared with 20 300 cm^{-1} for $[Ti(H_2O)_6]^{3+}$. As we mentioned earlier, we find that Δ values for $[M(H_2O)_6]^{n+}$, where M is a first-row transition element, are between 15 000 and 20 000 cm^{-1} for tripositive ions and around 10 000 cm^{-1} for dipositive ions.

We have limited our discussion of spectra to complexes formed by metal ions with d^1 or d^9 configurations because they are simple to understand and because the spectral peaks can be correlated with electron promotion between simple d orbitals. In other d^n ions (n = 2–8) the situation is more complicated because interaction between electrons have to be considered and various 'terms' arise for each configuration. Discussion of the spectra of complexes of such ions is deferred until the next chapter.

Our discussion has been concerned mainly with *d–d* peaks, but it must be emphasized that the spectra of most transition metal complexes also contain more intense 'charge-transfer' peaks in the ultraviolet region. A typical example is provided by the spectrum of $TiCl_3,3C_6H_5N$ (see *Figure 11.9*), where in addition to the *d–d* peak at 16 600 cm^{-1} there is a much

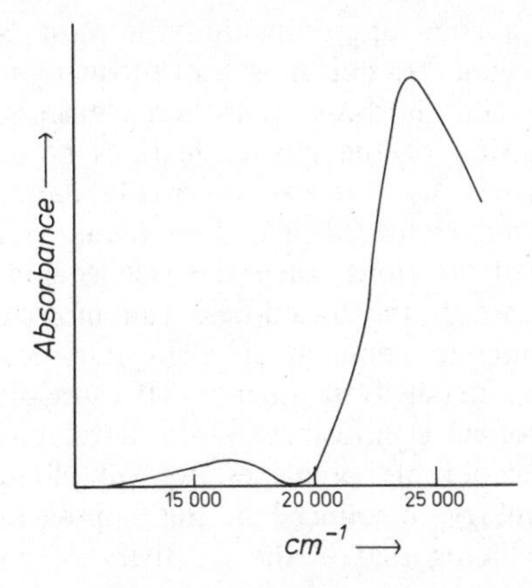

Figure 11.9 Spectrum of $TiCl_3,3py$

more intense peak at 24 300 cm^{-1}. This charge-transfer peak is probably associated with the transition of an electron from a non-bonding titanium *d* orbital to a pyridine antibonding π^* orbital. We cannot use the simple diagram of *Figure 11.7* to discuss such spectra, but must use a full molecular-orbital diagram; this would resemble *Figure 11.4* but also incorporate the ligand π and π^* orbitals.

We have discussed only the spectra of complexes formed by the first transition series, because here we normally find a clean-cut separation between *d–d* and charge-transfer peaks. With the heavier elements of the second and third transition series, Δ (ΔE) gets bigger by at least 30%, so that the *d–d* peaks occur at shorter wavelength and are often partly obscured by charge-transfer peaks.

11.4 Magnetic Susceptibility[12]

All atoms show diamagnetism (i.e. they tend to move from the strongest to the weakest part of an applied magnetic field), but atoms containing electrons with unpaired spins show paramagnetism as well. In effect the unpaired electron behaves as a little magnet with consequent attraction to the applied field. When an orbital contains two electrons, their spins must be opposed and their magnetic moments being equal and opposite will cancel out each other, leaving the atom with only diamagnetic characteristics. Thus an atom with an unpaired electron is paramagnetic, but the extent of the paramagnetism is reduced by the diamagnetism of any electron pairs present.

The magnetic moment associated with the unpaired electron can be treated as a vector quantity, and the magnitude of the moment of an atom is accordingly directly related to the number of unpaired electrons. For the first-row transition elements there is a simple relationship between the magnetic moment, μ (expressed in Bohr magnetons*, BM), and the number of unpaired electrons, n:

$$\mu = [n(n+2)]^{1/2}$$

This simple 'spin-only' formula is a good approximation for most complexes of the first-row transition elements, but it is less adequate for compounds of the heavier second- and third-row transition elements.

If we limit ourselves to a discussion of the grosser features of magnetic behaviour, and especially if we merely use the experimentally determined values of magnetic susceptibility (and hence calculated magnetic moments) to establish the number of unpaired electrons, then the valence-bond approach is generally adequate, although the ligand-field and molecular-orbital descriptions are just as simple to apply at an elementary level and they can be extended to more detailed treatments. Because the valence-bond approach has an historical significance in the development of the treatment of magnetic behaviour of complexes, we will discuss it briefly, especially since the terminology introduced in this approach is applicable to other discussions, including that of the reactivity of complexes (see Section 11.19).

In the valence-bond description of octahedral complexes formed by the first-row transition elements (see Section 11.2), use is made of the $3d^2 4s4p^3$ set of hybrid orbitals to describe the bonding between the metal ion and the six ligands. The remaining three $3d$ orbitals ($3d_{xy}$, $3d_{xz}$, $3d_{yz}$) are used to accommodate the metal ion $3d$ electrons. Hence Hund's rule is applied and the electrons are fed into these orbitals one at a time so as to give the maximum number of unpaired spins. There is no problem with ions of d^1, d^2 and d^3 configurations (e.g. Ti^{3+}, V^{3+} and Cr^{3+}), but difficulties arise with ions with four or more d electrons, because electron pairing becomes necessary unless other orbitals are made available. Experimentally it is found that the Cr^{2+} ion (d^4) forms two magnetically different types of complex. With the $[Cr(CN)_6]^{4-}$ ion, pairing indeed takes place, and the complex has a magnetic moment corresponding to two unpaired electrons; in other cases, such as $CrCl_2,4H_2O$, the complex has four unpaired electrons, and it is clear that the four electrons are in separate orbitals. To account for this behaviour it was suggested that while for 'spin-paired' complexes (two unpaired electrons) the Cr^{2+} ion uses the $3d^2 4s4p^3$ set of hybrid orbitals for bonding, in the 'spin-free' complexes (four unpaired electrons) the Cr^{2+} ion uses the $4s4p^3 4d^2$ hybrids, thus making the $3d$ orbitals available for the four d electrons of Cr^{2+}. The position is illustrated in *Figure 11.10*. Thus the valence-bond approach divides the complexes into two distinct categories, according to whether the bonding involves 'inner' ($3d^2 4s4p^3$) or 'outer' ($4s4p^3 4d^2$) hybrid orbitals; the complexes are referred to as inner (or inner-orbital) and outer (or outer-orbital) complexes, respectively.

***The Bohr magneton (BM) is defined as 1 BM = $eh/4\pi m$, where e is the electron charge, m the electron mass, and h is Planck's constant.**

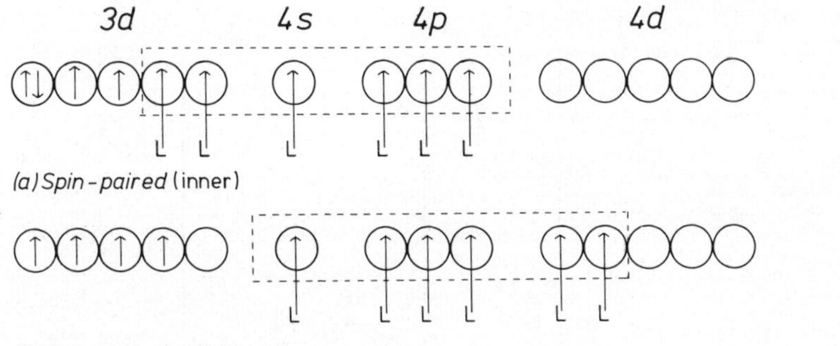

Figure 11.10 Inner and outer complexes of Cr^{2+}

A similar explanation can be given for the varying magnetic behaviour of complexes formed by d^5, d^6 and d^7 ions. *Table 11.4* summarizes the magnetic behaviour of octahedral complexes of the d^4–d^7 ions, and shows some examples of each. At this point the expressions low-spin and high-spin are introduced; they correspond to the valence-bond description of inner and outer orbitals.

While this description does account for the number of unpaired electrons, it introduces a rather artificial division of complexes into inner and outer, which implies significantly different bonding. This can be avoided in the ligand-field approach. We saw in Section 11.3 (*Figure 11.7*) that, under the influence of the field produced by the six octahedrally-disposed ligands, the five $3d$ orbitals split into two groups, the triplet t_{2g} (d_{xy}, d_{xz} and d_{yz}) and the doublet e_g ($d_{x^2-y^2}$ and d_{z^2}). The d electrons of the metal ion fill up the d levels according to Hund's rule. Once again there is no problem over d^1, d^2 and d^3 ions because each electron goes into a separate orbital of the t_{2g} set. However, with d^4 we are again in a dilemma. If the electron is placed in a t_{2g} level it must pair up. The alternative is to place the electron in one of the e_g orbitals even though it is of appreciably higher energy. The alternatives are illustrated in *Figure 11.11*.

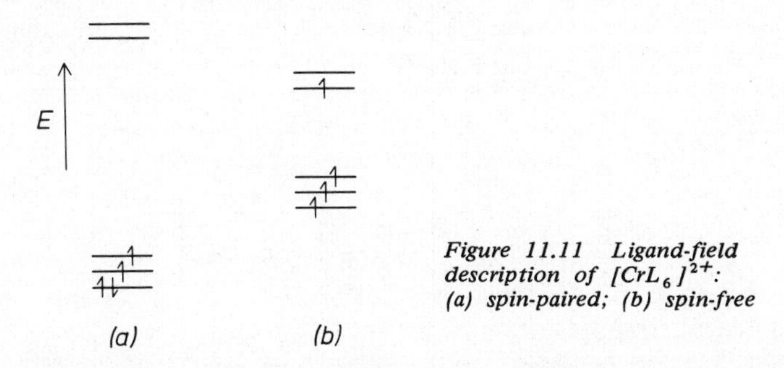

Figure 11.11 Ligand-field description of $[CrL_6]^{2+}$: (a) spin-paired; (b) spin-free

Of course, it can be readily seen that we are describing the location of the d electrons in essentially the same way as in the valence-bond approach, the first three electrons entering the d_{xy}, d_{xz} and d_{yz} orbitals (t_{2g}) and the next electron going into the $d_{x^2-y^2}$ or d_{z^2} orbital (e_g). The difference is that in the ligand-field approach we have not specified the nature of

Table 11.4 Magnetic behaviour of octahedral complexes of first row transition elements for d^4–d^7

No. of d *electrons*	*Element*	*Inner ≡ low spin*		*Outer ≡ high spin*	
		No. of unpaired electrons	*Examples*	*No. of unpaired electrons*	*Examples*
4	Cr(II)	2	$K_4[Cr(CN)_6]$	4	$CrCl_2,4H_2O$
	Mn(III)	2	$K_3[Mn(CN)_6]$	4	$Mn(acac)_3$
5	Mn(II)	1	$K_4[Mn(CN)_6]$	5	$[Mn(bipy)_3]Br_2$
	Fe(III)	1	$K_3[Fe(CN)_6]$	5	$Fe(acac)_3$
6	Fe(II)	0	$[Fe(bipy)_3](ClO_4)_2$	4	$[Fe(en)_3]Cl_2$
	Co(III)	0	$[Co(NH_3)_6]Cl_3$	4	$K_3[CoF_6]$
7	Co(II)	1	$[Co(diars)_3](ClO_4)_2$	3	$[Co(NH_3)_6]Cl_2$
	Ni(III)	1	$[Ni(diars)_2Cl_2]Cl$	3	–

the bonding in the complexes, but merely specified the relative energies of the d levels. When the ligand field is big, the energy gap Δ will be big, and the electrons will prefer to pair up in the t_{2g} orbitals; when the ligand field is small, however, Δ will be small, and the fourth electron will go into one of the e_g orbitals so as to give the maximum charge separation. This is demonstrated in *Figure 11.12*, which shows the relative energies of low- and high-spin configurations with a variation in the value of Δ. For small values of Δ the spin-free (high spin) configuration is much more stable than the spin-paired (low spin), but as Δ increases so the energy difference gets less, and at a certain critical value (Δ') the two configurations have the same energy. As Δ continues to increase, so the spin-paired state gets increasingly stable relative to the spin-free configurations.

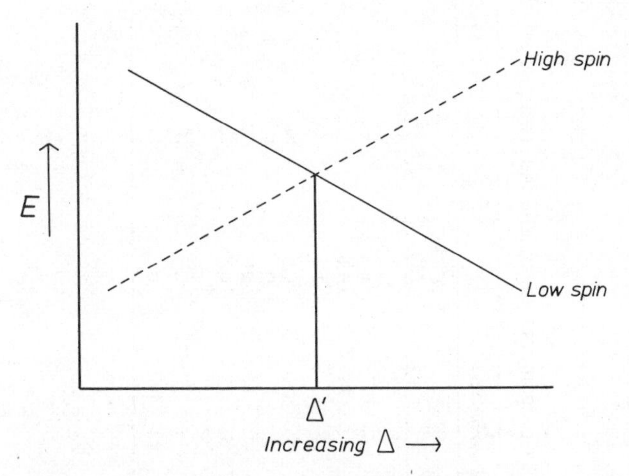

Figure 11.12 Energy relationships for octahedral spin-paired (low spin) and spin-free (high spin) complexes

Thus although there is a distinct division of complexes of a given ion according to the number of unpaired electrons, the ligand-field approach does not require a sharp distinction in bond type. Of course, in each case the bonding will involve all the orbitals that are suitable on grounds of energy and symmetry.

Table 11.5 lists the electron arrangements for the ions (d^0–d^{10}) forming octahedral complexes, and gives the crystal-field stabilization energy for each, i.e. the energy saved by having the majority of the electrons in the t_{2g} orbitals.

At this point we should also refer to the molecular-orbital approach, which has much in common with the ligand-field, except that the upper doublet (e_g) is no longer pure d but an antibonding pair of orbitals $\phi_1{}'$ and $\phi_2{}'$ derived from the $d_{x^2-y^2}$ and d_{z^2} orbitals. The metal electrons are then accommodated in either the non-bonding or the antibonding orbitals. When the bonding is weak then ΔE (the energy difference between the non-bonding d orbitals and the lowest-energy antibonding orbitals) will also be small, and we shall get the maximum number of unpaired electrons, i.e. high-spin complexes. When ΔE is large the electrons prefer to pair up in the non-bonding orbitals; for d^7 ions, the

Table 11.5 Ligand-field description of electron arrangements in octahedral complexes of first-row transition elements

Examples	*No. of* d *electrons*	*Weak field (small Δ)* t_{2g}	e_g	*CFSE (Δ)*	*Strong field (large Δ)* t_{2g}	e_g	*CFSE (Δ)*
Sc^{3+}	0			0			0
Ti^{3+}	1	↑		0.4	↑		0.4
V^{3+}	2	↑ ↑		0.8	↑ ↑		0.8
Cr^{3+}	3	↑ ↑ ↑		1.2	↑ ↑ ↑		1.2
Cr^{2+}	4	↑ ↑ ↑	↑	0.6	↑↓ ↑ ↑		1.6
Fe^{3+}	5	↑ ↑ ↑	↑ ↑	0	↑↓ ↑↓ ↑		2.0
Co^{3+}	6	↑↓ ↑ ↑	↑ ↑	0.4	↑↓ ↑↓ ↑↓		2.4
Co^{2+}	7	↑↓ ↑↓ ↑	↑ ↑	0.8	↑↓ ↑↓ ↑↓	↑	1.8
Ni^{2+}	8	↑↓ ↑↓ ↑↓	↑ ↑	1.2	↑↓ ↑↓ ↑↓	↑ ↑	1.2
Cu^{2+}	9	↑↓ ↑↓ ↑↓	↑↓ ↑	0.6	↑↓ ↑↓ ↑↓	↑↓ ↑	0.6
Zn^{2+}	10	↑↓ ↑↓ ↑↓	↑↓ ↑↓	0	↑↓ ↑↓ ↑↓	↑↓ ↑↓	0

seventh electron enters one of the antibonding orbitals and so weakens the bonding between the ion and the ligand.

For the remainder of this chapter we will confine our discussion of magnetic moments to their simple relationship to the number of unpaired electrons, but in Chapter 12 a somewhat fuller account of magnetochemistry will be given.

11.5 The Stability of Complex Compounds[13]

The stability of a complex compound is closely related to its electronic configuration, and depends on the type of bonding that is present. If the bonding is essentially of an ion–dipole type then we must expect that the strongest bonds will be formed by the smallest ions, since they produce electrostatic fields of the greatest intensity. Considering the hydrated bivalent ions, Mg^{2+}, Ca^{2+}, Sr^{2+} and Ba^{2+}, and plotting the absolute enthalpies of hydration against the reciprocal of the radius, we get the smooth curve of *Figure 11.13*, which shows the decreasing strength of the ion–dipole bonds with increasing metal ion radius.

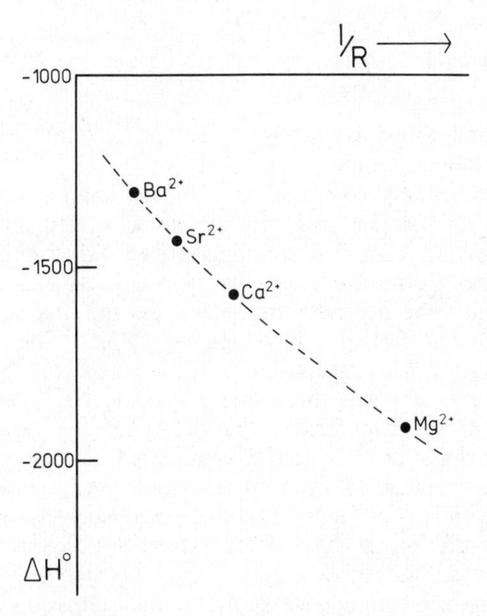

Figure 11.13 A plot of enthalpy of hydration (ΔH°) against reciprocal radius for the bivalent Group IIA ions

We might expect a similar smooth curve for the absolute enthalpies of hydration of the bivalent ions of the first transition series, since the ionic radii decrease from Ca^{2+} (99 pm) to Zn^{2+} (72 pm). However, a plot of the available experimental data (*Figure 11.14*) shows that the enthalpy of hydration does not change steadily as the ionic radius decreases, but gives a double-humped curve. It can be seen that hydrated ions other than Mn^{2+} are more stable than would be expected from considerations of

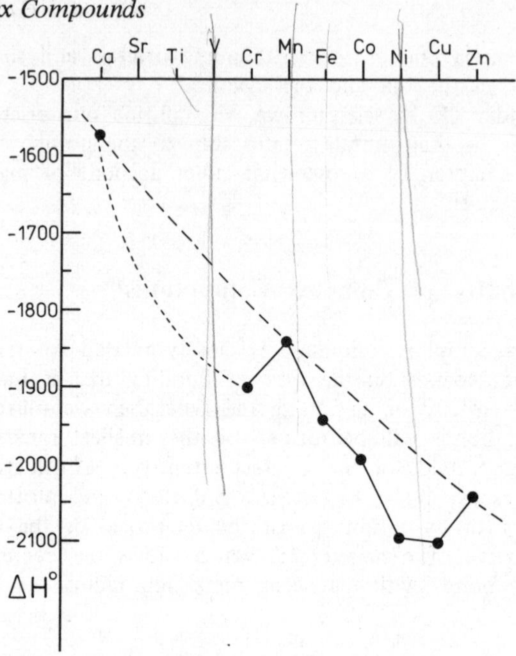

Figure 11.14 Enthalpies of hydration for M^{2+} *(Ca to Zn)*

ionic radii alone, and this extra stability may be accounted for in terms of crystal field stabilization energies.

The ions are octahedrally co-ordinated with six water molecules, giving $[M(H_2O)_6]^{2+}$, the d levels splitting into the usual t_{2g} triplet and e_g doublet. These hydrates are weak-field complexes (see *Table 11.5*), so that while the first three d electrons enter the t_{2g} levels and confer stability, the next two go into the e_g levels and make the ion less stable; at d^5 there is one electron in each d orbital and no CFSE. The sequence is repeated over the d^6–d^{10} configurations. Hence there will be the maximum CFSE at d^3 and d^8, and these ions (V^{2+} and Ni^{2+}) will accordingly be at the minima of the two dips in the curve. If the experimental values for the enthalpies of hydration are modified to allow for this CFSE (using Δ values obtained from spectroscopic measurements), then the smooth broken curve of *Figure 11.14* is obtained. A similar stabilization effect is observed for the hydrates of tervalent ions, with maximum stabilization for the d^3 ion, $[Cr(H_2O)_6]^{3+}$, and no stabilization for d^5, $[Fe(H_2O)_6]^{3+}$. The same principles apply to the formation of complexes formed by ligands other than water, of course, and interested readers may like to draw up a curve for, say, a series of hypothetical octahedral complexes $[ML_6]^{3+}$, in which the ions adopt spin-paired arrangements.

The 'stability' of complexes in aqueous solution can also be discussed in terms of ligand-field effects. Here we are concerned with a situation in which one or more of the water molecules co-ordinated to the metal ion is replaced by another ligand, i.e.

$$[M(H_2O)_6]^{2+} + L \rightleftharpoons [M(H_2O)_5L]^{2+} + H_2O$$

If the equilibrium constant (often called the 'stability constant' in this context) for this replacement reaction is K, then log K, which represents the free energy change of the reaction, is a measure of the stability of the $[M(H_2O)_5L]^{2+}$ ion. Many stability constants have been recorded and interested readers are referred to Bjerrum's summary[14]. We will restrict our discussion to ethylenediamine (en) complexes, in which each bidentate ethylenediamine ligand can replace two water molecules, in a series of reactions:

$$[M(H_2O)_6]^{2+} + en \rightleftarrows [M(H_2O)_4en]^{2+} + 2H_2O \quad (K_1)$$

$$[M(H_2O)_4en]^{2+} + en \rightleftarrows [M(H_2O)_2(en)_2]^{2+} + 2H_2O \quad (K_2)$$

$$[M(H_2O)_2(en)_2]^{2+} + en \rightleftarrows [M(en)_3]^{2+} + 2H_2O \quad (K_3)$$

The sum of the log K values (log K_1 + log K_2 + log K_3) gives a measure of the stability of $[M(en)_3]^{2+}$ complexes compared with $[M(H_2O)_6]^{2+}$. These values have been determined for the ions Mn^{2+} to Zn^{2+}, and their plot (*Figure 11.15*) shows the expected 'humped' curve; when allowance is made for the CFSE, the result is the smooth broken curve, showing the smooth increase for Mn^{2+} to Zn^{2+}.

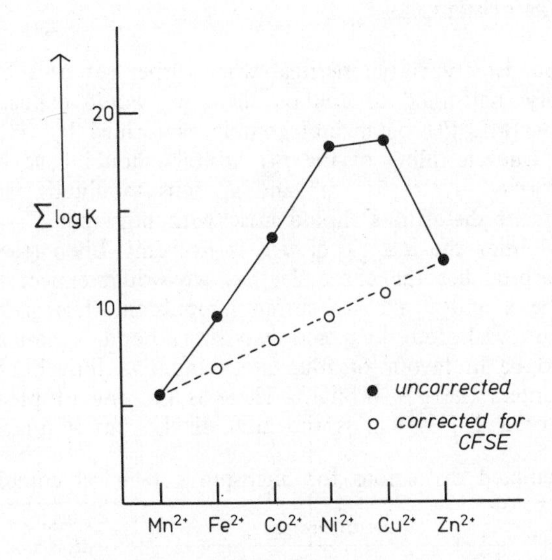

Figure 11.15 Plot of Σ log K for $[M(en)_3]^{2+}$ complexes

Similar considerations of CFSE have been used to explain analogous variations in the energy relationships of octahedral complexes of the first transition series. These include the complex cyanides of the M^{2+} ions and the acetylacetone complexes of the M^{3+} ions. Readers interested in the more general application of these ideas to variations in the thermodynamic properties of complexes are referred to an excellent article by George and McClure[15].

11.6 Distortion of Octahedral Complexes

Up until now we have assumed that in a complex ML_6 all the bond lengths are identical, and this is true for such complex ions as $[TiCl_6]^{2-}$, although we might expect some bond angles to deviate from 90° when all the ligands are not the same. Thus mutual repulsion between bulky groups is to be expected and when polydentate ligands are involved we might expect the bond angles to depend upon the size of the ring.

There is a further factor that must be considered, namely the repulsion between the ligand electron pairs and any d electrons on the central metal ion. We have already seen that the d electrons can be divided into two categories, one in which the electrons point along the axes towards the ligands (e_g), and one in which the electrons point between the axes (t_{2g}). If the e_g levels are unevenly filled then we should expect some of the ligands to experience a bigger repulsion than others, and the metal–ligand bond lengths should differ. If it is the t_{2g} levels that are unevenly filled we should expect a very much smaller effect, because the ligands are not influenced so much. Let us now consider in turn the high-spin and low-spin octahedral complexes and compare predicted and observed distortions.

11.6.1 HIGH-SPIN COMPLEXES

Since the charge density is symmetrical when either set of orbitals (t_{2g} and e_g) is empty, half-filled or doubly filled, we would expect no distortions to arise in $[ML_6]^{n+}$ octahedral complexes formed by d^0, d^3, d^5, d^8 and d^{10} ions. Uneven filling of the t_{2g} orbitals should have little effect, so complexes formed by d^1, d^2, d^6 and d^7 ions should be fairly regular. However, significant distortions should arise with high-spin d^4 and d^9 complexes, because either the $d_{x^2-y^2}$ or d_{z^2} is unevenly filled (see *Table 11.6*). If the $d_{x^2-y^2}$ orbital has the extra electron we would expect the four ligands along the x and y axes to suffer a repulsion, thus giving us a distorted octahedron with four long and two short bonds. Similarly, the distortion would be in favour of four short and two long bonds if the extra electron enters the d_{z^2} orbital. There is no way of predicting whether a 4/2 or a 2/4 system is the more likely, but in practice it is

Table 11.6 Predicted distortions for high-spin octahedral complexes

No. of d *electrons*	*Configuration*		*Predicted distortion*
	t_{2g}	e_g	
0	0	0	None
1	1	0	Slight
2	2	0	Slight
3	3	0	None
4	3	1	Considerable
5	3	2	None
6	4	2	Slight
7	5	2	Slight
8	6	2	None
9	6	3	Considerable
10	6	4	None

the 4 short/2 long arrangement that predominates. This distortion is referred to as a tetragonal distortion.

Compounds that permit the testing of our predictions are the complex fluorides, KMF_3, where M = Cr, Mn, Fe, Co, Ni or Cu. These do not contain the $[MF_3]^-$ ion, but are polymeric through M–F→M bridging, giving each metal a six-co-ordinate environment of fluorines. X-ray studies have shown that there is a regular octahedral arrangement of fluorines about Mn, Fe, Co and Ni, but that with both $KCrF_3$ and $KCuF_3$ there is considerable tetragonal distortion, with two M–F bonds much longer than the other four. A similar effect is observed for the hydrated chlorides ($MCl_2,2H_2O$) of Mn, Fe, Co and Cu, which have polymeric planar chains of MCl_2 units with two water molecules taking up the remaining two positions in an octahedral arrangement (2). As the data in *Table 11.7*

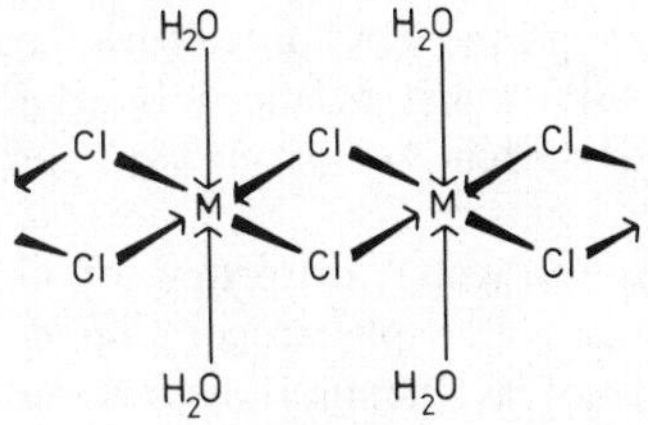

H_2O H_2O Cl Cl Cl M M Cl Cl Cl H_2O H_2O

(2)

show, the M–Cl distances are quite similar, except with $CuCl_2,2H_2O$, where two of the Cu–Cl bonds are much longer than the other two.

The simple dihalides and trihalides of the first-row transition elements are also six-co-ordinate (octahedral) through metal-halogen bridging, and structural data show that halides of d^4 (e.g. $CrCl_2$ and MnF_3) and d^9 (e.g. CuX_2) metals show typical tetragonal distortions with 4/2 structures.

A number of high-spin complexes of d^4 and d^9 ions also show similar distortions. Thus manganese(III) tropolonate, $Mn(O_2C_7H_5)_3$, has six oxygens surrounding the manganese, four at a distance of 194 pm and two at 213 pm (Adveef *et al.*[16]). A recent d^9 example is $[Cu(bipy)_3](ClO_4)_2$, which has four planar Cu–N bonds of length 203 pm, and two much longer axial bonds at 223 and 245 pm (Anderson[17]).

These tetragonal distortions observed for 'octahedral' high-spin complexes of d^4 and d^9 ions are illustrations of the general Jahn–Teller theorem, which says that if a non-linear molecule has a degenerate state, then there is at least one vibrational co-ordinate along which a distortion can occur so as to remove the degeneracy. Thus in an octahedral arrangement, the four ligands in the *xy* plane can move towards the metal ion, and simultaneously the two ligands on the *z* axis move away, so that the $d_{x^2-y^2}$

Table 11.7 Bond lengths in $MCl_2,2H_2O$ complexes

Metal	*M–O*/pm	*M–Cl*/pm
Mn	215	252; 259
Fe	208	249; 254
Co	204	245; 248
Cu	201	231; 298

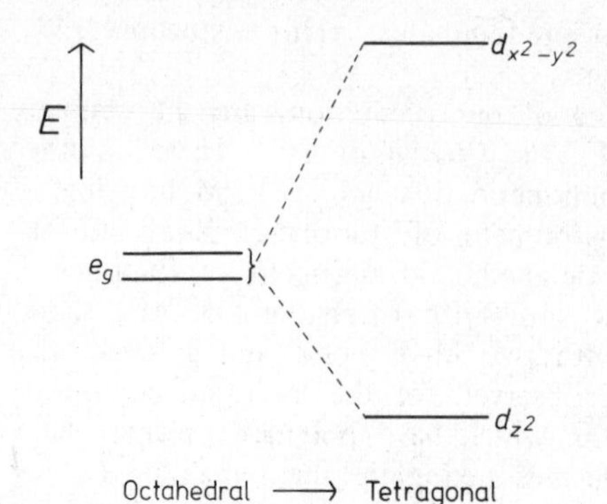

Figure 11.16 **Splitting of e_g doublet by tetragonal distortion of octahedral complex**

orbital becomes of higher energy than the d_{z^2} orbital. This distortion continues until the energy gained (by putting the odd e_g electron in the lower-energy of d_{z^2} orbital) is just balanced by the energy required to compress four bonds and stretch the other two. *Figure 11.16* illustrates this splitting.

We shall return to this tetragonal distortion of octahedral complexes when we come to consider square-planar compounds (see p.235), because the latter can be thought of as the limiting case in which the two z axis ligands are moved so far away from the metal as not to be exerting a significant ligand field on the metal ion.

Although we have interpreted the distortions in terms of ligand-field theory, the molecular-orbital and valence-bond approaches can be used. In the molecular-orbital description the fourth electron (for d^4) is placed in the $\phi_1{}'$ antibonding orbital which is concentrated along the z axis; accordingly the bonds between the metal ion and the two ligands along this axis will be weakened. The valence-bond description is one of two sets of hybrids, dsp^2 for the four short bonds (using $3d_{x^2-y^2}$, $4s$, $4p_x$ and $4p_y$) and pd for the two long bonds (using $4p_z$ and $4d_{z^2}$).

11.6.2 LOW-SPIN COMPLEXES

As the data in *Table 11.8* show, it is only with the d^7 configuration, with the odd electron in the e_g level, that we should expect a significant distortion. A recent study of $[NiCl_2(diars)_2]Cl$ (3) and the cobalt(III)

(3)

analogue (Bernstein *et al.*[18]) show the effect quite clearly. Thus while M–As bond lengths (cf. *Table 11.9*) are essentially the same, the Ni–Cl distance is 17 pm greater than the Co–Cl distance.

Table 11.8 Predicted distortions for low-spin octahedral complexes

No. of d *electrons*	*Configuration* t_{2g}	e_g	*Predicted distortion*
4	4	0	Slight
5	5	0	Slight
6	6	0	None
7	6	1	Considerable

Table 11.9 Bond lengths in the $[MCl_2(diars)_2]Cl$ complexes

M	*M–As*/pm	*M–Cl*/pm
Co	233	226
Ni	234	243

A previous study was made of the related *trans*-$M(diars)_2I_2$ complexes of the d^8 divalent metals, Ni, Pd and Pt. Up until now we have assumed that all d^8 complexes adopt a $(t_{2g})^6(e_g)^2$ configuration, with one electron in each of the e_g orbitals. While this is generally true, the diarsine complexes are diamagnetic. On the ligand-field approach we consider these electrons to be paired up in one of the split e_g levels with the two electrons in the lower level. Accordingly we predict that there should be a considerable distortion from the ideal octahedral configuration in these compounds, and the crystal structure determinations confirm this (cf. *Table 11.10*). The M–I distances are considerably longer than

Table 11.10 Bond lengths in $M(diars)_2I_2$ complexes

M	*M–I*/pm	*M–As*/pm
Ni	321	229
Pd	340	239
Pt	350	238

expected for single bonds (the Pd–I distance in a typical *trans* square-planar complex is 265 pm), and it would seem that the pair of electrons is occupying the d_{z^2} orbital and strongly repelling the two iodine atoms.

11.7 Co-ordination Numbers Other Than 6

The three methods outlined earlier for octahedral complexes, namely valence-bond, molecular-orbital and ligand-field, can be used to describe the bonding in complexes of other co-ordination numbers. In Chapter 8 we discussed in some detail the application of the valence-bond and molecular-orbital methods to such simple covalent molecules such as $BeCl_2$ (linear), BCl_3 (trigonal-planar) and CH_4 (tetrahedral), and very little modification is needed when these two theories are applied to complex compounds with the same symmetries, except that the bonding description may involve *d* orbitals as well as *s* and *p* orbitals.

The known complexes of co-ordination number 2 are formed by ions with d^0 or d^{10} configurations, so that the ligand-field approach is of no value, and this is also the position with the majority of compounds of co-ordination number 3. For co-ordination number 4, the ligand-field description will be discussed at some length and compared with the valence-bond one. In recent years a considerable number of complex compounds of co-ordination numbers 5, 7 and 8 has been fully characterized and the ligand-field approach to bonding is increasing in importance, although for many purposes the simple valence-bond approach is adequate.

11.8 Co-ordination Number 2

The only well-characterized complex compounds of co-ordination number 2 are those formed by the d^{10} ions, Cu^+, Ag^+ and Au^+, or d^0 ions such as $[UO_2]^{2+}$. In every case the compounds have the expected linear bonding arrangement.

The best known examples are probably the anionic cyanide complexes of univalent silver and gold, $[M(CN)_2]^-$, in which the two cyanide groups bond through carbon to the silver or gold ions. The bonding in these

$$\bar{N}C \longrightarrow \overset{+}{M} \longleftarrow \bar{C}N$$

(4)

anions is described most simply in valence-bond terms (4), with each cyanide ion donating a pair of electrons into vacant *sp* hybrid orbitals of the metal ion. Thus for Ag^+ we write

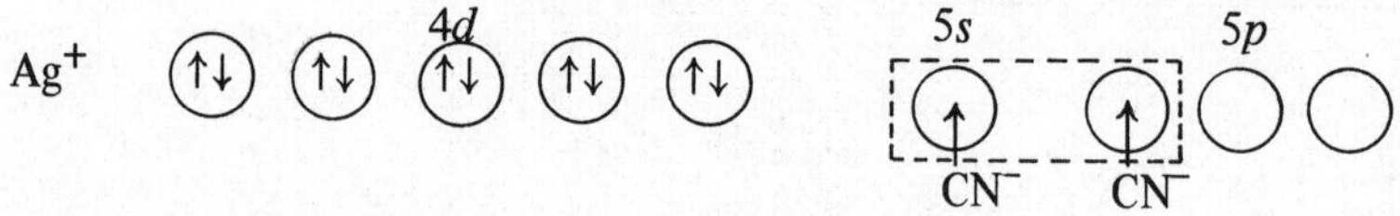

	4*d*					5*s*	5*p*		
Ag^+	(↑↓)	(↑↓)	(↑↓)	(↑↓)	(↑↓)	(↑)	(↑)	( )	( )
						CN^-	CN^-		

There is a similar description for $[Au(CN)_2]^-$, with the Au^+ ion using the 6*s*6*p* combination of orbitals. Of course, a molecular-orbital can be used, and as with $BeCl_2$ the linear arrangement of bonds can be described in terms of two three-centre bonding molecular orbitals, a similar combination of atomic orbitals and related energy diagram being concerned (p.90).

Another recently characterized linear anion is $[CuCl_2]^-$ (Newton *et al.*[19]). The ammine cations of silver(I) and gold(I), $[M(NH_3)_2]^+$, probably have the same linear bonding arrangement, although caution is necessary in assigning structures to species that have not been fully characterized. Thus since $[Ag(CN)_2]^-$ and $[Au(CN)_2]^-$ have linear structures it might be assumed that the analogous copper(I) anion, $[Cu(CN)_2]^-$, would also be linear; in practice this anion [in $KCu(CN)_2$] is polymeric with spirals of linked copper atoms with a co-ordination number of 3 (see Section 11.9).

The simple cyanides and thiocyanates of univalent silver also have polymeric structures, which in this case result in two-co-ordinate linear structures (5) and (6). It is interesting to note that while the cyanide gives a

$\rightarrow Ag-C\equiv N\rightarrow Ag-C\equiv N\rightarrow Ag-$

(5)

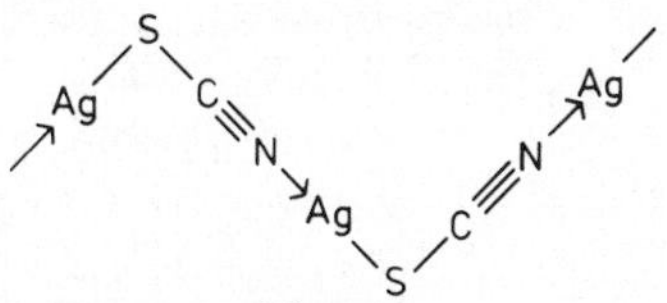

(6)

linear chain structure, the thiocyanate structure contains a zig-zag chain because of the 'tetrahedral' arrangement of electron pairs about the sulphur atoms (see *Chapter 10,* p.172).

Recently, the structures of a number of simple 1 : 1 adducts of gold(I) halides have been determined, and linear arrangements of Au–X and Au←L bonds confirmed. The adducts studied include $AuBr,AsPh_3$ (Einstein and Restivo[20]), and $(AuCl)_2,PhSCH_2CH_2SPh$ (7) (Drew and Riedl[21]). The linear arrangement is also found in the simple compounds of divalent mercury,

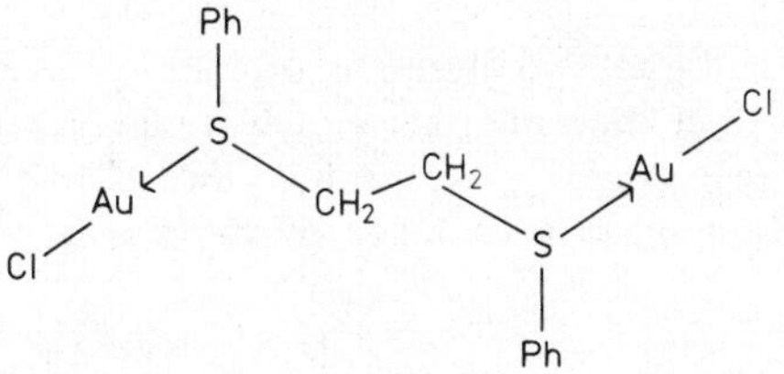

(7)

such as the halides (HgX_2, X = Cl, Br or I) and alkyl halides (RHgX, X = Cl or Br). Another recent example is $PhCH_2-Hg-SCPh_3$ (Bach *et al.*[22]).

It is clear that compounds of co-ordination number two are prevalent amongst d^{10} elements, particularly the bigger and heavier ones, and in a very simple way we can consider that the d^{10} charge clouds of the heavier atoms are deformed relatively easily on the approach of the ligands. Thus the two ligands would tend to approach from opposite directions, say along the z axis, and the d^{10} charge cloud would be distorted so as to increase the charge density in the xy plane. This increased charge density in the xy plane helps to prevent the close approach of other ligands. Orgel[23] has described this charge redistribution in terms of the hybridization of the s and d_{z^2} orbitals (cf. *Figure 11.17*). Thus:

$$\psi_1 = \phi_s + \phi_{d_{z^2}}$$

$$\psi_2 = \phi_s - \phi_{d_{z^2}}$$

ψ_2 is occupied by the two electrons originally considered to be in the d_{z^2} orbital. It must be emphasized, however, that this hybridization description is merely a convenient way of looking at the charge cloud reorientation and does not describe the bonding between the metal and the ligands.

We can make use of the unoccupied orbital (ψ_1), however, to produce a simple molecular-orbital scheme for the σ bonding, by combining it

(a) $\psi_1 = s + d_{z^2}$

(b) $\psi_2 = s - d_{z^2}$

Figure 11.17 sd *Hybridization*

with the p_z orbital and the two ligand σ orbitals. *Figure 11.18* shows the resultant energy level diagram; this closely resembles the diagram for the $BeCl_2$ molecule (*Figure 8.7*), except that the *sd* hybrid replaces the *sp* orbital, and the bonding axis is taken as the *z* axis.

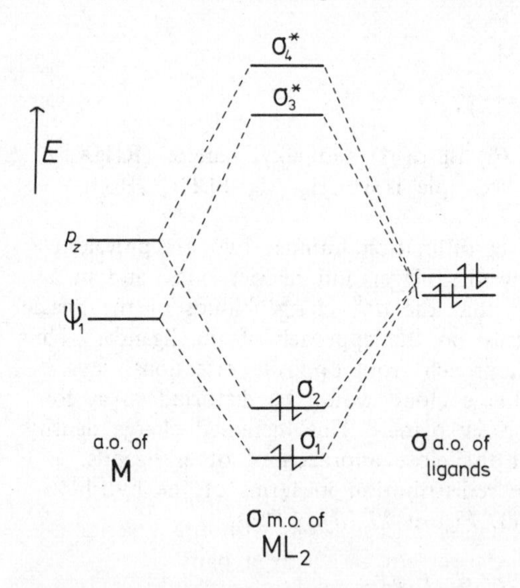

Figure 11.18 Molecular-orbital energy diagram for linear ML_2 *complexes*

We would expect a collinear arrangement of bonds in the complex oxo cations $[VO_2]^+$, $[MoO_2]^{2+}$ and $[UO_2]^{2+}$, in which the metal atom can be considered formally as having a d^0 configuration. Thus $[VO_2]^+$ can be written as $V^{5+}(O^{2-})_2$, with each doubly charged oxide ion donating two electrons into vacant *sp* hybrid orbitals of V^{5+}. However, a more satisfying valence-bond description is shown in (8), with σ bonds arising

$$O{=}\overset{+}{V}{=}O$$

(8)

from *sp* or *pd* hybrid orbitals, and π bonds resulting from the overlap of singly filled *d* orbitals of V^+ with singly filled *p* orbitals on the oxygen atoms:

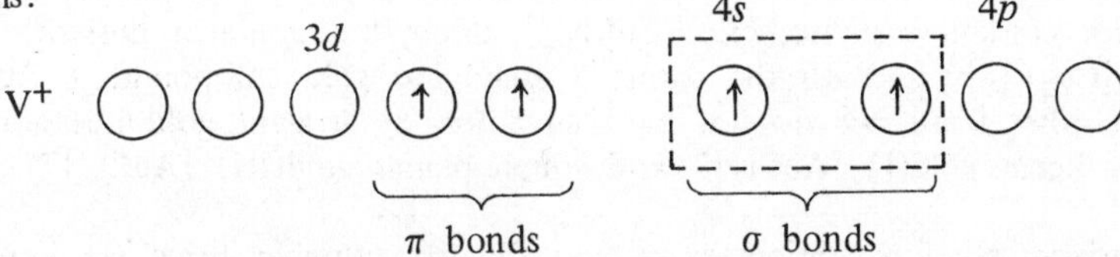

The molecular-orbital approach to the bonding in such oxo cations uses the *s* and d_{z^2} atomic orbitals for σ bonding along the *z* axis and the d_{xz} and d_{yz} orbitals for π bonding. For clarity, the σ- and π-bonding schemes are shown separately in *Figures 11.19(a)* and *11.19(b)*. The two bonding π molecular orbitals will be similar to those proposed for CO_2 (see p.148) and be distributed over all three atoms.

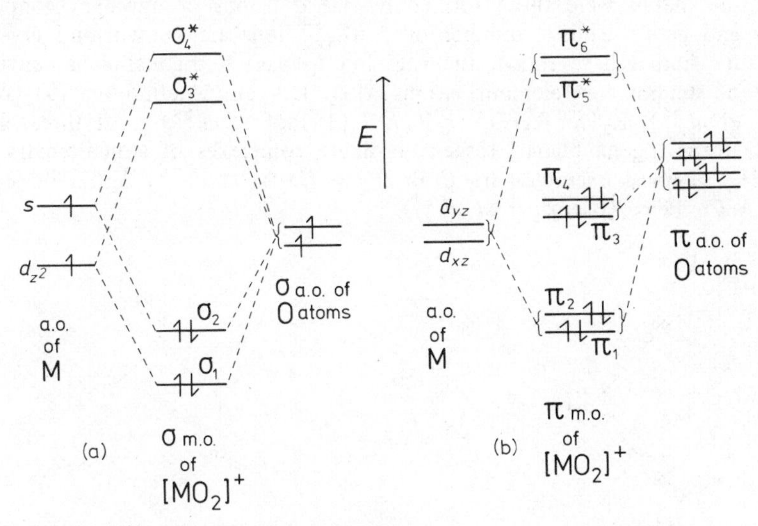

Figure 11.19 Molecular-orbital energy diagrams for $[MO_2]^+$*: (a)* σ*; (b)* π

Non-linear complex compounds of co-ordination number two are unusual. We discussed one such example, $[ICl_2]^+$, in the previous chapter (p.181), where we saw that the structure was influenced by the two lone pairs of electrons, which resulted in an angular $[ICl_2]^+$ ion, with the two lone pairs making up a 'tetrahedral' arrangement of electron pairs.

11.9 Co-ordination Number 3

Simple molecules such as BCl_3, in which the central atom is linked to three bonding pairs of electrons and no lone pairs, have trigonal-planar structures (see Chapter 10, p.90). However, as we have noted already, stoichiometry is not an infallible guide to molecular structure, and many apparently simple (MX_3) molecules may be more complex. Thus $AlCl_3$ is dimeric both in the vapour phase and in solution, with a bridging pair of chlorine atoms giving each aluminium an environment of four chlorines, while the solid is still more polymeric with chlorine-sharing, resulting in six-co-ordinate aluminium. Similarly, a salt such as $KCuCl_3$ does not contain a monomeric $[CuCl_3]^-$ anion, but a planar dimeric anion $[Cu_2Cl_6]^{2-}$. With the formally analogous gold compound, $CsAuCl_3$, on the other hand, we do not have an anion of divalent gold but a mixture of linear gold(I) $[AuCl_2]^-$ and square-planar gold(III) $[AuCl_4]^-$ anions.

Comparatively few complexes of co-ordination number three are known, although there is little doubt that a diligent search would reveal others. As far as the Main Group elements are concerned, there are species such as $[SnCl_3]^-$, $[H_3O]^+$ and $[Me_2S]^+$ which have pyramidal configurations (see Chapter 10), and adducts such as $BeMe_2,NMe_3$; the latter molecule is monomeric in the vapour phase, and presumably has a trigonal-planar structure. Until recently the only known three-co-ordinate complexes of transition metals were those formed by the d^{10} ions of univalent copper, silver and gold. Thus a number of $[CuL_3]^+$ ions are known, and crystal-structure studies have shown the Cu^+ ion to have a trigonal-planar environment of sulphur (or selenium) atoms when L = ethylenethiourea (9) (Weinmyer *et al.*[24]), R_3PS, R_3PSe or R_3AsS (Tielhof *et al.*[25]). All three elements form trigonal-planar, three-co-ordinate complexes of stoichiometry MX,2L. Typical examples are $CuBr,2PPh_3$ (Davis *et al.*[26]) $AgI,2PPhMe_2$ and $AuCl,2PPh_3$ (Baenziger *et al.*[27]).

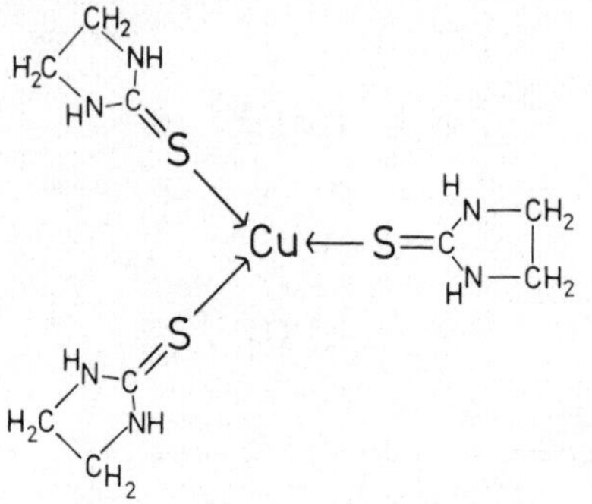

(9)

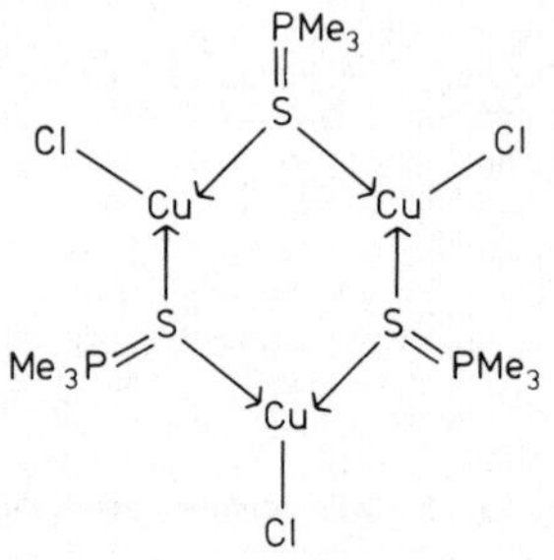

(10)

As we saw in the last section, $KCu(CN)_2$, does not contain the linear $[Cu(CN)_2]^-$ ion, but has a polymeric chain structure in which each copper atom is three-co-ordinate. A similar situation exists with $CuCl,SPMe_3$ (Tielhof *et al.*[28]), which has the trimeric structure shown in (10). Each copper atom has a trigonal-planar environment, and the sulphur atoms are pyramidal, with the six-membered ring adopting a chair symmetry.

It is interesting to note that the majority of the three-co-ordinate complexes are formed with ligands containing second- or third-row donor atoms, namely phosphorus, sulphur and selenium, linked to other bulky groups. Although these donor atoms are noted for their ability to form π bonds, it seems likely that this is not a major factor in establishing three-co-ordinatior but rather that the ligands are bulky so that not more than three of them can be readily accommodated around each metal. Evidence in support of this comes from complexes formed by the phosphorus donor ligand (11) in which the attached groups have been constrained away from the metal. Thus Cu, Ag and Au all form tetrahedral $[ML_4]ClO_4$ complexes.

(11) (12)

Steric requirements also appear to play a major role in determining the structure of compounds incorporating bulky $-N(SiMe_3)_2$ groups. Thus a number of transition metal compounds have been prepared with stoichiometry $M[N(SiMe_3)_2]_3$, and in the titanium, vanadium, chromium and iron compounds it has been shown that each transition metal has a trigonal-planar arrangement of nitrogens (12) (Bradley[29], Hursthouse and Rodesiler[30]). Closely related trigonal-planar complexes include the high-spin cobalt(II) compound $Co[N(SiMe_3)_2]_2,PPh_3$ and the nickel(I) complex $Ni[N(SiMe_3)_2],2PPh_3$ (Bradley *et al.*[31]).

The trigonal-planar stereochemistry can be discussed along the same lines as those proposed for linear complexes. Thus as the three ligands approach the metal atom they will tend to be as far apart from one another as possible, say in the *xy* plane. The consequent deformation of the d^{10} charge cloud gives an increased charge density along the *z* axis, which may be described by saying that the d_{z^2} electrons now occupy an *sd* hybrid orbital (that given by ψ_1 on p.229). The bonding can be described on a simple molecular-orbital basis by using the other *sd* hybrid (ψ_2) and the p_x and p_y orbitals of the metal together with an orbital from each ligand. *Figure 11.20* shows the simple molecular-orbital diagram, and it may be seen that this closely resembles the diagram given for BCl_3 (*Figure 8.9*) except that an *sd* hybrid replaces the *s* orbital used for boron.

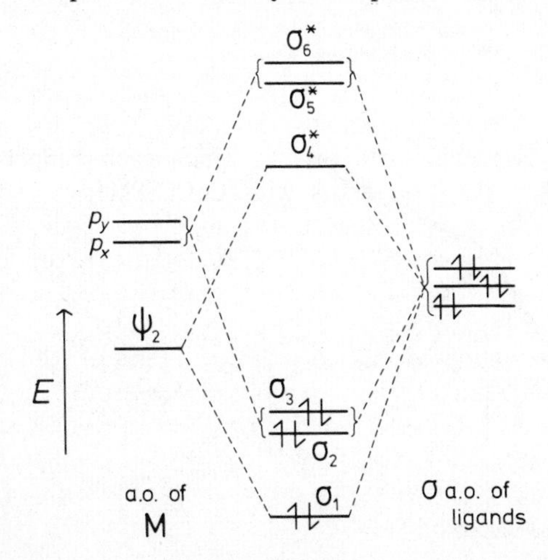

Figure 11.20 Molecular-orbital energy diagram for trigonal-planar ML_3 complexes

11.10 Co-ordination Number 4

The principal stereochemistries observed experimentally are tetrahedral and square-planar, although various distorted versions of these two structures may be found when all the ligands are not the same or if the *d* levels are filled unevenly. If we limit our discussion for the moment to undistorted complexes, then the valence-bond description of tetrahedral bonding involves sp^3 or sd^3 (d_{xy}, d_{xz}, d_{yz}) hybridization schemes, or a mixture of the two; square-planar hybridization arises from the combination of $d_{x^2-y^2}$, s, p_x and p_y atomic orbitals, the resulting hybrids having axes lying in the *xy* plane.

The tetrahedral arrangement is found for complexes of divalent beryllium, zinc, cadmium and mercury, and of tervalent boron, aluminium and gallium. These elements have no *d* orbitals comparable in energy with the valence shell *s* and *p* orbitals, so that if the co-ordination is restricted to four then a tetrahedral arrangement is to be expected. (With some of these metals the co-ordination number can be increased to 5 or 6 through the incorporation of the high-energy *d* orbitals into the bonding scheme, e.g. $AlH_3,2NMe_3$.) If, however, *d* levels are energetically available, then the four-co-ordinate complex may be either tetrahedral or square-planar. It is not possible to predict from valence-bond calculations which configuration will have the lowest energy for a given metal ion and ligands, and most quantitative work has been based on either the molecular-orbital or the ligand-field methods.

The molecular-orbital method uses the same orbitals as the valence-bond approach, but compounds them with ligand orbitals to give molecular orbitals. The orbital symmetries are such that the $d_{x^2-y^2}$ and *s* orbitals of the metal can combine with orbitals on all four ligands, but the p_x and p_y orbitals can only overlap with the ligands orientated along the *x* and *y* axes, respectively.

We shall describe the ligand-field theory of four-co-ordination in rather more detail since it has been used extensively to describe these complexes. *Figure 11.21* illustrates a tetrahedral bond configuration, and shows that

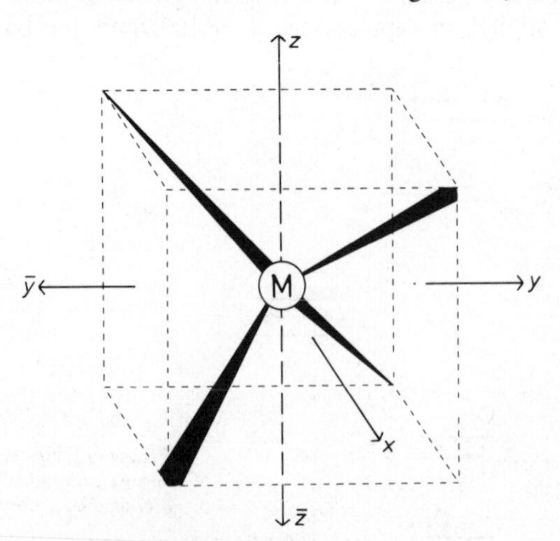

Figure 11.21 Tetrahedral bond configurations

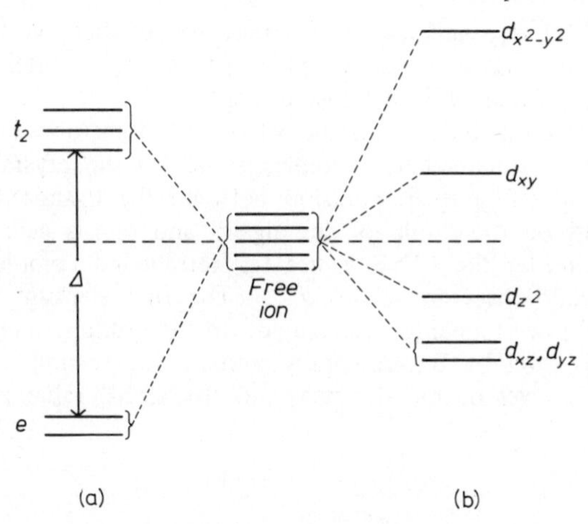

Figure 11.22 d-*Orbital splitting in (a) tetrahedral ligand field; (b) square-planar ligand field (not to scale)*

the x, y and z axes bisect the angles between pairs of ligands. If we now apply the argument already outlined on p.210 for octahedral complexes, we see that the d_{z^2} and $d_{x^2-y^2}$ orbitals of the metal ion are further away from the ligand electrons than are the electrons in the d_{xy}, d_{yz} and d_{xz} orbitals. In this case, therefore, the effect of the ligand field is to split the d levels, giving a stabilized doublet level, e, of lower energy than the undistorted d orbitals, and less stable triplet levels, t_2, of higher energy [see *Figure 11.22(a)*]. The field produced at the metal ion by four tetrahedrally arranged ligands is, however, smaller than that produced by six of the same ligands at the same distance in an octahedral configuration; thus, if Δ is the splitting for octahedral co-ordination the value for tetrahedral co-ordination is considered to be about $4\Delta/9$. We saw on p.211 that the t_{2g} and e_g levels in an octahedral field have energies of $-2\Delta/5$ and $+3\Delta/5$, respectively, relative to a weighted mean zero. The corresponding values for the same ligands in a tetrahedral arrangement (where the doublet e level now has the lowest energy) will therefore be

$$e = -3/5 \times 4\Delta/9 = -0.27\Delta$$

and

$$t_2 = +2/5 \times 4\Delta/9 = +0.18\Delta$$

As we saw earlier, the extent to which the ligands stabilize the d orbitals is called the 'crystal field stabilization energy', abbreviated CFSE.

The splitting produced by a square-planar arrangement of ligands is shown in *Figure 11.22(b)*. The $d_{x^2-y^2}$ level is the least stable and is of much higher energy than the others. The d_{xy} orbital has the next highest energy, since the axes of its lobes all lie in the plane of the ligands. The d_{xz} and d_{yz} orbitals are degenerate, since they must be influenced by the ligand field to the same extent, but the experts differ on the energy of these orbitals relative to that of the d_{z^2} orbital. We shall use the scheme described by Basolo and Pearson where the d_{z^2} orbital is given an energy higher than that of d_{xz} and d_{yz}. This may be justified on the ground that the 'collar' of charge in the xy plane of the d_{z^2} orbital (see p.210) gives a greater repulsion with the ligand field.

The principal factors that determine whether a transition metal ion will form tetrahedral or square-planar complexes are (a) the crystal field stabilization energy, and (b) mutual repulsion between the ligands – which, in turn, will depend on the 'bulk' of the ligand and on its electronegativity. We shall first consider the CFSE values for tetrahedral complexes of metal ions with configurations from d^1 to d^{10}. The first electron will occupy the d orbital of lowest energy, i.e. one of the e orbitals, and in doing so will stabilize this level by 0.27Δ energy units. The second d electron enters the second level of the doublet, and the CFSE value goes up to

$$2 \times 0.27\Delta = 0.54\Delta$$

There are two alternatives for the third electron; it may go either into the e levels, in which case it must 'pair' with an electron already in residence, to give a total CFSE of 0.81Δ, or it may go into one of the higher t_2 levels, giving a state with three unpaired electrons and a CFSE value of $(0.54 - 0.18)\Delta = 0.36\Delta$. There is, therefore, a competition between the additional crystal field stabilization produced by 'spin-pairing', and the operation of Hund's rules, similar to that discussed earlier for octahedral complexes. The choice may be determined by the strength of the ligand field; if the field is weak the CFSE value is small, and the third electron goes into the t_2 orbital to give a 'spin-free' arrangement, whereas strong ligand fields favour 'spin-pairing' in the e level. The argument outlined here can be extended to other d electron configurations ($d^4 \rightarrow d^{10}$) and to square-planar four-co-ordination. The relevant CFSE values are shown in *Table 11.11*.

Table 11.11 shows clearly that if the CFSE were the only factor to be considered, then all four-co-ordinate complexes would be square-planar, except the d^0, d^5 (spin-free) and d^{10} ones, where the CFSE is zero. However, the mutual repulsion between the ligands must also be taken into account, and this repulsion will be significant when the ligands are either electronegative or bulky. This steric factor will be particularly important when the field is weak and the CFSE correspondingly small, but repulsion will be rather less important in complexes formed by transition elements of the second and third rows where Δ is larger (compare tetrahedral $[NiCl_4]^{2-}$ and square planar $[PtCl_4]^{2-}$). Tetrahedral complexes are known for all divalent elements of the first transition series, although up till now none with a spin-paired configuration has been characterized. Such spin-pairing is only likely to occur for large values of Δ (produced either by ligands giving a strong field or with metals of the second and third transition series), and we shall see that when Δ is large we get square-planar rather than tetrahedral complexes. Another factor to note is the tendency for compounds (especially of the heavier elements) to polymerize.

We now discuss some examples of four-co-ordination for metal ions of different d electron configurations.

Table 11.11 Crystal field stabilization energies (in Δ units) for tetrahedral and square-planar complexes

Number of d electrons	Weak field			Strong field		
	Tetrahedral	*Square-planar*	*Difference (square-planar) – (tetrahedral)*	*Tetrahedral*	*Square-planar*[*]	*Difference (square-planar) – (tetrahedral)*
0	0	0	0	0	0	0
1	0.27	0.51	0.24	0.27	0.51	0.24
2	0.54	1.02	0.48	0.54	1.02	0.48
3	0.36	1.45	1.09	0.81	1.45	0.64
4	0.18	1.22	1.04	1.08	1.96	0.88
5	0	0	0	0.90	2.47	1.57
6	0.27	0.51	0.24	0.72	2.90	2.18
7	0.54	1.02	0.48	0.54	2.67	2.13
8	0.36	1.45	1.09	0.36	2.44	2.12
9	0.18	1.22	1.04	0.18	1.22	1.04
10	0	0	0	0	0	0

*Assuming that even in a strong field the d_{z^2} and the d_{xz} and d_{yz} levels are sufficiently close for one electron to go into each orbital before pairing occurs. Slightly different values will be obtained if the electrons are allowed to pair up in the d_{xz} and d_{yz} orbitals before entering the d_{z^2} orbital

11.10.1 d^0, d^5 (SPIN-FREE) AND d^{10}

Here the CFSE is zero because the orbitals are either empty, all singly filled or all doubly filled, e.g. if each orbital contains one electron, the CFSE value is

$$[2 \times (-0.27\Delta) + 3 \times (+0.18\Delta)] \quad = \quad 0$$

In each case the ligands will take up the tetrahedral position, which minimizes the repulsion between them. Tetrahedral arrangements are thus found in $TiCl_4$, $[FeCl_4]^-$ and $[ZnX_4]^{2-}$, although the latter two are somewhat distorted because of the influence of the cations in the crystal lattice.

11.10.2 d^1, d^6

With either one or six *d* electrons, the CFSE is very small so we should expect the repulsion factor to be the dominating one, and indeed a tetrahedral configuration is found for the d^1 compound VCl_4; NbI_4, however, is polymeric with octahedrally co-ordinated niobium atoms forming long chains through shared edges. Four-co-ordinate complexes of divalent iron (d^6) have the expected tetrahedral configuration. The anion in $[NMe_4][FeCl_4]$ is a somewhat flattened tetrahedron (Lauher and Ibers[32]), but this distortion is the same as in the analogous zinc salt and may be attributed to crystal packing requirements. Another tetrahedral complex is $Fe(SPMe_2NPMe_2S)_2$ (13) in which the iron has a somewhat distorted tetrahedral environment of four sulphur atoms (Churchill *et al.*[33]).

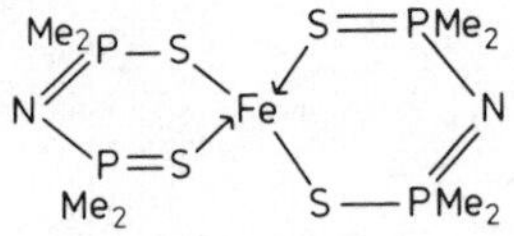

(13)

11.10.3 d^2, d^7

The CFSE is still rather small, and it is unlikely to be the deciding factor for first row transition elements. No d^2 complex has been properly characterized up until now, but the tetrahedral configuration is certainly present in many divalent cobalt complexes (d^7 configuration). Thus the tetrahedral arrangement is found both in anions, $[CoX_4]^{2-}$ (X = Cl, Br, I or NCS), and in neutral compounds of the type CoX_2,2L (with X = halogen, and L = pyridine, *p*-toluidine or triphenylphosphine oxide).

Low-spin, four-co-ordinate complexes of divalent cobalt are invariably square-planar, two typical examples being the maleonitrile-dithiolate complexes $[NR_4]_2[Co(S_2C_2(CN)_2)_2]$ (14) and the dithioacetylacetone complex, $Co(sacsac)_2$ (15) (Beckett and Hoskins[34]); in each case the cobalt atom is linked to four sulphur atoms which form a square-planar environment.

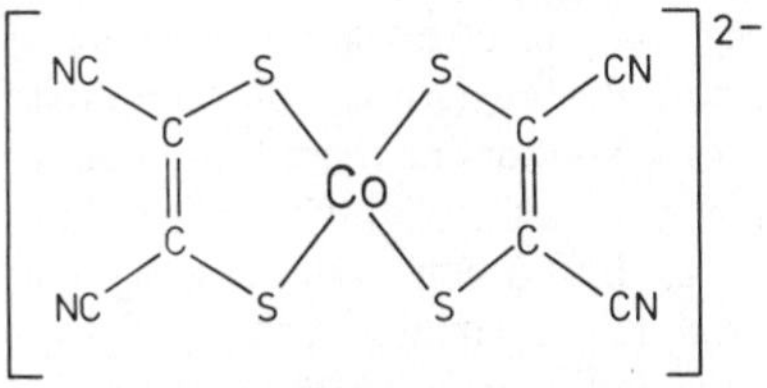

(14)

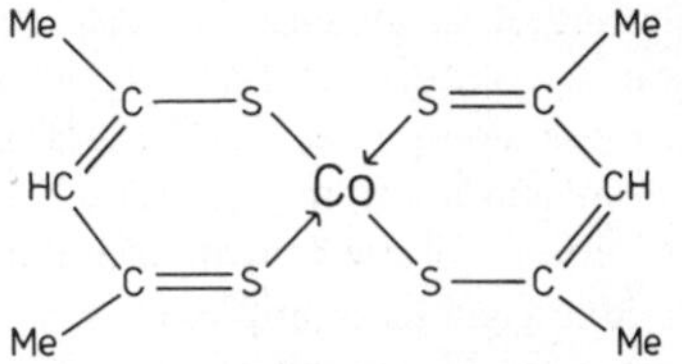

(15)

11.10.4 d^3, d^4

In view of the CFSE differences for square-planar and tetrahedral complexes of d^3 and d^4 ions, we should expect all complexes to be square-planar, although very little experimental evidence is available. A recent determination of the structure of the chromium(II) complex $Cr[N(SiMe_3)_2]_2$,2thf (Bradley *et al.*[35]) shows the chromium atom to have a square-planar environment of two nitrogens and two oxygens; thus even though the bulky groups favour a tetrahedral arrangement, the CFSE is more significant and the square-planar distribution is favoured.

With the heavier elements of the second and third transition series, the complexes of ML_4 stoichiometry are invariably polymeric. Thus $TcCl_4$ (d^3) has an octahedral arrangement (16) through sharing of pairs of chlorines between neighbouring technetium atoms and $[ReCl_4]^-$ is a dimer.

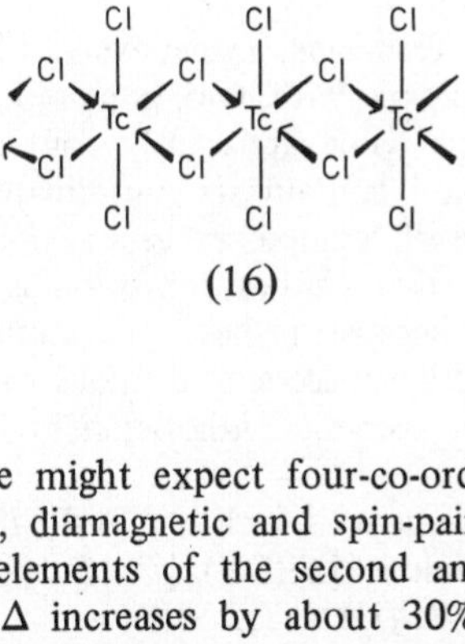

(16)

11.10.5 d^8

On grounds of CFSE we might expect four-co-ordinate complexes of d^8 ions to be square-planar, diamagnetic and spin-paired. This is especially likely with the heavier elements of the second and third row, because for a given complex, ML_4, Δ increases by about 30% on going from a first-row element to a second-row element. In practice we find that all four-co-ordinate complexes of divalent palladium and platinum are square-planar, even for such anions as $[PdCl_4]^{2-}$ and $[PtCl_4]^{2-}$ in which the chloride ligands are electronegative and produce a comparatively small ligand field. Hence the stereochemical problems associated with palladium-(II) and platinum(II) complexes are concerned with such matters as *cis–trans* isomerism in complexes of the type MX_2,2L and the mechanism of their reactions (see the *trans* effect, p.282).

With divalent nickel, however, we have a much smaller metal ion, and factors such as repulsions between bulky and electronegative ligands play a much more important role. The position is by no means clear cut, but in general we find that square-planar complexes are more common, with tetrahedral arrangements frequently arising with electronegative and bulky ligands. Before crystallographers established the structures of a number of these complexes, there was a simple rough guideline based on colour and magnetic moment. Thus nickel(II) complexes of stoichiometry NiL_4 are either yellow-brown and diamagnetic, or blue-green and paramagnetic; the diamagnetic complexes were considered to be square-planar and the paramagnetic complexes tetrahedral. If we look again at the simple ligand-field splitting diagrams for square-planar and tetrahedral nickel complexes (*Figure 11.23*) we can see that the electron arrangements give rise

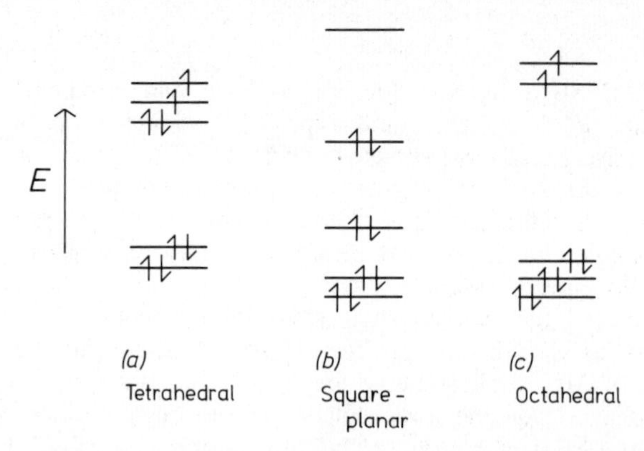

Figure 11.23 *d-Electron arrangements for (a) tetrahedral, (b) square-planar and (c) oc hedral* d^8 *complexes*

to zero and two unpaired electrons, respectively. However, it must not be assumed that all the complexes with two unpaired electrons have a tetrahedral configuration because some apparently four-co-ordinate nickel(II) complexes are polymeric and contain six-co-ordinate nickel. As *Figure 11.23*(*c*) shows, octahedral d^8 complexes can also have two unpaired electrons. One such complex that caused confusion is $Ni(acac)_2$, which was believed to be tetrahedral because it has a magnetic moment corresponding to two unpaired electrons; a subsequent crystal-structure determination showed the complex to be trimeric with octahedrally co-ordinated nickel atoms.

The square-planar arrangement is well established in many nickel(II) complexes, including the anions $[Ni(CN)_4]^{2-}$ and $[Ni(S_2C_2(CN)_2)_2]^{2-}$, and bis(ethylmethylglyoxime)nickel(II) (17).

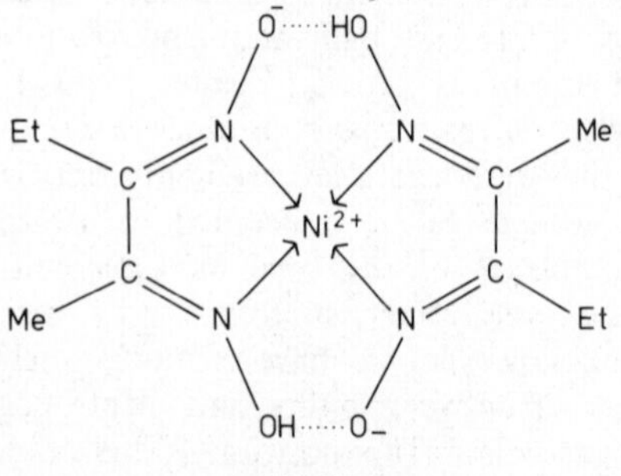

(17)

Quite a number of tetrahedral nickel(II) complexes have been characterized in recent years, including the anions $[NiCl_4]^{2-}$, $[NiBr_4]^{2-}$ and $[NiI_4]^{2-}$, which involve electronegative halide ligands that provide relatively small ligand fields. Some complexes involving bulky tertiary phosphine ligands are also tetrahedral, as for instance with $NiCl_2,2PPh_3$ and $[NiBr_3(PBu^t_3)]^-$ (Alyea *et al.*[36]), but this is by no means always the case, since $NiBr_2,2PEt_3$ has a *trans* square-planar arrangement. For one particularly interesting compound, $NiBr_2,2PPh_2$Benzyl, which is a green paramagnetic complex (μ = 2.7 BM), the crystal structure has revealed the presence of two tetrahedral and one square-planar nickel species in each unit cell. Hence with the bulky ligands there is a fine balance between square-planar and tetrahedral structures.

11.10.6 d^9

Bivalent copper resembles nickel in that CFSE considerations result in most complexes being square-planar. The only tetrahedral copper(II) complexes fully characterized are the $[CuBr_4]^{2-}$ and $[CuCl_4]^{2-}$ anions, where it seems that the repulsion between the electronegative halide ions is sufficient to overcome the loss of CFSE, particularly since the ligand field is comparatively small. However, once again the factors are finely balanced, and while a flattened tetrahedral arrangement is found in salts formed with the large cations Cs^+, $[Me_4N]^+$ and $[Et_3NH]^+$, the ammonium salt contains a square-planar anion (Cu–Cl = 230 pm) linked into a tetragonally distorted octahedral structure through two longer *trans* Cu–Cl links (Cu–Cl = 279 pm). The salt $[PhCH_2CH_2NH_2Me][CuCl_4]$ exists in two forms, a green one with a square-planar anion (Cu–Cl = 225 and 228 pm), and a yellow one in which the anion is a flattened tetrahedron with Cu–Cl = 219 and 223 pm (Harlow *et al.*[37]).

Another set of complex anions that can be either tetrahedral or square-planar are those found in compounds of stoichiometry $MCuCl_3$. These complexes do not contain monomeric $[CuCl_3]^-$ ions, but the dimeric $[Cu_2Cl_6]^{2-}$ anions. With the large cations such as $[Ph_4P]^+$ and $[Ph_4As]^+$, the $[Cu_2Cl_6]^{2-}$ anions (18) adopt a flattened tetrahedral, chlorine-bridged structure with Cu–Cl terminal bonds around 220 pm, and longer (230 pm) bridging bonds (Textor *et al.*[38], Willett and Chow[39]). With smaller cations such as Li^+, K^+ and NH_4^+ the dimeric anions have a planar arrangement (19), being linked through longer chlorine bridges (~310 pm) to provide a distorted octahedral structure. Thus the units are stacked so that the Cu atom in one unit comes between Cl atoms of the units immediately above and below.

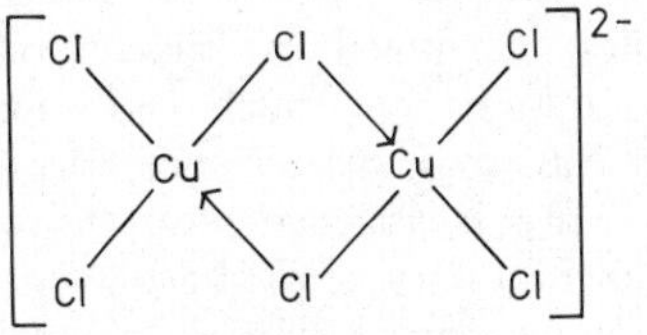

(18)

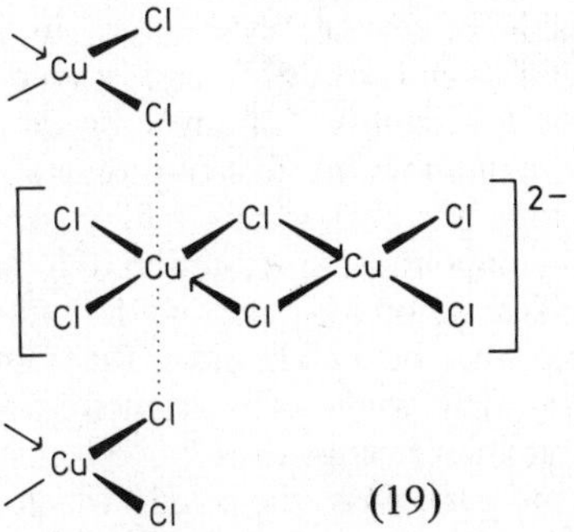

(19)

It is worth commenting further on this relationship between square-planar and distorted octahedral complexes. Earlier we pointed out that the d^9 configuration led to an uneven filling of the e_g orbitals with two electrons in the d_{z^2} orbital repelling the two z axis ligands, so producing a 4/2 arrangement. A somewhat more sophisticated interpretation invoked the Jahn–Teller theorem with a splitting of the e_g levels; at the same time the t_{2g} levels split into a doublet and a singlet, and in the limiting case in which the two z axis ligands are removed so far away as to exert no influence we get the ligand-field diagram for square-planar complexes. *Figure 11.24* shows this relationship between the splitting in square-planar and tetragonally distorted octahedral complexes.

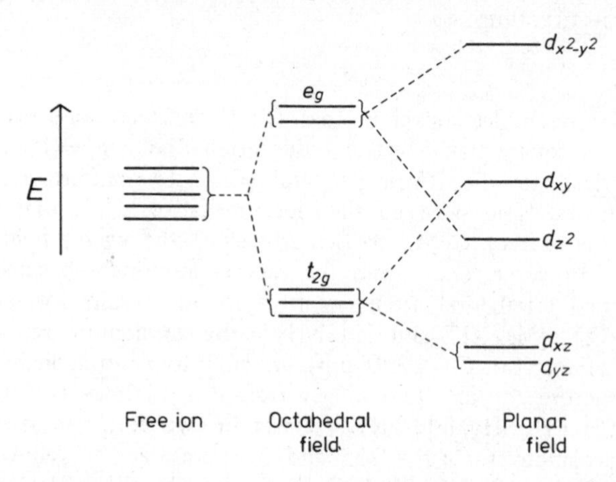

Figure 11.24 d-*Orbital splitting for octahedral and square-planar fields*

The observed distortion of the tetrahedral $[CuCl_4]^{2-}$ and $[Cu_2Cl_6]^{2-}$ anions is to be expected, since the t_2 levels (which point towards the ligands) are unevenly filled.

11.11 Co-ordination Number 5

In recent years considerable progress has been made in the characterization of complexes of co-ordination number five, and quite a number of crystal-structure determinations has been carried out. Excellent summaries are available in several reviews and in particular by those of Muetterties and Schunn[40] and Wood[41], and readers interested in full details are referred to these accounts. It must be emphasized, however, that even though quite a number of five-co-ordinate molecules has now been characterized, compounds showing this odd co-ordination number are much less common than compounds of co-ordination numbers four and six. Hence care must be taken before a co-ordination number of five is assigned, and the evidence must be much more than simple stoichiometry. We have seen previously that while PCl_5 is monomeric in the gas phase with a trigonal-bipyramidal arrangement of chlorines about the phosphorus atom, this is not the case with the solid, which consists of tetrahedral $[PCl_4]^+$ and octahedral $[PCl_6]^-$ ions. Some solid pentahalides, such as $NbCl_5$, $TaCl_5$ and $MoCl_5$, contain dimeric species (20). Other molecules that illustrate

(20)

the dangers of assigning structures on stoichiometry alone include the salts Cs_3MnCl_5 (Goodyear and Kennedy[42]) and Cs_3CoCl_5, which contain Cl^- and tetrahedral $[MCl_4]^{2-}$ ions, rather than $[MCl_5]^{3-}$ ions.

There are two basic configurations that can be adopted by complex compounds of co-ordination number five, the trigonal-bipyramidal and the square-pyramidal (see *Figure 11.25*). We pointed out in Chapter 10 that

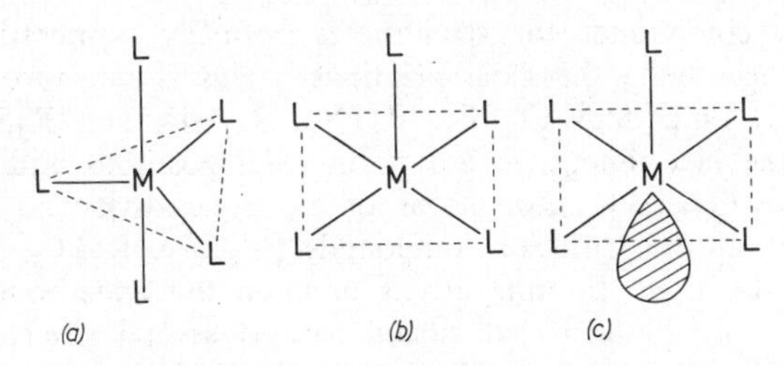

***Figure 11.25** Configurations for complexes of co-ordination number 5: (a) trigonal-bipyramidal; (b) square-pyramidal; (c) octahedral*

both structures could be described in terms of dsp^3 hybridization, with the d_{z^2} orbital being involved in the combination describing the trigonal-bipyramidal (TBP) arrangement; for the square-pyramidal arrangement (SP) the $d_{x^2-y^2}$ orbital is used in the hybridization scheme. For those readers who prefer to use a hybridization approach for a discussion of molecular shape, it may help if we describe the TBP structure in terms of two different sets of hybrids, sp^2 (giving the three trigonal-planar bonds) and dp (giving the two linear axial bonds). In the same way the SP structure can be viewed as $d_{x^2-y^2}sp^2$ hybrids giving the four basal bonds and p_z giving the apical bond.

All the evidence suggests that there is very little to choose between the alternative structures, there being a very fine balance determined by the relative influence of various factors including:

(i) electronegativity of the ligands;
(ii) steric requirements of the ligands;
(iii) mutual repulsion of bonding pairs and lone pairs of electrons;
(iv) mutual repulsion of bonding pairs and any *d*-shell electrons;
(v) CFSE

For the moment let us restrict our discussion to compounds of the non-transition elements so that we do not have to worry about the complications associated with the presence of *d* electrons (e.g. factors iv and v). In Chapter 10 we discussed the compounds of the Group V elements and saw that simple MX_5 compounds normally adopt a TBP configuration, although $SbPh_5$ is a notable exception. The TBP arrangement is also found for simple mono- and di-substituted compounds of the Group V elements, namely MX_4Y and MX_3Y_2; the various isomers that are possible are shown in *Figure 11.26*.

Figure 11.26 Isomers for trigonal-bipyramidal MX_4Y and MX_3Y_2

With the MX_4Y compounds the odd Y group normally occupies an axial position if it is more electronegative than X, as with $PFCl_4$ and $SbFCl_4$; when Y is very bulky, however, it is found in an equatorial position, typical examples being $MePF_4$ and Et_2NPF_4.

With MX_3Y_2 compounds the structure is normally symmetrical with the two Y groups occupying the axial positions. This is not too unexpected for the difluoro compounds Cl_3PF_2, R_3PF_2, R_3AsF_2 and R_3SbF_2, since here we have the electronegative groups in axial positions and bulky groups in equatorial positions. Considerations of electronegativity do not appear to be important in the trifluoro compounds (Cl_2PF_3, R_2PF_3 and Ph_2AsF_3), however, since the three fluorine atoms bond in the equatorial plane.

In Chapter 10 we noticed that silicon formed several five-co-ordinate compounds with TBP structures and that in Me_3SnOH the three methyl groups formed a trigonal plane about tin, with the OH group bridging neighbouring tins to give a polymeric TBP arrangement. Trimethyl-lead acetate has a similar structure (21) (Sheldrick and Taylor)[43] to Me_3SnOH, with acetate groups acting as bridges. The TBP arrangement is also found for two five-co-ordinate tin complexes that have been characterized recently, namely $[SnCl_3Me_2]^-$ (22) (Buttenshaw *et al.*[44]) and the compound (23) in which the chelating dibenzoylmethane ligand is bonded to the Ph_3Sn grouping (Bancroft *et al.*[45]). In the anion the axial positions are occupied by the

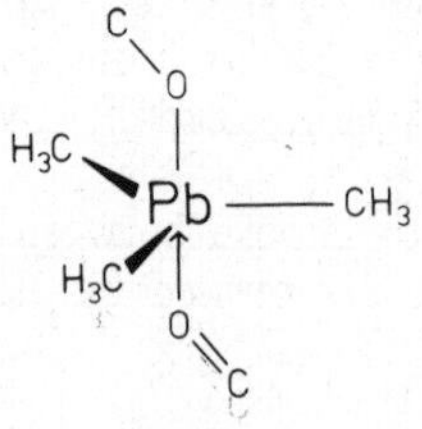

(21)

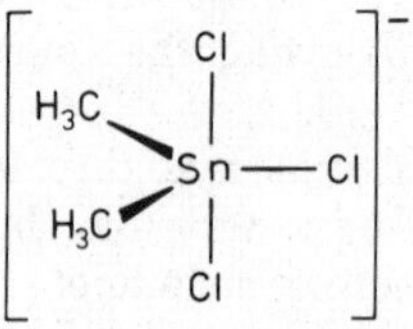

(22)

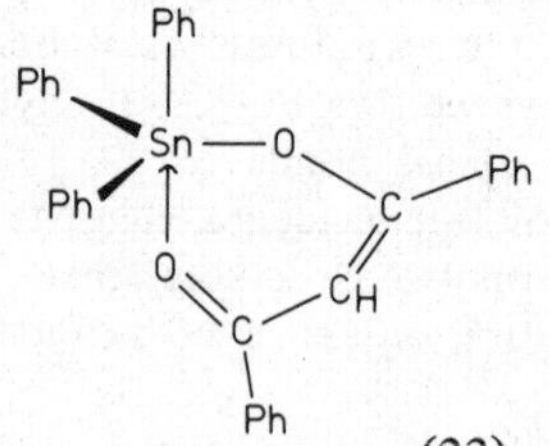

(23)

more electronegative Cl atoms, with bulky CH_3 groups in equatorial positions; the axial Sn–Cl bonds are significantly longer than the equatorial Sn–Cl bond (257 pm and 241 pm, respectively).

When we come to consider known five-co-ordinate compounds of the transition elements, then the position is much more complicated because of the possible influence of *d* electrons. Bonding can still be described

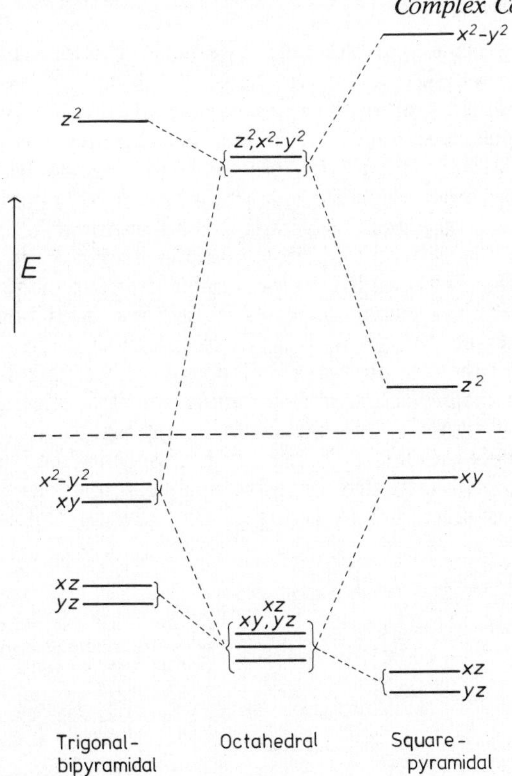

Figure 11.27 Ligand-field splitting diagram for five-co-ordinate complexes

in valence-bond terms (dsp^3 or sp^3d hybrid orbitals) but crystal-field and molecular-orbital descriptions are more useful now that the range of fully characterized five-co-ordinate complexes is so extensive. For many purposes we can use the simple crystal-field splitting diagram (*Figure 11.27*). The observed magnetic properties can be accounted for on the basis of these diagrams, and allowance made for the stability that a particular configuration gains through CFSE. *Table 11.12* lists CFSE values for both TBP and SP structures calculated on the basis of *Figure 11.27*, the

Table 11.12 Five-co-ordinate complex CFSE (in Δ units) for TBP and SP arrangements

No. of d *electrons*	*TBP/* Δ	*SP/* Δ	Δ *difference (SP − TBP)*
0	0	0	0
1	0.27	0.46	0.19
2	0.54	0.92	0.38
3	0.62	1.01	0.39
4	0.70	0.92	0.22
5	0.97	1.38	0.41
6	1.24	1.84	0.60
7	1.32	1.93	0.61
8	1.40	1.84	0.44
9	0.70	0.92	0.22
10	0	0	0

assumption being made that the four lowest-energy *d* orbitals are filled singly, and then doubly, the highest-energy *d* orbitals not being occupied at all until the other four orbitals are completely filled. This is the commonest arrangement, but we shall see that spin-free complexes are occasionally found for d^5 and d^8 configurations. It can be seen that for all configurations (other than d^0, spin-free d^5, and d^{10}) the CFSE associated with an SP arrangement is greater than that for a TBP ligand distribution, although the difference is not anything like as great as that we noted between the two possible four-co-ordinate arrangements. Moreover, CFSE for the SP arrangement decreases if the four basal bonds are depressed so that the L_{apical}–M–L_{basal} angle increases from 90 to 100° or more. Thus we can sum up the CFSE position by saying that it is unlikely to be a major factor in determining which configuration is adopted by a given set of ligands. In practice there appears to be very little difference in energy between the TBP and SP arrangements, and one structure is readily transformed into the other by small deformations of bond angles; this relationship is illustrated in *Figure 11.28*. Readers interested in a more

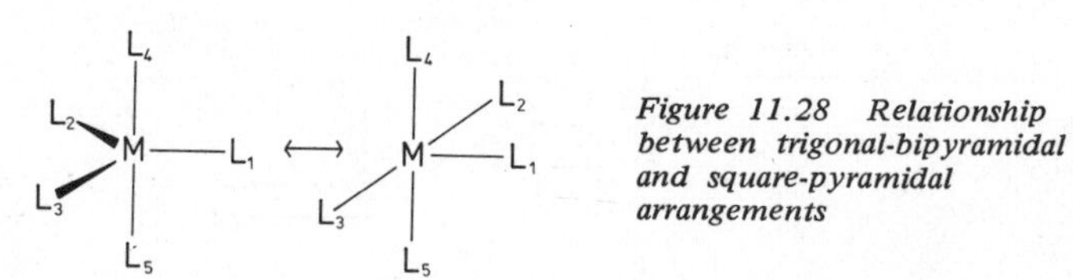

Figure 11.28 Relationship between trigonal-bipyramidal and square-pyramidal arrangements

detailed theoretical discussion are referred to a recent paper by Rossi and Hoffmann[46] that provides a full molecular-orbital description of both structures and their inter-relation.

In the following discussion a representative selection of structures is listed in *Tables 11.13–11.17*. The majority of the compounds are referred to in Wood's review[41] and from this readers can obtain original references to most papers prior to 1972; only more recent references are provided. Two other reviews of considerable interest are that by Morassi and co-workers[47], which provides a very detailed discussion of five-co-ordinate compounds of divalent iron, cobalt and nickel, and that by Hoskins and Whillans[48] which weighs up the factors that may determine the final configuration.

The first impression readers will get on glancing at *Table 11.13* is that the TBP arrangement is more common than the SP one, and although this is true it must be emphasized that some structures are appreciably distorted from the ideal one. From our knowledge of ML_5 compounds of the main group elements (see Chapter 10) we would certainly predict a preference for a TBP-based structure unless the *d* electron arrangement played a significant role, i.e. through CFSE or particular π-bonding requirements of the ligands. If we start with d^5 (spin-free) and d^{10} configurations, then the CFSE will be zero for both TBP and SP arrangements, and purely on the basis of electron-pair repulsions we would expect to find the TBP structure; this is indeed the case for the $[Fe(N_3)_5]^{2-}$ and $[CdCl_5]^{3-}$ anions. It is perhaps a little surprising to find that the $[MnCl_5]^{3-}$ anion has an SP arrangement, since the Cl^- ligands do not produce a very big field, and in any case the CFSE difference between the two structures (i.e. SP − TBP) is not marked

Table 11.13 Stereochemistry of five-co-ordinate complexes of the transition elements (ML_5)

Configuration *(d^n, n =)*	*Trigonal-bipyramid* *(TBP)*	*Square-pyramid* *(SP)*
1	$MoCl_5$ (vapour)	–
4	–	$[MnCl_5]^{2-}$
5	$[Fe(N_3)_5]^{2-}$	–
7	$[Co(2\text{-}picNO)_5]^{2+}$	–
8	$[MnCO)_5]^-$	
	$Fe(CO)_5$	
	$[Co(CNMe)_5]^+$	
	$[Ni\{P(OCH)_3(CH_2)_3\}_5]^{2+}$	$[Ni(OAsMe_3)_5]^{2+}$
	$[Ni(CN)_5]^{3-}$	$[Ni(CN)_5]^{3-}$
9	$[CuBr_5]^{3-}$	–
	$[CuCl_5]^{3-}$	
10	$[CdCl_5]^{3-}$	–

for the d^4 configuration. It is noteworthy, however, that the ion is appreciably distorted from the SP, with the apical Mn–Cl bond (257 pm) very much longer than the axial bonds (236 and 224 pm). Such a distortion would be predicted since the d_{z^2} orbital contains a single electron while the $d_{x^2-y^2}$ orbital is empty, so that the apical ligand will suffer a repulsion not experienced by the equatorial ligands. A cautionary note must be struck, too, because the crystal structure was carried out on the bipyridinium salt, and it is possible that packing effects may influence the anion to some extent.

The nickel complexes (d^8) are particularly interesting because both TBP and SP bonding distributions are observed, and both spin-paired and spin-free complexes have been isolated. The $[Cr(en)_3][Ni(CN)_5]$ salt is spin-paired and has a structure in which the unit cell contains two different anions, one being SP and the other mid-way between TBP and SP; this is a clear indication of the similarity in the energy of the TBP and SP arrangements. The other nickel complex worthy of special mention is $[Ni(OAsMe_3)_5]^{2+}$ which has a spin-free configuration, presumably because the field produced by the five trimethylarsine oxide ligands is relatively small. Despite this, the complex is said to adopt a distorted SP structure.

The second group of five-co-ordinate complexes to be discussed is that in which one donor atom differs from the other four, as in complexes of stoichiometry MLL'_4 and MLB_2 (B = bidentate ligand). As the selection of examples in *Table 11.14* shows, both stoichiometries give rise to both stereochemistries.

With the TBP arrangement, the 'odd' ligand is found in one of the axial positions for MLL'_4 complexes and in an equatorial position for the MLB_2 complexes. Thus as *Figure 11.29* shows, the most symmetrical structure is found in each case. With the SP complexes the odd ligand is invariably found in the apical position [see *Figure 11.29*(*c*)], and it should be noted that all of the SP complexes formed by d^0, d^1 or d^2 metal ions contain a multiply bonded apical ligand (either =O or ≡N); the apical position is favoured by such ligands since π bonding is more favourable along the z axis. This is particularly so since the four basal ligands are not in the same plane as the metal atom but are considerably depressed, the L_{apical}-ML_{basal} bond angle normally being between 100 and 106°.

Table 11.14 Stereochemistry of five-co-ordinate complexes of the transition elements (MLL'_4 and MLB_2)*

Configuration (d^n, n =)	*TBP*	*SP*
0	$TiCl_4,NMe_3$ (axial N)	$[TiOCl_4]^{2-}$ (apical O) $MoOF_4$ (apical O)[49]
1	–	$VO(acac)_2$ (apical O) $VO(bzac)_2$ (apical O) $VO(S_2CNEt_2)_2$ (apical O)[50] $[VO(NCS)_4]^{2-}$ (apical O)
2	–	$[ReOBr_4]^-$ (apical O) $ReN(S_2CNEt_2)_2$ (apical N) $[RuNCl_4]^-$ (apical N) $OsO(O_2C_2H_4)_2$ (apical O)[51]
5	–	$MnI(OPPh_3)_4$ (apical I)[52]
7	$[Co(dpe)_2Cl]^+$ (green) (equat. Cl)[53]	$[Co(dpe)_2Cl]^+$ (red) (apical Cl)[53]
8	$Fe(CO)_4,SbPh_3$ (axial Sb)[54] $[Fe(CO)_4(CN)]^-$ (axial CN)[55] $Co(CO)_4H$ (axial H)	$NiBr(2\text{-Meim})_4$ (apical Br) $Ni(S_2PEt_2)_2$,quin (apical N)
9	$[Cu(bipy)_2I]^+$ (equat. I) $[Cu(phen)_2(CN)]^+$ (equat. CN)[56]	$Cu(acac)_2$,quin (apical N)
10	$Zn(S_2CNEt_2)_2$,py (equat. N) $Zn(acac)_2,H_2O$ (equat. H_2O)	–

*B = bidentate ligand

Another point worth noting is that although the complexes are listed in the table as TBP and SP, considerable distortions from the ideal structure are sometimes observed, and in one or two instances the complex could well be regarded as arising from either basic structure. A classic example is $Zn(acac)_2,H_2O$ which we have listed as TBP [see *Figure 11.28(b)*], as reported by one set of workers; however, another crystal-structure determination led to a description as intermediate, and perhaps slightly closer

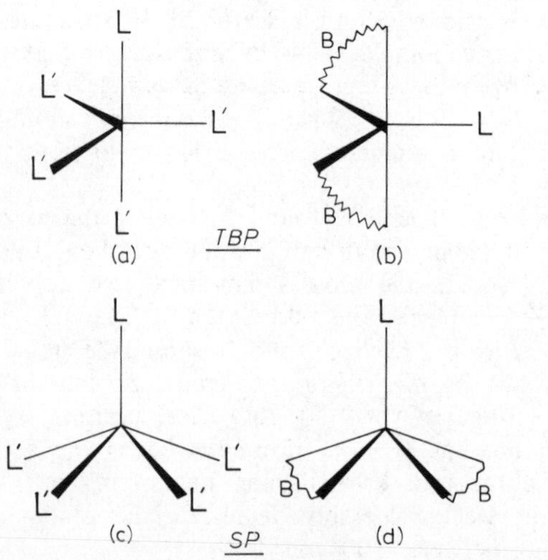

Figure 11.29 Configurations for MLL'_4 and MLB_2 complexes

to SP. As we have seen previously, the TBP and SP structures can be changed one into the other by comparatively small bond angle changes; in this case a small movement of the axial bonds (TBP) away from L, and of the equatorial bonds towards L, produces an SP structure with L occupying the apical position.

The close relationship between the two structures is further illustrated by $[Co(dpe)_2Cl][SnCl_3]$, which exists in two forms, a green one in which the $[Co(dpe)_2Cl]^+$ cation has a TBP structure with an equatorial Cl, and a red form in which the cation is SP with an apical Cl; both forms are spin-paired.

A final remark on the compounds listed in *Table 11.13* concerns $MnI(OPPh_3)_4$, which has a d^5 (spin-free) distorted SP arrangement. CFSE plays no part here, and the only reason for this particular distribution of ligands would appear to be their bulk.

If we turn now to complexes of stoichiometry $ML_3L'_2$, $ML_3L'L''$ or ML_3B, it can be seen that all the complexes listed in *Table 11.15* show a basically TBP arrangement of ligands, except for the three complexes formed by the heavier elements (Mo and Au). However, it must be emphasized that some of the TBP complexes, particularly those of cobalt, nickel and copper, show quite large distortions towards an SP arrangement.

The three basic TBP arrangements that are possible for $ML_3L'_2$ complexes are illustrated in *Figure 11.30*. The particular configuration adopted by a given complex can be rationalized on the basis that the axial positions are normally occupied by the ligands that are highest in the spectrochemical series (see p.211), i.e. $CN^- > PR_3 > NMe_3 > Cl^- > Br^- > I^-$. Thus the symmetrical *trans*-TBP structure is adopted by the trimethylamine complexes formed by the trihalides of titanium, vanadium and chromium, and by $NiBr_3,2PPhMe_2$. With the phosphine complexes formed by divalent cobalt or nickel the axial positions are occupied by cyano groups in

Table 11.15 Stereochemistry of five-co-ordination complexes of the transition elements ($ML_3L'_2$, $ML_3L'L''$ and ML_3B)

Configuration (d^n, n =)	*TBP*	*SP*
1	$TiBr_3,2NMe_3$ (axial N,N) $TiCl_3,2NMe_3$	$MoOCl_3,SPPh_3$ (apical Cl)[57]
2	$VBr_3,2NMe_3$ (axial N,N) $VCl_3,2NMe_3$ (axial N,N)	–
3	$CrCl_3,2NMe_3$ (axial N,N)	–
5	$MnCl_2,(2\text{-Meim})_3$ (axial Cl,N)[58]	–
7	$CoBr_2,3PHPh_2$ (axial P,P) $NiBr_3,2PPhMe_2$ (axial P,P)	–
8	$NiBr_2,3PMe_3$ (axial P,P)[59] $Ni(CN)_2,3PPhMe_2$ (axial CN,CN) $Ni(CN)_2,3PPh(OEt)_2$ (axial CN,CN) $NiI_2,3PHPh_2$ (axial P,P) $NiI_2,3P(OMe)_3$ (axial P,P)[60]	$AuBr_3$,dmp (apical N)[61] $AuCl_3$,dmp (apical N)[61]
9	$CuCl_2(dimim)_3$ (axial Cl,N)[62] $[Cu(NH_3)_2(NCS)_3]^-$ (axial N,N)	–

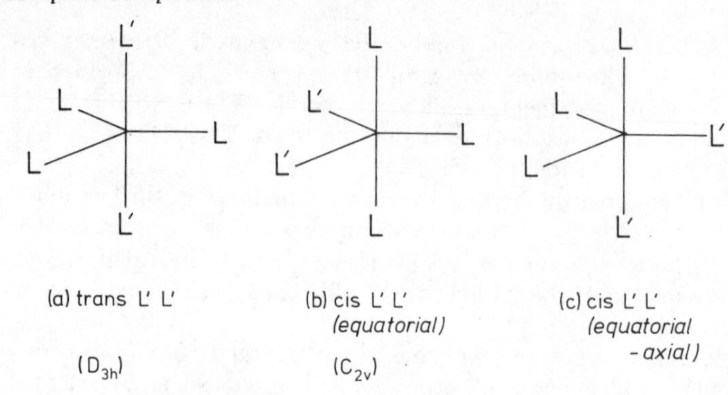

Figure 11.30 Configurations for $ML_3L'_2$ complexes

$Ni(CN)_2,3PPhMe_2$ and $Ni(CN)_2,3PPh(OEt)_2$, which accordingly assume the *trans*-TBP arrangement. With the analogous bromo and iodo complexes, however, the axial positions are occupied by phosphine ligands [see *Figure 11.30(b)*].

The trimethylamine adducts are an interesting spin-free series (d^1, d^2 and d^3). The first two electrons enter the lower doublet (d_{xz} and d_{yz}) and would be expected to have little or no influence on the stereochemistry since the electrons do not occupy the regions of space pointing towards the ligands. With $CrCl_3,2NMe_3$, however, we have a third electron which enters either the $d_{x^2-y^2}$ or d_{xy} level (see *Figure 11.27*). This uneven filling of the two orbitals in the equatorial plane should mean that the three equatorial chlorine atoms will be exposed to different repulsion forces so that some distortion is to be expected, and this has been confirmed by a crystal-structure determination, although the distortion is in the bond angles rather than bond lengths, with two ClCCl angles around 124° and the other close to 112°.

It is also interesting to look at the electronic spectra of $TiBr_3,2NMe_3$ and $TiCl_3,2NMe_3$ (see *Figure 11.31*), each of which shows two peaks, one in the near-infrared region (around 5000 cm^{-1}) and the other in the

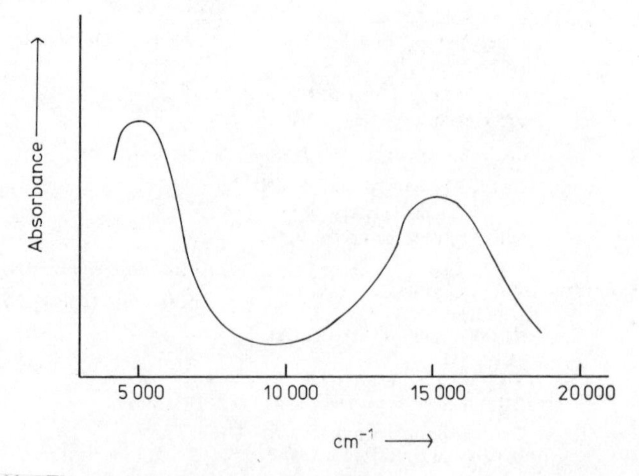

Figure 11.31 Electronic spectrum of $TiCl_3,2NMe_3$

visible region (around 15 000 cm^{-1}). The two peaks are easily explained on the basis of the splitting diagram of *Figure 11.26*, since the single electron can be promoted to either the doublet ($d_{x^2-y^2}$, d_{xy}) or singlet (d_{z^2}) level. As chlorine ligands produce a somewhat larger field than bromine ligands, there is a bigger separation of *d* levels for $TiCl_3$,$2NMe_3$ and the two peaks are found at slightly higher energies.

Quite a number of five-co-ordinate complexes have been prepared in which the halides or isothiocyanates of divalent cobalt, nickel, copper, zinc and cadmium are complexed by tridentate ligands, the complexes having the stoichiometry MX_2T. The listing of these compounds in *Table 11.16*

Table 11.16 Stereochemistry of five-co-ordinate complexes of the transition metals (ML_2T and $MLL'T$)*

Configuration (d^n, n =)	*TBP*	*SP*
7	$CoCl_2$,terpy (axial N,N) $CoCl_2$,Me_5dien (axial N,Cl) $CoCl_2$,Et_4dien (axial N,Cl) $Co(NCS)_2$,nnpO (axial N,NCS)[63] $Co(NCS)_2$,nnp (axial N,NCS)[64] $Co(NCS)_2$,nnpMe (axial N,NCS)[64]	$CoCl_2$,paphy (apical Cl)
8	$NiCl_2$,terpy (axial N,N) $NiBr_2$,Me_2dmpa (axial N,N)	$NiBr_2$,triars (apical Br) $NiBr_2$,dppea (apical Br) NiI_2,dsp (apical S)
9	$CuCl_2$,terpy (axial N,N) $CuBr(N_3)$,Et_4dien (axial N,N_3)[65]	–
10	$ZnCl_2$,terpy (axial N,N) CdX_2,terpy (axial N,N)[66] [X = $Mn(CO)_5$]	–

*T = tridentate ligand

as either TBP or SP is somewhat arbitrary, since the structures are dominated by the steric requirements of the ligands and are considerably distorted from the ideal. Despite this reservation the division is useful, and on this basis we find two types of arrangement for each structure (see *Figure 11.32*). In every case in which some of the donor atoms are incorporated into heterocyclic ring systems the ligands take up a *trans* arrangement of the donor atoms [either TBP (*a*) or SP (*c*) in *Figure 11.32*], and this has been described as a 'straight-jacket' effect. It can be seen that comparatively small angular changes in respect of the M–X bonds will convert a TBP structure into an SP one, and vice versa. The *trans* arrangement is also found for the triars and dppea ligands in basically SP structures.

The remaining ligands assume a *cis* arrangement of donor atoms, as shown for TBP (*b*) and SP (*d*) in *Figure 11.32.* Once again the two structures are in principle easily converted one into the other, so the observed distortions are not unexpected.

All the complexes listed are high-spin, apart from $NiBr_2$,triars, although the magnetic moment of $Co(NCS)_2$,nnp is very temperature-dependent, and at room temperature is only 3.58 BM whereas that of $Co(NCS)_2$,nnpMe is 4.5 BM; thus five-co-ordinate complexes of divalent cobalt with a N_4P

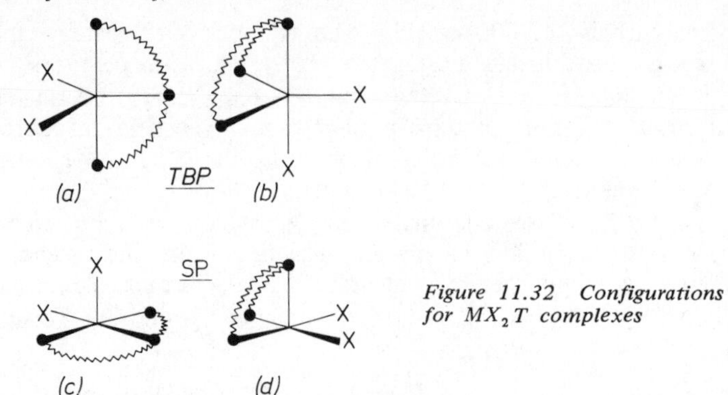

Figure 11.32 Configurations for MX_2T complexes

donor set clearly fall into the cross-over range between high- and low-spin complexes (Orlandini *et al.*[64]).

Over the past ten years quite a number of five-co-ordinate complexes have been prepared with stoichiometry ML(Tet), where Tet represents a tetradentate ligand (see *Table 11.17*). These complexes fall broadly into the two listed structural categories, TBP and SP, but once again it must be emphasized that the structures are rarely ideal. Moreover, with these complexes the five-co-ordination is frequently a consequence of the bulky nature of the tetradentate ligands that prevents the approach of a sixth (monodentate) ligand, that might otherwise complete an octahedral environment – especially for SP complexes. The detailed stereochemistry of ML(Tet) complexes appears to be determined by the steric requirements of the tetradentate ligand rather than such factors as the CFSE. Thus the ligands fall into two broad categories, that may be described colloquially as 'tophat' (or 'tripod') and 'chain'; these are illustrated in *Figure 11.33.* All the TBP complexes incorporate a 'tophat' ligand with the

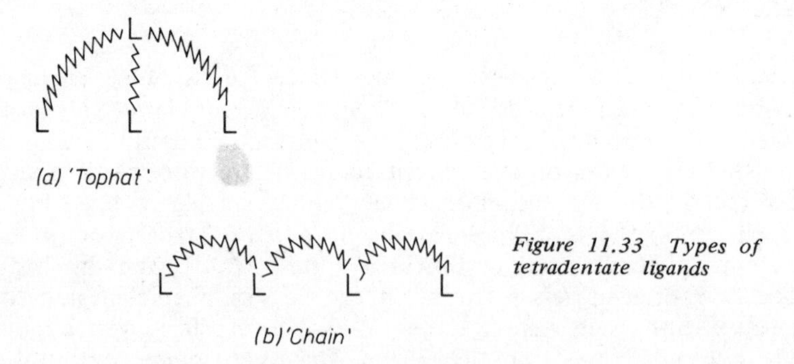

Figure 11.33 Types of tetradentate ligands

fifth monodentate ligand occupying an axial position [see *Figure 11.34(a)*]. The SP complexes are usually formed with the 'chain-type' ligands, which may occupy the four basal positions [*Figure 11.34(b)*], as in $[Co(PNNP)I]^+$, or occupy the apical position and three of the four basal positions [*Figure 11.34(c)*], as in $[Pd(tpas)Cl]^+$.

Readers will also note that both high- and low-spin complexes can be prepared for either TBP or SP configurations. It can be seen that in almost every instance low-spin complexes are formed when three or more of the donor atoms come from the second or subsequent row of the

Table 11.17 Stereochemistry of five-co-ordinate complexes of the transition elements (MLTet)*

Configuration d^n (n =)	TBP				SP			
	Examples	*Donor set*	*Spin*		*Examples*	*Donor set*	*Spin*	
			High	*Low*			*High*	*Low*
5	[Mn(Me_6tren)Br]$^{+}$	N_4Br	✓		—			
6	[Fe(Me_6tren)Br]$^{+}$	N_4Br	✓		—			
7	[Co(Me_6tren)Br]$^{+}$	N_4Br	✓		[Co(PNNP)I]$^{+}$ (apical I)	N_2P_2I		✓
	[Co(tpa)Br]$^{+}$	N_4Br	✓		Co(disalen)py (apical)	N_3O_2		✓
	[Co(n_2p_2)I]$^{+}$ [67]	N_2P_2I	✓					
	[Co(TPN)Cl]$^{+}$	NP_3Cl		✓				
	[Co(TPN)I]$^{+}$	NP_3I		✓				
	[Co(QP)Cl]$^{+}$	P_4Cl		✓				
					Ni(dacoDA)H_2O	N_2O_3	✓	
8	[Ni(Me_6tren)Br]$^{+}$	N_4Br	✓		[Ni(dapip)Cl]$^{+}$ (apical Cl)	N_4Cl	✓	
	[Ni(n_2p_2)I]$^{+}$	N_2P_2I		✓	[Ni(PNNP)I]$^{+}$ (apical I)	N_2P_2I		✓
	[Ni(TPN)I]$^{+}$	NP_3I		✓	[Ni(bdpPme)I]$^{+}$ (apical O)	NOP_2I		✓
	[Ni(TAP)CN]$^{+}$	PAs_3CN		✓				
	[Ni(TSP)Cl]$^{+}$	SP_3Cl		✓				
	[Ni(SAs_3)Br]$^{+}$	SAs_3Br		✓	[Pd(tpas)Cl]$^{+}$ (apical As)	As_4Cl		✓
	[Ni(n_2p_2)I]$^{+}$	N_2P_2I		✓				
	[Pt(QAS)I]$^{+}$	As_4I		✓				
9	[Cu(Me_6tren)Br]$^{+}$	N_4Br			[Cu(trien)SCN]$^{+}$ (apical SCN)	N_4S		
	[Cu(tren)NCS]$^{+}$	N_5			[Cu(bPyaenMeOH)Br]$^{+}$			
					(apical N)	N_4Br		
					[Cu(bPyDAH)NCS]$^{+}$			
					Cu(disalen)H_2O (apical H_2O)	N_2O_3		
10	[Zn(Me_6tren)Br]$^{+}$	N_4Cl			[Zn(trien)I]$^{+}$ (apical I)	N_4I		
	[Zn(tren)Cl]$^{+}$	N_4Cl			Zn(disalen)H_2O (apical H_2O)	N_2O_3		

*Tet = tetradentate ligand

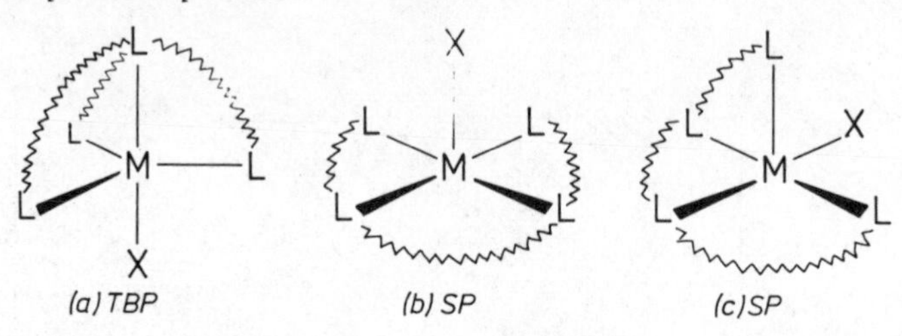

Figure 11.34 Configuration of MX(Tet) complexes

periodic table (i.e. P, As, S, Br, I), and that high-spin complexes result when the majority of the donor atoms are nitrogen. It is a fine balance, however, and with the N_2P_2I set of donors (with the ligand n_2p_2) cobalt gives a high-spin TBP complex and a nickel a low-spin one. Readers wishing to go into this matter in more detail are referred to the review by Morassi and co-workers[47].

Finally it may be noted that five-co-ordinate complexes can be made in which the metal atom is associated with a single pentadentate ligand, although such a ligand has to be chosen carefully so as to avoid excessive intraligand repulsions. Very little detailed structural work has been carried out on this type of complex, although a recent study (Hoskins and Whillans[69]) of the d^8 high-spin nickel complex cation $[Ni(tepen)]^{2+}$ has shown it to have the SP structure shown in (24).

(24)

11.12 Co-ordination Number 7[70]

Once again it must be emphasized that the assignment of co-ordination number cannot be based solely on the stoichiometry of a complex, and that many complexes that appear to be seven-co-ordinate frequently turn out to be otherwise; thus $(NH_4)_3SiF_7$ does not contain $[SiF_7]^{3-}$ ions, but a mixture of F^- and $[SiF_6]^{2-}$ ions. Nevertheless, since the appearance of the third edition of this book some ten years ago a considerable number of genuine seven-co-ordinate complexes has been fully characterized by X-ray crystallography or by electron or neutron diffraction. It is apparent that most of the complexes can be properly described in terms of one of the ideal geometries, pentagonal bipyramid (PBP), capped octahedron (CO) and capped trigonal prism (CTP), which are illustrated in *Figure 11.35.* It is perhaps worth digressing slightly in respect of the CTP structure to point out that the related trigonal prism is a possible alternative to the octahedron for six-co-ordinate complexes, and while it is unusual it has

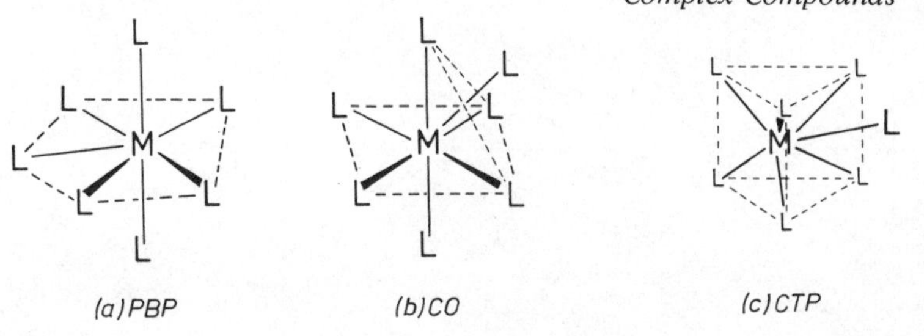

Figure 11.35 Stereochemistry of seven-co-ordinate complexes

been well-established for a number of transition metal complexes, particularly those formed with bidentate dithiolato ligands; interested readers are referred to a review by Eisenberg[71].

Although over 200 seven-co-ordinate structures have now been determined, it is almost impossible to predict the structure of a new seven-co-ordinate complex even when it is closely related to a complex whose structure is known. Thus $[ZrF_7]^{3-}$ has a PBP arrangement, while the isoelectronic ion $[NbF_7]^{2-}$ takes up a CTP structure.

Some of the simpler examples of seven-co-ordinate complexes, namely those involving seven monodentate ligands, are listed in *Table 11.18*, and

Table 11.18 Stereochemistry of seven-co-ordinate complexes of transition elements with mondentate ligands

General formula	*Examples*		
	PBP	*CO*	*CTP*
ML_7	$[ZrF_7]^{3-}$ $[V(CN)_7]^{4-}$ $[Mo(CN)_7]^{5-}$ ReF_7	–	$[NbF_7]^{2-}$ $[Mo(CNR)_7]^{2+}$
ML_6L'	$[NbOF_6]^{3-}$	–	$[Mo(CNR)_6I]^+$
$ML_5L'_2$	$[UO_2F_5]^{2-}$	$[UCl(tmpo)_6]^{3+}$	–
$ML_4L'_3$	–	$MoX_4,3PR_3$ $[W(CO)_4Br_3]^-$	–

it can be seen that several PBP and CTP structures are known in which all seven ligands are identical, but CO structures of this type have not been observed up until now. It is interesting to note that rhenium(VII) fluoride has the same PBP structure as IF_7, there being some puckering of the skirt of five equatorial fluorines. In seven-co-ordinate uranyl complexes such as $[UO_2F_5]^{3-}$ (25) and $[UO_2(NCS)_5]^{3-}$, the two uranium–oxygen multiple bonds occupy the axial positions with the other five ligands in the equatorial plane. Readers will recall our previous discussion of the linear nature of the $[UO_2]^{2+}$ ion (p.230).

Since the equatorial sites are crowded in the PBP arrangement, we would not predict this stereochemistry for complexes that contain the larger halides or bulky groups such as trialkyl- or triaryl-phosphines, and in such instances the capped octahedral arrangement becomes more favourable; the carbonyl halides of molybdenum and tungsten, and their complexes, also adopt the CO arrangement.

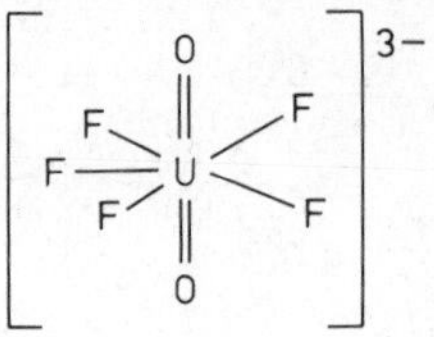

(25) 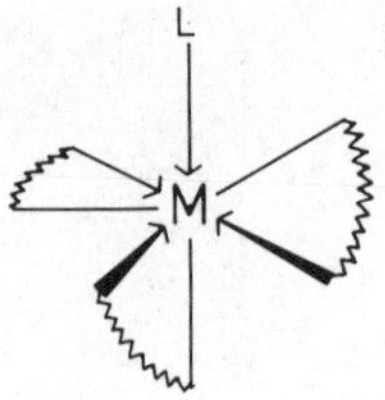 (26) 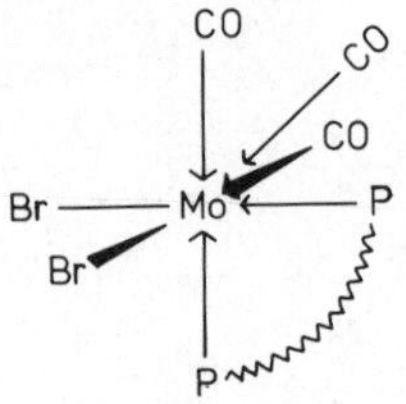 (27)

When bidentate ligands are involved, the PBP arrangement is the commonest, except when the complexes contain carbonyl or bulky groups, in which case the CO structure dominates. The PBP structure is especially favoured when the bidentate groups have small 'bites', as with ligands such as nitrate and oxalate, and here the equatorial positions are occupied; only when a third bidentate group is present in MB_3L (26) does it bridge the equational and axial positions.

CO and CTP structures are also found for complexes involving bidentate groups, particularly those of stoichiometry ML_5B in which some of the monodentate ligands are carbonyl groups. A typical example is $Mo(CO)_3(dpe)Br_2$, which has the structure shown in (27).

In view of the complexity of the structures a detailed discussion of bonding schemes would be out of place, but it is worth noting that the ligand-field splitting diagram for the PBP arrangement is essentially the same as for the TBP arrangement in five-co-ordinate compounds (cf. *Figure 11.27*); since there are five ligands in the equatorial plane the d_{xy}, $d_{x^2-y^2}$ doublet will be appreciably closer to the d_{z^2} orbital than in the five-co-ordinate case. On the basis of such a diagram we can predict that for d^1 complexes there should again be two *d–d* peaks in the electronic spectra, and this is observed for such complexes of tervalent titanium as $Ti_2(ox)_3,10H_2O$ which has a bridged structure (28) (Drew *et al.*[72]) in which each titanium atom has a PBP environment; two *d–d* transitions are observed at 9400 and 12 300 cm^{-1}.

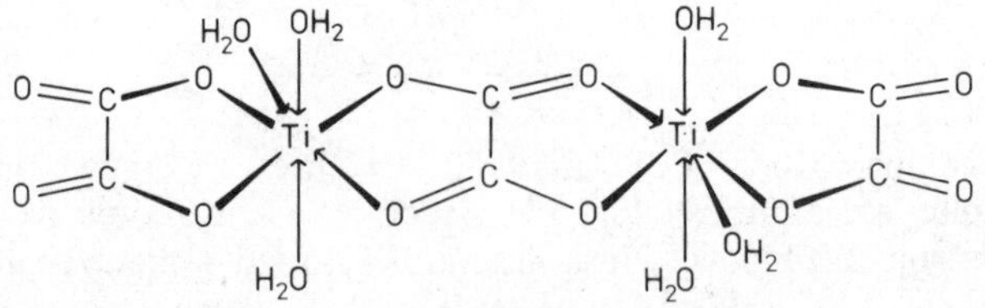

(28)

11.13 Co-ordination Number 8[73]

Generally speaking we should expect eight co-ordination to be most common for complexes formed between large cations and small ligands, since overcrowding will be at a minimum and the steric strain less. Moreover, high co-ordination numbers will be most likely when the cations contain comparatively few *d* electrons as there will then be less repulsion with the ligand electrons. A final general point is that when a donor–acceptor bond is formed there is transfer of electron density from the donor ligand to the acceptor metal, so that the formation of eight such bonds results in a considerable charge build-up on the central metal; this build-up is not so noticeable if the metal acceptor is in a high oxidation state.

All of these considerations explain why eight co-ordination is mainly found for complexes of metal ions of the second and third transition series, the lanthanides and the actinides, which have d^0, d^1 or d^2 electronic configurations and a formal charge of 3+ or more. However, in recent years several eight-co-ordinate complexes have been characterized for first-row elements, as some of the examples given in *Table 11.19* show.

The molecular geometries for eight-co-ordinate complexes are not simple, and once again it must be emphasized that in many cases the actual structure may be somewhat distorted from the ideal; the four basic geometries (hexagonal-bipyramidal, cubic, dodecahedral and square-antiprism) are illustrated schematically in *Figure 11.36.*

The hexagonal-bipyramidal arrangement is found only with uranyl complexes, for which there is an axial O=U=O skeleton (z axis) with the remaining six donor atoms in the equatorial plane. The six equatorial ligands may be somewhat crowded, which frequently results in a small amount of puckering. Two classes of eight-co-ordinate uranyl complexes are observed, namely $UO_2L_2B_2$ and UO_2B_3, where L and B are monodentate and bidentate ligands, respectively. In the first type, B is normally the nitrate ion behaving in a bidentate manner, and L may be H_2O, $(EtO)_3PO$ or ethyl carbamate[83], e.g. (29). With the UO_2B_3 complexes the bidentate ligands are normally 'small-bite' oxo species such as acetate, nitrate or carbonate (Graziani *et al.*[84]), e.g. (30), although the $[UO_2(S_2CNEt_2)_3]^-$ anion has also been characterized.

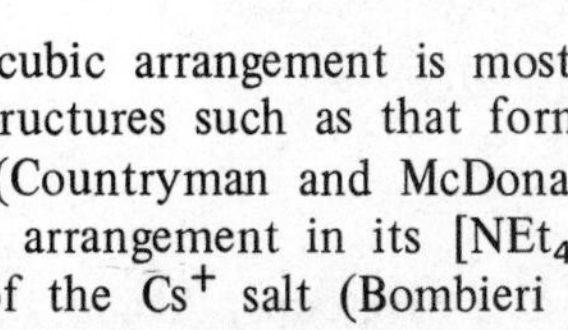

(29)

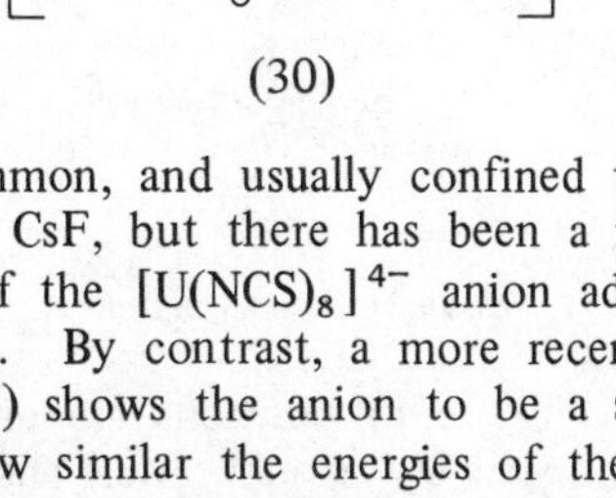

(30)

The cubic arrangement is most uncommon, and usually confined to ionic structures such as that formed by CsF, but there has been a recent report (Countryman and McDonald[85]) of the $[U(NCS)_8]^{4-}$ anion adopting a cubic arrangement in its $[NEt_4]^+$ salt. By contrast, a more recent study of the Cs^+ salt (Bombieri *et al.*[77]) shows the anion to be a square-antiprism. This vividly demonstrates how similar the energies of the various configurations must be, and a look at the examples listed in *Table 11.19* confirms this. Thus the $[Mo(CN)_8]^{3-}$ anion is dodecahedral in the $[NBu^n_4]^+$ salt but square-antiprismatic in the Na^+ salt, and while the octafluoride anions $[MF_8]^{n-}$ are generally square-antiprismatic, $[ZrF_8]^{4-}$ is dodecahedral.

We should, perhaps, look in a little more detail at the dodecahedral arrangement since it is rather more common than the square-antiprismatic one. The schematic dodecahedral arrangement is elaborated in *Figure 11.37,*

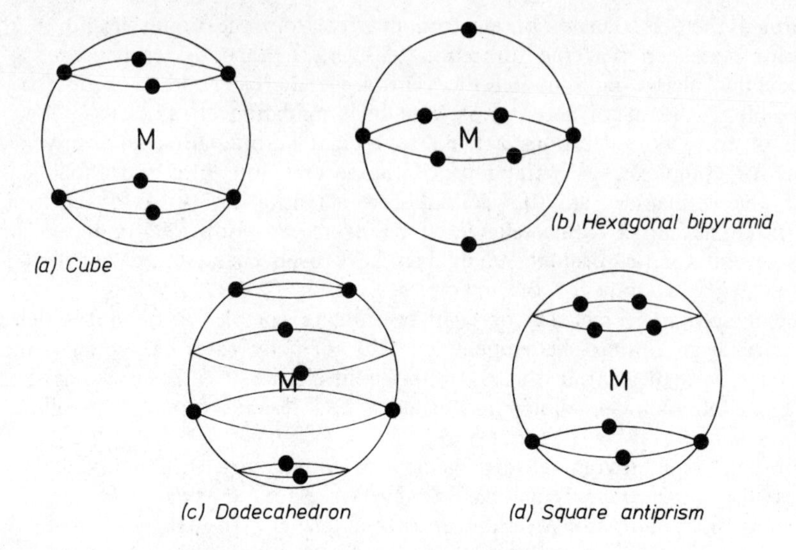

Figure 11.36 Stereochemistry of eight-co-ordinate complexes

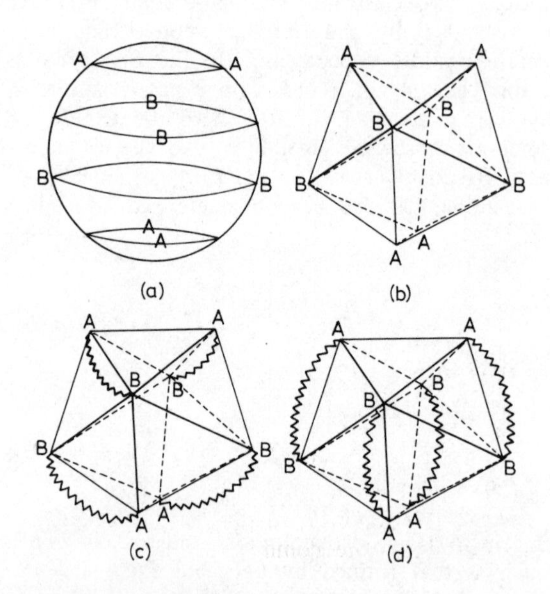

Figure 11.37 Dodecahedral complexes

and the eight donor positions are divided into two groups, A and B, which provide respectively an elongated and a flattened tetrahedron about the central metal.

In the diamagnetic molybdenum(IV) complex $Mo(CN)_4(CNMe)_4$, the CN groups occupy the A positions, and spectroscopic studies (Parish and Simms[86]) have shown this to be the case with a whole series of similar complexes of quadrivalent molybdenum and tungsten of general formula $M(CN)_4(CNR)_4$. In the diarsine complex cation $[MoCl_4(diars)_2]^+$, the four arsenic atoms occupy the A positions.

Table 11.19 Eight-co-ordinate complexes of the transition elements* (dodecahedral and square antiprism)

General formula	*Dodecahedral*	*Square antiprism*
ML_8	–	$[La(pyO)_8]^{3+}$ [74]
	$[ZrF_8]^{4-}$	$[TaF_8]^{3-}$; $[UF_8]^{4-}$; $[ReF_8]^{2-}$
	$[M(CN)_8]^{4-}$ (M = Mo, W)	–
	$[Mo(CN)_8]^{3-}$ ($Bu^n_4N^+$ salt)[75]	$[Mo(CN)_8]^{3-}$ (Na^+ salt)[76]
		$[W(CN)_8]^{3-}$ (Na^+ salt)[76]
	–	$[U(NCS)_8]^{4-}$ (Cs^+ salt)[77]
$ML_6L'_2$	$[UCl_2(Me_2SO)_6]^{2+}$ [78]	–
$ML_4L'_4$	$Mo(CN)_4(CNMe)_4$ [79]	–
ML_4B_2	$TiCl_4$,2diars	
	$[MoCl_4(diars)_2]^+$ [80]	–
ML_2B_3	–	$M(acac)_3,2H_2O$ [76] (M = La, Pr, Nd, Sm)
MB_4	$Ti(NO_3)_4$	$M(acac)_4$ (M = Ce, Zr, Th)
	$[Zr(ox)_4]^{4-}$	
	$Th(S_2CNEt_2)_4$	
	$V(S_2CCH_2Ph)_4$ [81]	
	$Mo(S_2CR)_4$ [82]	

*References in Kepert (ref. 73) unless otherwise given

The dodecahedral arrangement is especially favoured in complexes involving four bidentate ligands with small 'bite', and each of the ligands chelates one A and one B position, the groups adopting the arrangement of *Figure 11.37(c)*; $Th(S_2CNEt_2)_4$ is the odd one out with the chelation still spanning A and B sites but in a slightly different way [*Figure 11.37 (d)*].

The bidentate acetylacetone ligand gives square-antiprismatic structures in $[M(acac)_4]$ complexes, with the acac groups bridging the edges of the square faces as in (31). A similar structure is found with a number of

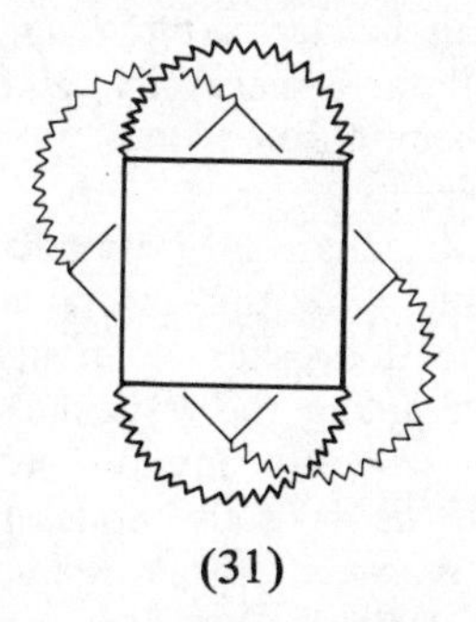

(31)

$M(acac)_3,2H_2O$ complexes (see *Table 11.19*), with the two water molecules taking the place of one chelate group, but with other analogous lanthanide complexes there is appreciable distortion towards the dodecahedral arrangement.

The bonding in eight-co-ordinate complexes (see Lippard[87]) can be discussed in terms of valence-bond, molecular-orbital or ligand-field approaches, but the apparent restriction of such complexes to elements with few *d* electrons limits the use of the ligand-field method. The *d*-orbital splittings for dodecahedral and square-antiprismatic arrangements are shown in *Figure 11.38*, and it can be seen that in each case there is a single d orbital appreciably lower in energy than the rest; for this reason the d^2 $Mo(CN)_4(CNR)_4$ complexes are diamagnetic with the two electrons paired in the d_{xy} orbital. The spectra of $NbCl_4,B_2$ complexes have been examined and show the three predicted peaks.

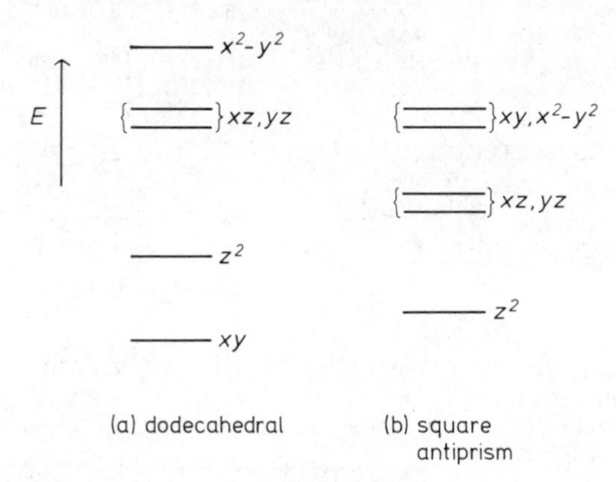

Figure 11.38 d-*Orbital splitting for eight-co-ordinate complexes*

In the valence-bond approach, the hybridization schemes used are:

cube	sp^3d^3f (using d_{xy}, d_{xz} and d_{yz})
hexagonal bipyramid	axial – sp or sd (using p_z or d_{z^2})
	equatorial – sp^2d^2f or p^2d^3f (using p_x, p_y; d_{z^2}, $d_{x^2-y^2}$, d_{xy})
dodecahedron	sp^3d^4 (using d_{z^2}, d_{xy}, d_{xz}, d_{yz})
square antiprism	sp^3d^4 (using $d_{x^2-y^2}$, d_{xy}, d_{xz}, d_{yz})

It can be seen that for an hexagonal-bipyramidal or cubic arrangement the participation of an f orbital is necessary, and up until now complexes with these structures have been formed only with uranium, which has $5f$ orbitals energetically available.

For both dodecahedral and square-antiprismatic arrangements there is a d orbital unused in σ bonding, and this orbital may be involved in π bonding with the ligands. In the dodecahedral arrangement the available $d_{x^2-y^2}$ orbital may be used for π bonding the B ligands only [see *Figure 11.37(b)*] but for the square-antiprism structure any of the ligands may π bond with the available d_{z^2} orbital. If we take the dodecahedral structure, then for d^0 ions the $d_{x^2-y^2}$ orbital is vacant, so it can accept electrons from π donors, and we would expect the better π donors to occupy the B sites if there were no other significant factors. This is indeed the case with $TiCl_4$,2diars, in which the Cl atoms occupy the B position. With d^1 or d^2 ions, on the other hand, we should anticipate the B positions being occupied by π acceptors, to accept electrons from the singly or doubly filled orbital. This is the case with $M(CN)_4(CNR)_4$ complexes, for which the CNR groups occupy the B positions, but not with $[MoCl_4(diars)_2]^+$ which has the same structure as $TiCl_4$,2diars, with Cl atoms in the B positions. Thus with the diarsine complexes it seems that the steric requirements of the bulky chelate ligands may be more important than a small amount of π bonding.

Readers interested in the valence-bond approach to eight co-ordination are referred to Lippard's review[87].

11.14 Co-ordination Numbers 9, 10 and 12[74]

Such high co-ordination numbers are comparatively rare and the best characterized complexes are those formed by the lanthanide and actinide elements.

For co-ordination number nine, the two ideal geometries are the monocapped square antiprism (with capping on one square face) and the tricapped trigonal prism (with capping on each of the three square faces). The monocapped square antiprism is very uncommon, but in a somewhat distorted form it has been reported recently for $[Pr(terpy)Cl(H_2O)_5]^{2+}$ [see *Figure 11.39(a)*]. Four water ligands occupy the corners of one square face, with the remaining water molecule and the chlorine occupying opposite corners of the other square face; the terpy ligand spans the remaining two corners with the centre nitrogen above the centre of the face.

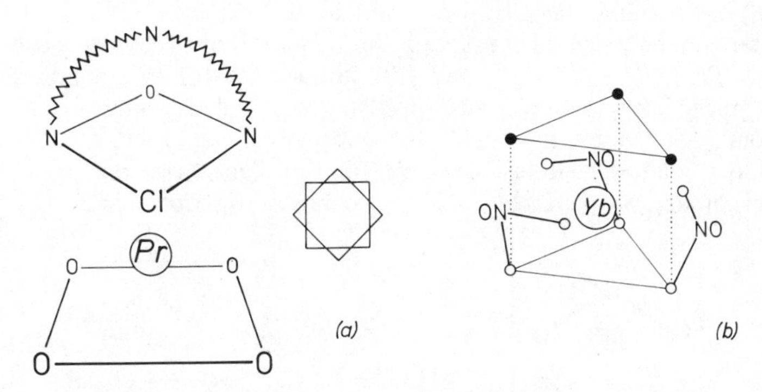

Figure 11.39 Structure of nine-co-ordinate compounds: (a) $[Pr(terpy)Cl(H_2O)_5]$; (b) $Yb(NO_3)_3(dmso)_3$

The tricapped trigonal-prismatic structure is found much more frequently, typical examples being the hydrated cations $[M(H_2O)_9]^{3+}$ formed by the larger lanthanide ions, and the polymeric trifluorides, MF_3. This structure has been established recently (Bhandary *et al.*[88]) for the ytterbium complex $Yb(NO_3)_3(dmso)_3$ [see *Figure 11.39(b)*], in which the three monodentate dmso ligands use three corners of one of the triangular faces; the three bidentate groups each link one corner of the other triangular face with the centre of a square face.

It is interesting to note that we get nine co-ordination with Yb^{3+}, which is one of the smaller lanthanide ions, but that with the larger lanthanum ion La^{3+}, an extra dmso ligand can be accommodated, giving a ten-co-ordinate complex $La(NO_3)_3(dmso)_4$[84]. Complexes of this co-ordination number have structures based on the bicapping of either of the eight-co-ordinate polyhedra, the square-antiprism and the dodecahedron. Calculations by

Al-Karaghouli and Wood[89] based on ligand repulsions suggest that the bicapped square antiprism is the lowest energy form.

Twelve co-ordination is even less common, but one authentic example is $[Ce(NO_3)_6]^{3-}$, in which the six bidentate nitrate groups establish an icosahedron around the cerium. Readers interested in the stereochemistry of complexes of co-ordination number ten and higher are referred to an excellent review by Muetterties and Wright[90].

11.15 Metal–Metal Bonding: Cluster Compounds

Quite a number of elements form compounds in which there is one or more direct element–element bonds. With the *p* block elements, for instance, we get simple σ bonding in compounds such as $Me_3Sn{-}SnMe_3$ and with elemental phosphorus we get P_4 units. In Chapter 13 we shall discuss some boron compounds which contain groups or clusters of boron atoms.

Analogous metal–metal bonds are formed by quite a few transition elements, and the structures and bonding of such compounds have been the subject of much research over the past 10–15 years. Single σ M–M bonds are well established in molecules such as $Mn_2(CO)_{10}$ (see Section 11.16) and multiple bonding can also be found in such ions as $[Re_2Cl_8]^{2-}$ (32). The latter ion contains two eclipsed $ReCl_4$ units linked by an extremely short Re–Re bond (224 pm), and it is thought that the σ bonding may be strengthened by two π bonds (involving d_{xz} and d_{yz} orbitals) and a delta bond, formed by the side-by-side overlap of d_{xy} orbitals. The multiple bonding is illustrated in *Figure 11.40.* Readers interested in such multiple bonding are referred to a recent review by Cotton[91].

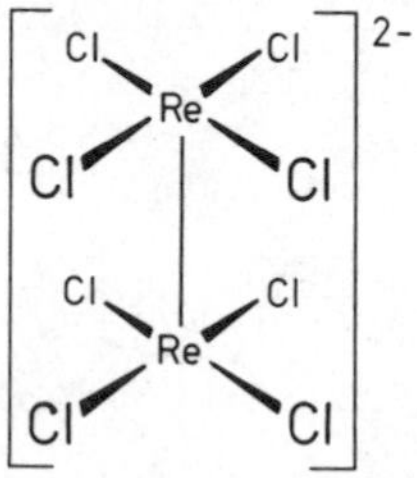

(32)

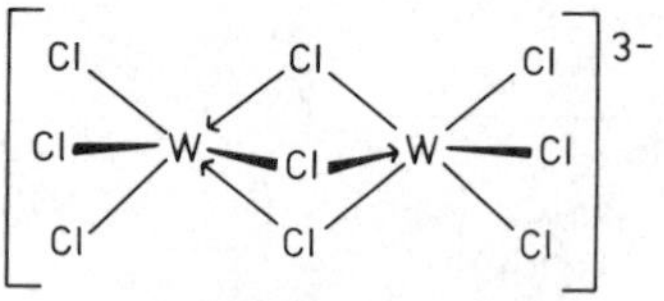

(33)

Metal–metal bonding can be encouraged if there is dimer or polymer formation through bridging groups, such that the neighbouring metal atoms are brought sufficiently close together for orbital overlap and spin pairing. A typical example is the anion $[W_2Cl_9]^{3-}$ (33).

Several transition elements form compounds containing 'clusters' of metal atoms, the commonest clusters being M_3 (triangular) and M_6 (octahedral). The M_3 triangular cluster is mainly found in derivatives of rhenium(III) halides, such as $[Re_3Cl_{12}]^{3-}$, whose structure is shown in *Figure 11.41.* The Re–Re distances in the triangular cluster are around 248 pm, and the Re_3 unit is also held together by three chlorine bridges in the same plane; each Re also is bonded to three terminal chlorines. In solid $ReCl_3$ itself, the planar Re_3Cl_3 unit remains intact, with bridging through two of each

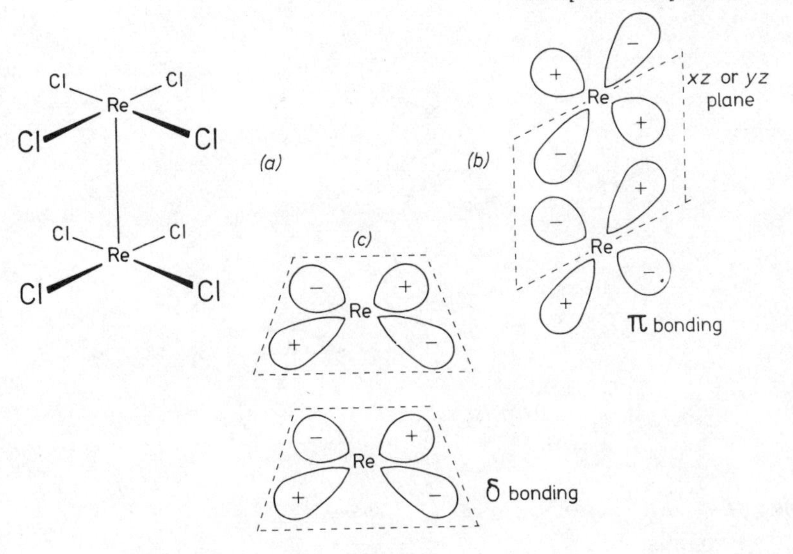

Figure 11.40 Multiple bonding in $[Re_2Cl_8]^{2-}$

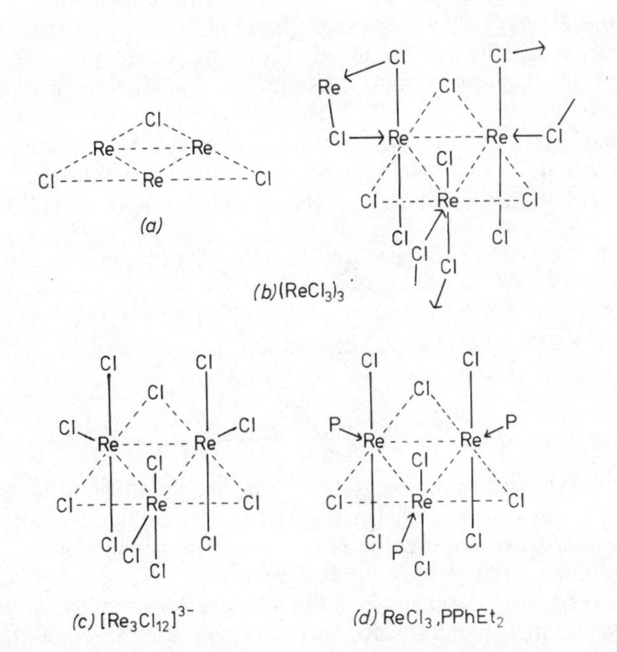

Figure 11.41 Structure of rhenium compounds based on the Re_3Cl_3 *triangular unit: (a) basic unit; (b)* $(ReCl_3)_3$*; (c)* $[Re_3Cl_{12}]^{3-}$*; (d)* $ReCl_3,PPhEt_2$

set of three terminal chlorines to neighbouring units. These bridges are readily broken down by ligands other than chloride (Walton[92]), and compounds such as $ReCl_3,L$ are formed; the crystal structure of $ReCl_3,PPhEt_2$ shows the molecule to have the basic trimer structure (cf. $[Re_3Cl_{12}]^{3-}$) with the phosphine ligands occupying terminal positions in the Re_3Cl_3 plane.

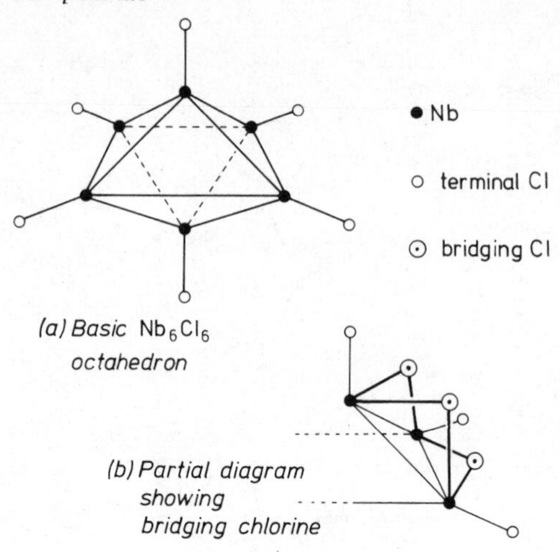

Figure 11.42 Nb_6 *cluster in* $[Nb_6Cl_{18}]^{n-}$ *anions*

Octahedral M_6 clusters are formed particularly by lower halides of niobium, tantalum, molybdenum and tungsten. The niobium and tantalum halides form $[M_6X_{12}]^{n+}$ clusters (see *Figure 11.42*) in which the twelve halide atoms bridge the twelve edges of the M_6 octahedron. The $[Nb_6Cl_{12}]^{n+}$ ion has been studied (Koknat and McCarley[93]) in particular detail and established for n = 2, 3 and 4. In the series of derived anions, $[Nb_6Cl_{18}]^{4-}$, $[Nb_6Cl_{18}]^{3-}$ and $[Nb_6Cl_{18}]^{2-}$, a chloride ion has been co-ordinated to each niobium atom. Crystal-structure determinations show (see *Table 11.20*) that as the $[Nb_6Cl_{18}]^{4-}$ ion is oxidized so

Table 11.20 Nb–Nb bond distances in $[Nb_6Cl_{18}]^{n-}$ ions

Ion	*Nb–Nb*/ pm	*Nb–Cl(bridge)*/ pm	*Nb–Cl(terminal)*/ pm
$[Nb_6Cl_{18}]^{4-}$	292	248	261
$[Nb_6Cl_{18}]^{3-}$	297	243	252
$[Nb_6Cl_{18}]^{2-}$	302	242	246

the Nb–Nb distance increases and the terminal Nb–Cl bond shortens. The increase in the Nb–Nb distance is attributed to bond weakening through the successive removal of electrons from the multicentre bonding molecular orbitals that hold together the Nb_6 octahedron. In all the anions the terminal Nb–Cl bonds are longer than would be expected for a single bond (~240 pm), and this can be accounted for by the difficulty of accommodating the chlorine; as the Nb_6 octahedron expands, so there is more room for this chlorine and the Nb–Cl distance decreases.

The dihalides of molybdenum and tungsten, MX_2, are best considered as $[M_6X_8]X_4$, in which there is a M_6 octahedron surrounded by eight X atoms, each one at the centre of a face and so bridging three M atoms. This is illustrated in *Figure 11.43* for $MoCl_2$, where the eight 'face-bridging' chlorines are shown as being at the corners of a cube which embraces the Mo_6 octahedron. Two of the remaining chlorines are linked terminally

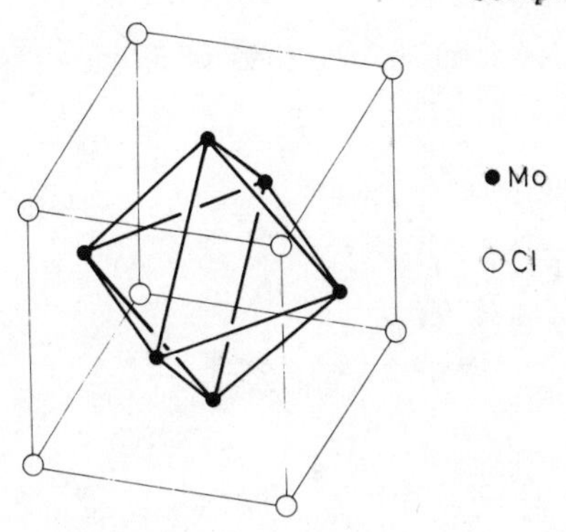

Figure 11.43 Structure of Mo_6Cl_8 cluster unit

and the others bridge neighbouring units. The stereochemistry is simplified if we group together the basic cube/octahedron (C/O) Mo_6Cl_8 unit, and illustrate the four bridging and two terminal chlorines arranged octahedrally around this unit. Mo_6Cl_{12} undergoes a number of reactions with mondentate ligands, and non-electrolyte $(Mo_6Cl_{12})L_2$ complexes can be prepared in which the bridging between units has been broken to leave the basic (Mo_6Cl_8) unit linked terminally to the two ligands and four chlorines (see *Figure 11.44*). Further reaction results in one or two of the terminal

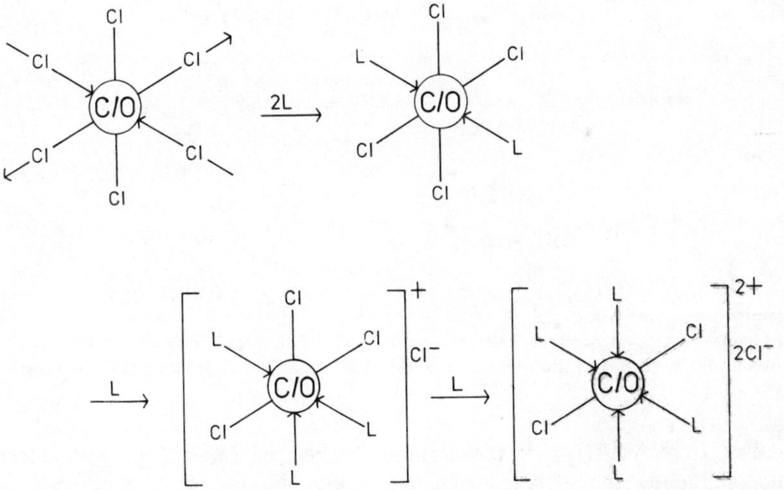

Figure 11.44 Structural interpretation of the reactions of $(MoCl_2)_6$ with donor ligands

Mo–Cl bonds being ionized to give the compounds $[(Mo_6Cl_8)Cl_3L_3]^+Cl^-$ and $[(Mo_6Cl_8)Cl_2L_4]^{2+}2Cl^-$.

Metal cluster compounds are being characterized with increasing frequency and heteronuclear clusters are now well established. One interesting recent example (Churchill and Bezman[94]) is the compound $Cu_4Ir_2(PPh_3)_2(C{\equiv}CPh)_8$, which contains an octahedral cluster (34) of four copper and two iridium atoms.

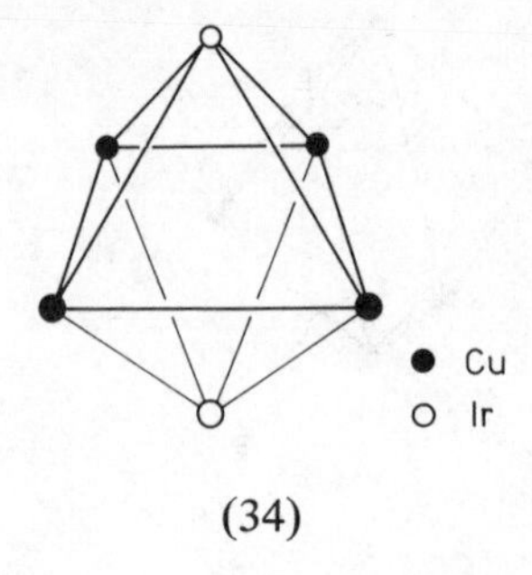

(34)

11.16 Carbonyls

Table 11.21 lists the reasonably well-characterized carbonyls containing up to four metal atoms; it does not give anionic or cationic species, although we shall refer briefly to some of these later in this section.

Table 11.21 Transition metal carbonyls*

Group V	*Group VI*	*Group VII*	*Group VIII*		
$V(CO)_6$	$Cr(CO)_6$	–	$Fe(CO)_5$	–	$Ni(CO)_4$
		$Mn_2(CO)_{10}$	$Fe(CO)_9$	$Co_2(CO)_8$	
			$Fe_3(CO)_{12}$	–	
				$Co_4(CO)_{12}$	
–	$Mo(CO)_6$	–	$Ru(CO)_5$	–	–
		$Tc_2(CO)_{10}$	–	$Rh_2(CO)_8$†	
			$Ru_3(CO)_{12}$	–	
				$Rh_4(CO)_{12}$	
–	$W(CO)_6$	–	$Os(CO)_5$	–	–
		$Re_2(CO)_{10}$	$Os_2(CO)_9$	$Ir_2(CO)_8$†	
			$Os_3(CO)_{12}$	–	
				$Ir_4(CO)_{12}$	

*List restricted to neutral carbonyls containing four or less metal atoms.
†Isolated by matrix technique and characterized by infrared spectra (Hanlan and Ozin[95])

Apart from $V(CO)_6$, all the carbonyls listed in *Table 11.21* are diamagnetic, and obey the so-called *effective atomic number* (EAN) or *noble gas rule*, which requires the number of electrons on the metal plus those donated by the co-ligands to equal the number of electrons held by the appropriate noble gas atom. Thus for the first-row elements we are aiming at the configuration of krypton, which has 36 electrons; by assuming that each CO group contributes two electrons we can see how the krypton configuration is attained (*Table 11.22*). $V(CO)_6$, the odd one out, is one electron short of the required 36, and we might expect it to dimerize in order to achieve the noble gas configuration; nevertheless it remains as a paramagnetic monomer, presumably because dimerization (and consequent

Table 11.22 Mononuclear metal carbonyls and effective atomic number (EAN)

Carbonyl	*No. of electrons from metal*	*No. of electrons from CO groups*	*Total no. of electrons*
$V(CO)_6$	23	12	35
$Cr(CO)_6$	24	12	36
$Fe(CO)_5$	26	10	36
$Ni(CO)_4$	28	8	36

seven co-ordination), would impose an unacceptable steric strain. The carbonyl can accept an electron to form the anion $[V(CO)_6]^-$, which does have the krypton configuration.

Although $V(CO)_6$ does not dimerize, the other hypothetical monomers $Mn(CO)_5$ and $Co(CO)_4$ form metal–metal bonds and so achieve the 36-electron configuration about each metal; we shall discuss their structure shortly. The necessary additional electron can also be acquired through covalent bonding with such atoms as hydrogen and chlorine to give the appropriate carbonyl hydride or carbonyl chloride, e.g. $HMn(CO)_5$, $HCo(CO)_4$ and $ClMn(CO)_5$.

If for the moment we restrict our discussion to mononuclear neutral carbonyls, we find that they have linear M–C–O systems with bonding of the CO group to the metal through the carbon atom. The carbonyls have the expected shapes, namely octahedral for $M(CO)_6$, trigonal-bipyramidal for $M(CO)_5$ and tetrahedral for $Ni(CO)_4$; these shapes are those predicted from simple electron-pair repulsion theory, and in valence-bond terms the M–C σ bonding can be described in terms of d^2sp^3, dsp^3 and sp^3 hybrid orbitals (see p.204). Rather better bonding descriptions for the σ bonds make use of the molecular orbital schemes previously discussed (see p.207 for octahedral and p.234 for tetrahedral molecules). The bonding between the metal and the CO groups cannot be simply σ, and arise solely from the donation of a σ lone pair from each carbon to the metal, because the carbon monoxide molecule is a notoriously bad Lewis base which does not form donor–acceptor molecules with good Lewis acids such as BF_3. A clue to the origin of the strength of the metal–carbon bonds is provided by the oxidation state of the metals. Thus in all the examples given the metals are formally in the zero oxidation state, and have substantial numbers of electrons in doubly filled d orbitals, and these electrons can be involved in 'back-bonding' through d_π–p_π interactions. This was discussed in Section 11.2, where we saw that three d orbitals (d_{xy}, d_{xz} and d_{yz}) may π bond with ligands that have orbitals of the correct symmetry, size and energy. For octahedral $[M(CO)_6]$ carbonyls any one of these d orbitals can be involved in π overlap with orbitals of the four ligands in the same plane (see *Figure 11.45*). As there are *three d* orbitals for bonding to *six* carbonyl groups, the maximum average π bonding can be only 0.5, giving each M–C a bond order of 1.5.

This concept of delocalized π bonding has been somewhat oversimplified and strictly speaking all π orbitals of suitable symmetry should be incorporated into the full molecular-orbital scheme; this would involve the p_π orbitals of both carbon and oxygen atoms. Readers may be helped by a more restricted consideration of the orbitals involved in just one M–C–O

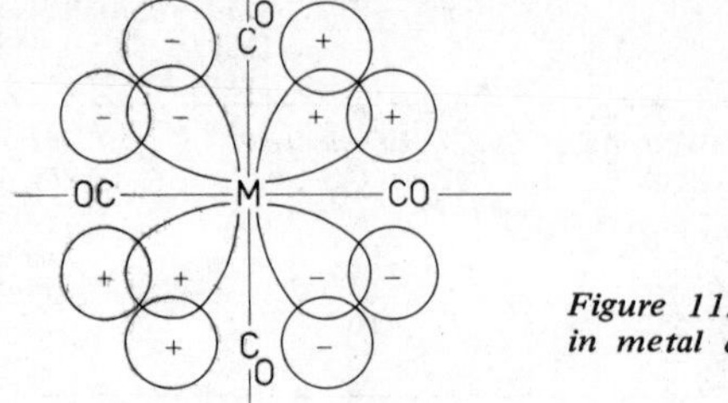

Figure 11.45 π Bonding in metal carbonyls

linkage, which is illustrated in *Figure 11.46.* This shows that the M–C d_π–p_π bonding can be extended to the oxygen atom which has a suitable doubly filled p_π orbital.

The back-bonding from the metal to the CO group not only strengthens the M–C bond, but also disperses the electron density that would otherwise build up as a result of the donation of electron pairs from the carbonyl group.

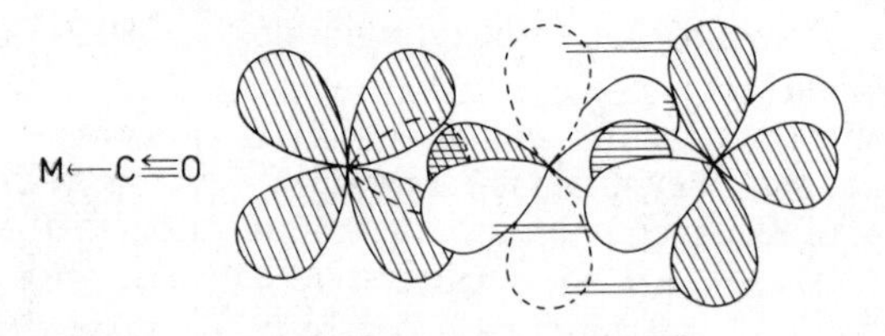

Figure 11.46 π Bonding in the MCO linkage

The binuclear carbonyls typified by $Mn_2(CO)_{10}$, $Fe_2(CO)_9$ and $Co_2(CO)_8$ all contain metal–metal bonds. With $Mn_2(CO)_{12}$ and the technetium and rhenium analogues, each atom achieves an octahedral environment by bonding with five CO groups, with the M–M bond occupying the sixth position; to reduce the steric repulsion between the carbonyl groups a staggered configuration (35) is adopted. Since $Fe_2(CO)_9$ is diamagnetic, it is clear that there is an Fe–Fe bond, but this is now supplemented by bridging of the iron atoms (Fe–Fe = 252 pm) by three carbonyl groups as in (36) (Cotton and Troup[96]), the stereochemistry being similar to that of $[W_2Cl_9]^{3-}$

(35)

(36)

(37)

which we discussed in the last section. $Co_2(CO)_8$ appears to exist in two forms, one which has a σ Co–Co bond linking two $Co(CO)_4$ groups, and a second [see (37)] in which two carbonyl groups bridge the cobalt atoms; once again the diamagnetism of the carbonyl confirms spin-pairing (i.e. bonding) between the two cobalt atoms.

The presence of bridging carbonyls has been established by X-ray crystal-structure determinations, of course, but normally infrared spectra can provide quick and fairly conclusive evidence for their presence. Thus while terminal M–CO groups in neutral carbonyls usually have CO stretching frequencies in the region between 1900 and 2150 cm^{-1}, a bridging CO group shows a frequency in the 1800 cm^{-1} region; the lower frequency of the bridging CO group reflects the lower CO bond order.

$Fe_3(CO)_{12}$ has been a difficult carbonyl to deal with since it crystallizes in thin plates, but a recent investigation by Cotton[97] has shown quite clearly that the structure contains a triangular Fe_3 cluster (38), with two carbonyl groups bridging between one pair of Fe atoms.

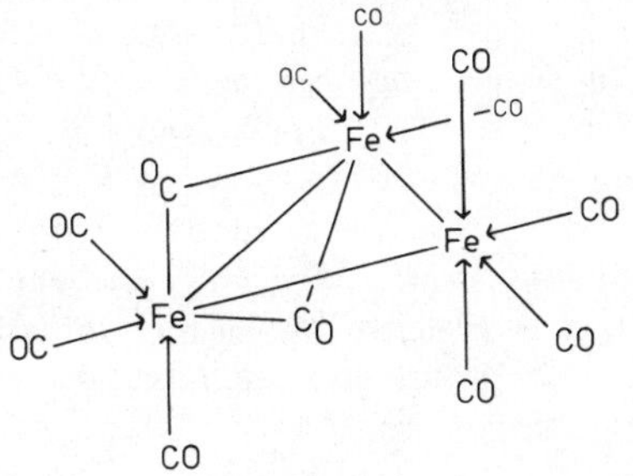

(38)

A detailed discussion of the more polymeric carbonyls would be out of place here, but it should be noted that they contain metal clusters, sometimes with bridging carbonyls. Thus $Rh_6(CO)_{16}$ contains an octahedral cluster of Rh atoms with two terminal CO groups on each Rh, and four bridging CO groups, each linking three Rh atoms. With $Os_6(CO)_{18}$, on the other hand, the Os_6 cluster adopts a bicapped tetrahedral structure (39) with three terminal CO groups associated with each Os atom (Mason *et al.*[98]).

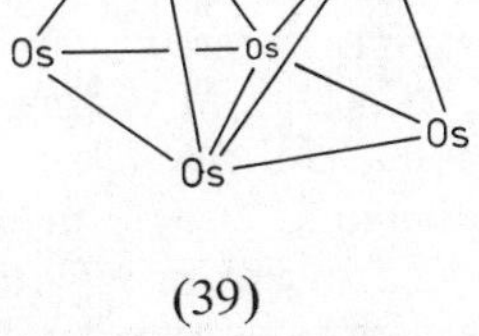

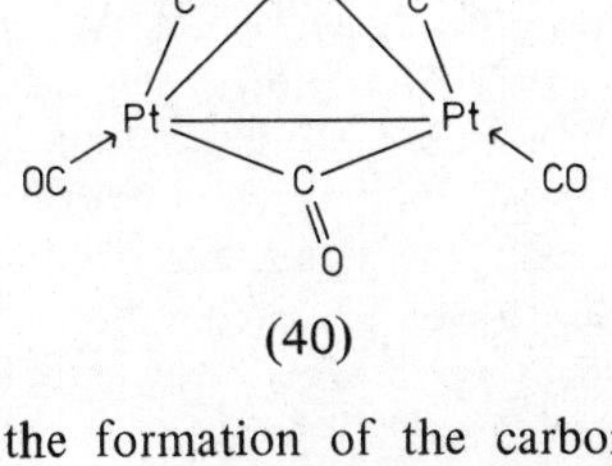

(39)

(40)

Earlier in this section we mentioned the formation of the carbonylate anion $[V(CO)_6]^-$, and it should be noted that similar anions can be formed by many of the carbonyls, e.g. $[Mn(CO)_5]^-$ and $[Co(CO)_4]^-$. Such species can also be formed by elements that do not form simple carbonyls. Thus

$[Nb(CO)_6]^-$ and $[Ta(CO)_6]^-$ have been prepared by reduction of the pentachlorides by sodium in diglyme at 100 °C under a high pressure of carbon monoxide, and by a similar reduction process a series of carbonylate anions of platinum have been prepared (Calabrese *et al.*[99]). These platinum complexes, which have the general formula $[Pt_3(CO)_6]_n^{2-}$, with n = 2, 3, 4 and 5, contain triangular clusters (40) of Pt_3 atoms with a Pt–Pt distance of 266 pm; there are three bridging and three terminal carbonyl groups. These planar units stack in an almost eclipsed form with the Pt–Pt distances between planes being 310 pm.

The CO groups in carbonyls can be replaced by typical ligands such as triarylphosphines or by π-donating groups such as cyclopentadiene (see next section) and olefins, etc. Thus donor ligands that provide a single σ lone pair will replace CO groups on a 1 : 1 basis so that $Mo(CO)_6$, for instance, will give $Mo(CO)_5L$, $Mo(CO)_4L_2$ and $Mo(CO)_3L_3$ complexes with quite a few monodentate ligands, typical examples being Ph_3P and py. Bidentate ligands such as diars replace two CO groups at a time to give complexes of the type $Mo(CO)_4$diars, $Fe(CO)_3$diars and $Ni(CO)_2$diars. These substituted carbonyls are particularly well characterized for ligands such as phosphines and arsines that can behave like CO groups as π-electron acceptors, such that the central metal can lighten its electron load through d_π-d_π bonding.

A range of compounds can be prepared by the reaction of metal carbonyls with π-electron donors such as cyclopentadiene (Cp), benzene and various alkenes and alkynes (Pauson[100]). Thus with cyclopentadiene compounds we get one Cp group providing five electrons, so that the substitution complexes take the form $CpV(CO)_4$, $[CpCr(CO)_3]_2$, $CpMn(CO)_3$, $[CpFe(CO)_2]_2$, $CpCo(CO)_2$ and $[CpNi(CO)]_2$. To achieve the noble gas configuration a metal–metal bond is necessary for those complexes listed as dimers.

Readers interested in the role played by metal clusters in the structures of metal carbonyls and their derivatives are referred to an excellent review by King[101].

11.17 Cyclopentadiene Complexes

The first example of this very interesting group of complexes to be discovered was 'ferrocene', an iron compound, $FeCp_2$, where Cp is the cyclopentadiene group (C_5H_5). A large number of similar compounds of general formula MCp_2, where M is one of the first-row transition elements Ti to Ni, has now been made, and several of these compounds have been oxidized to the cation, $[MCp_2]^+$, in which M is formally tervalent. These compounds are listed in *Table 11.23.* Several of the heavier transition metals, e.g. iridium and ruthenium, also form cyclopentadiene derivatives similar to those in *Table 11.23*, and many other related compounds such as Cp_2TaBr_3 are known; readers interested in detailed accounts of these compounds are referred to review articles by Wilkinson and Cotton[102] and Fischer and Fritz[103].

Cyclopentadiene compounds are soluble in the common organic solvents, and all the Cp_2M compounds (except Cp_2Ti) have melting points close to

Table 11.23 Cyclopentadiene complexes formed by the first-row transition metals

Element	*Compound*	*M.p. /°C*	*Colour*	*Magnetic moment /BM*	≡ *Unpaired electrons*
Cu(II)	–	–	–	–	–
Ni(II)	Cp_2Ni	173	Green	2.86	2
Ni(III)	$[Cp_2Ni]^+$	–	Yellow	1.75	1
Co(II)	Cp_2Co	173	Purple	1.76	1
Co(III)	$[Cp_2Co]^+$	–	Yellow	Diamagnetic	0
Fe(II)	Cp_2Fe	173	Orange	Diamagnetic	0
Fe(III)	$[Cp_2Fe]^+$	–	Blue	2.26	1
Mn(II)	Cp_2Mn	173	Pink	5.9	5
Cr(II)	Cp_2Cr	173	Scarlet	2.84	2
Cr(III)	$[Cp_2Cr]^+$	–	Green	3.81	3
V(II)	Cp_2V	168	Purple	3.82	3
V(III)	$[Cp_2V]^+$	–	Purple	2.86	2
Ti(II)	Cp_2Ti	>130	Green	Diamagnetic	0
Ti(III)	$[Cp_2Ti]^+$	–	Green	2.3	1

173 °C. The relatively simple infrared spectra of all these compounds are very similar to each other, and indicate a high degree of symmetry. The visible and ultraviolet absorption spectra, on the other hand, are quite dissimilar, indicating very different electronic configurations. 'Ferrocene' was first thought to have a simple σ bonded structure $(C_5H_5)–Fe–(C_5H_5)$, but this formulation is inconsistent with the properties of the compound. An X-ray investigation shows that the molecule has the hitherto unknown 'sandwich' structure, where the iron atom is placed between the two cyclopentadiene rings in a pentagonal-antiprismatical arrangement (*Figure 11.47*).

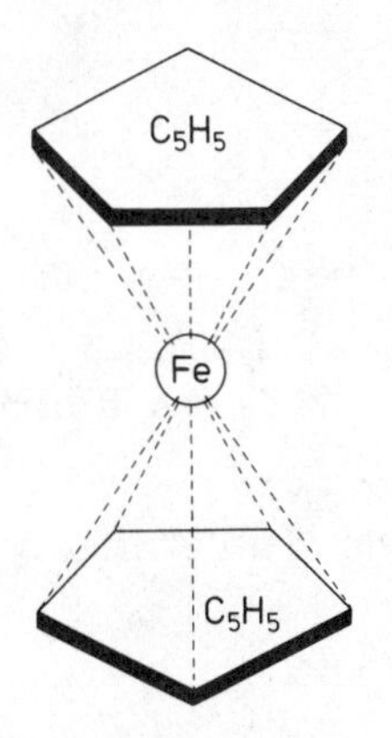

Figure 11.47 The 'sandwich' structure of ferrocene

While the arrangement of the atoms in ferrocene is known, there is a healthy measure of controversy about the nature of the bonding in the molecule, and quite a large number of theories has been advanced. It is not possible in a book of this size to discuss all these theories, some of which are extremely mathematical, and we shall limit ourselves to an elementary discussion in which we merely indicate some of the ideas behind the more quantitative theories.

We have to account for the bonding between the cyclopentadiene groups (C_5H_5) and the iron atom, and this bonding can be described as a pairing of an electron of each ring with one of the unpaired *d* electrons of the iron atom – so forming a single bond between the iron atom and each ring.

There are five atomic π-orbitals in each cyclopentadiene ring, one on each carbon atom, and five molecular orbitals can be compounded from them. (We recall that in benzene, p.100, there are six atomic π-orbitals which give six molecular orbitals, three bonding and three anti-bonding). The lowest-energy molecular orbital (ψ_1) of cyclopentadiene resembles the corresponding benzene orbital, with a continuous 'streamer' above and below the ring, extending over all five carbon atoms. The molecular orbitals of the next lowest energy (ψ_2 and ψ_3) are degenerate, and resemble the 'split streamers' of benzene, with a node in either the *xz* or *yz* plane. (We take the line joining the centroids of the cyclopentadiene rings and passing through the iron atoms as the *z* axis; the *x* and *y* axes lie in a plane containing the iron atom parallel to the cyclopentadiene rings.) There are three electrons in the two orbitals ψ_2 and ψ_3, so that one of them must contain an unpaired electron (see *Figure 11.48*).

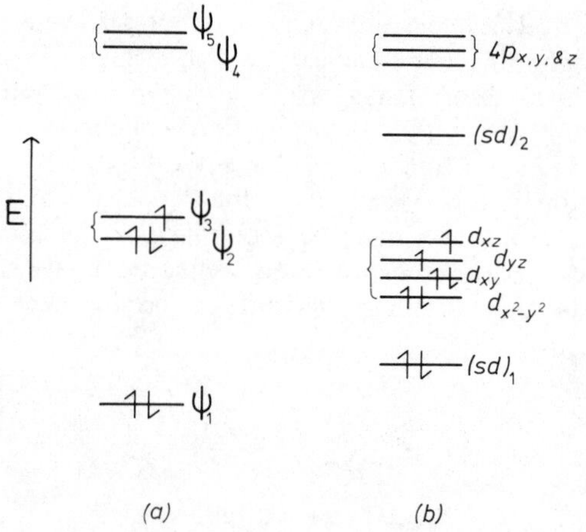

Figure 11.48 Relative energy levels for (a) π molecular orbitals of the C_5H_5 rings; (b) atomic orbitals (modified) of the iron atom. Degenerate levels, bracketed together, are separated for clarity

The electronic configuration of the ground state of iron is

$$\text{Fe(argon core)}(3d)^2(3d)^1(3d)^1(3d)^1(3d)^1(4s)^2$$

Moffitt suggested that under the influence of the approaching cyclopentadiene groups, the 4*s* and the $3d_{z^2}$ orbitals hybridize to give two non-equivalent *sd* hybrid orbitals (see earlier discussion on *sd* hybridization, p.229). One of these hybrid orbitals, ψ_2, denoted $(sd)_1$, has a lower energy than the $3d_{z^2}$ orbital, since the charge density is concentrated more in the *xy* plane, and repulsion with the electrons of the cyclopentadiene is diminished; the other orbital, $(sd)_2$, is of higher energy, approximately that of the $4p$ orbitals, with an enhanced charge density around the z axis. Thus the modified configuration of the iron atom is written

$$\text{Fe(argon core)}(sd)_1{}^2(3d)^2(3d)^2(3d)^1(3d)^1(sd)_2{}^0(4p_x)^0(4p_y)^0(4p_z)^0$$

We now get overlap between a molecular orbital of each ring with an appropriate d orbital of the iron atom (d_{xz} for one ring and d_{yz} for the other); one such overlap is illustrated in *Figure 11.49.* In some respects this bonding resembles that between olefins and metal ions (see Section 11.18). However, this representation of bonding is oversimplified, and in the more sophisticated molecular-orbital treatment the two bonds embrace the iron atom and

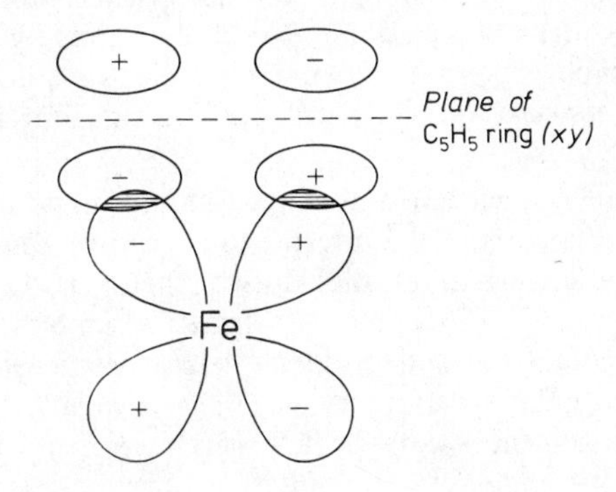

Figure 11.49 Overlap of Fe d_{xz} *orbital with a* π *molecular orbital (shown in cross section) of cyclopentadiene*

both cyclopentadiene rings. Thus Moffitt obtained two delocalized bonding orbitals by combining the two iron d orbitals with a π molecular orbital (ψ_2 or ψ_3) from each ring; two anti-bonding orbitals are obtained at the same time (see *Figure 11.50*). Other more elaborate treatments have been made

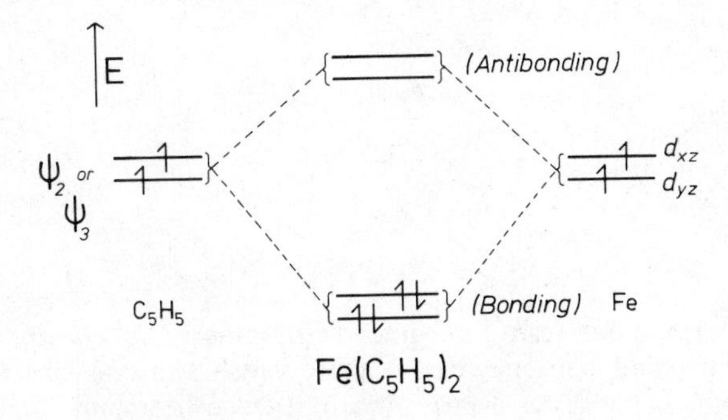

Figure 11.50 π *Molecular-orbital energy diagram for ferrocene*

which allow for the participation of the $4p$ orbitals of iron and the π-bonding electrons of the rings in the bonding. All of the π electrons in the rings cannot take part completely in the bonding, however, since the aromatic character of the cyclopentadiene rings is largely retained in the ferrocene-type compounds. Moreover, magnetic anisotropy measurements on Cp_2Fe indicate a single d_π–p_π bond between Fe and the rings.

We can use the Moffitt scheme (*Figure 11.48*) to account for the magnetic properties of some of the compounds. Thus the extra electron in Cp_2C (compared with Cp_2Fe) will go into either the $4p$ or the $(sd)_2$ orbitals, while in the corresponding nickel compound the two additional electrons will either singly occupy $4p$ orbitals or singly occupy a $4p$ and the $(sd)_2$ orbital. The chromium compound, with two electrons fewer than ferrocene has the configuration $(sd)_1{}^2(3d)^2(3d)^2(3d)^1(3d)^1$, with two unpaired electrons. The scheme is not entirely satisfactory for vanadium and titanium, however, which have three unpaired and no unpaired electrons, respectively, since to explain the three unpaired electrons in the vanadium compound it is necessary to assume that the $(sd)_1$ orbital has approximately the same energy as the $3d_{xy}$ and the $3d_{x^2-y^2}$ orbitals; this is not unreasonable, because the charge density in all three orbitals is largely concentrated in the xy plane, but it would predict two unpaired electrons for the titanium compound. More elaborate treatments proposed to account for this are outside the scope of this book.

We have not so far mentioned the manganese compound, but it is apparent that the magnetic moment, corresponding to five unpaired electrons, does not fit into the pattern of the Moffitt theory, which would predict one unpaired electron. The compound closely resembles the corresponding magnesium cyclopentadiene, and evidently has an ionic structure containing $(C_5H_5)^-$ anions and Mn^{2+} cations (which have five unpaired electrons). The sandwich arrangement would be a natural one for these ions.

All of the cyclopentadienyl compounds we have discussed have had 'sandwich' structures, with the two Cp rings parallel to one another and sandwiching the metal. In complexes of stoichiometry Cp_2MX_2, such as Cp_2NbCl_2 and Cp_2ZrCl_2, the Cp rings are tilted towards one another so as to form an angle of around 50°; the two X groups are in a plane at right angles to that containing the normals to the rings [see (41)]. In a recent paper, Prout and co-workers[104] review the structures of such compounds and comment on possible modes of bonding.

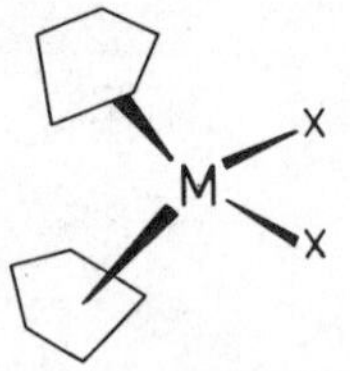

(41)

Sandwich compounds can be formed with other ring systems besides cyclopentadiene, and benzene for instance, which has one electron more than cyclopentadiene, gives diamagnetic dibenzenechromium, $Cr(C_6H_6)_2$, which has the configuration of ferrocene.

11.18 Alkene and Alkyne Complexes

Ethene, propene and a number of other olefins form well-defined complexes with Pd(II), Pt(II), Cu(I) and Ag(I). Divalent platinum, for instance, forms several complexes with ethene: $K[Pt(C_2H_4)Cl_3)],H_2O$, $[Pt(C_2H_4)Cl_2]_2$

and $Pt(C_2H_4)_2Cl_2$. The first of these, the so-called Zeise's salt, which has been known for almost 150 years, has the structure (42). The dimeric complex has a chlorine–chlorine bridged structure (43) and the compound

(42)

with two ethene groups per platinum atom has the *trans* configuration (44). In each molecule the platinum atom can be regarded as having a square-planar distribution, with one link pointing to the middle of the C=C bond and the C–C axis perpendicular to the $PtCl_3$ plane.

(43) (44)

The nature of the bonding in these alkene complexes has been long discussed and many suggestions have been advanced. X-ray studies have firmly ruled out the possibility of non-equivalent bonding of the carbon atoms to platinum, and it is known that the ethene moiety remains essentially intact. There appears to be some weakening of the carbon–carbon double bond in such complexes, however, since the infrared spectra show an appreciable lowering (80–150 cm^{-1}) of the carbon–carbon stretching frequency.

The simplest interpretation of the bonding is based on the so-called Dewar-Chatt model, in which the ethene molecule donates its electron pair of the π bond into a vacant orbital (dsp^2) of the platinum. This σ donation is considered to be supplemented by back-donation of electrons from a doubly filled d orbital of platinum into the antibonding π^* orbital of the ethene molecule. The overlap is somewhat improved if the platinum donor π electrons are placed in a dp hybrid orbital rather than a pure d orbital. This bonding scheme is illustrated in *Figure 11.51*.

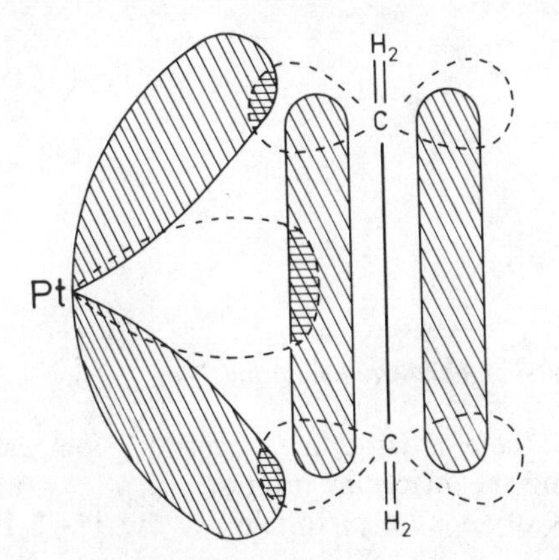

Figure 11.51 The orbital diagram for double bonding of the olefin complexes of divalent platinum

While this Dewar–Chatt model nicely accounts for the stability of olefin complexes it is oversimplified, and in recent years there has been considerable discussion about the relative amounts of σ and π bonding in the Pt–C_2H_4 linkage. Thus calorimetric studies (Partenheimer and Durham[105]) have led to the conclusion that σ bonding is much more important, and this is in agreement with detailed theoretical calculations (Rösch *et al.*[106]) of Zeise's anion, which suggests that π bonding does not exceed 25%. Of course, the amount of π bonding should be of much greater importance with better π acceptors such as C_2F_4.

Up until now these complexes have been discussed on the basis of the ethene molecule being planar, but neutron diffraction studies (Hamilton *et al.*[107]) have shown that the four C–H bonds are bent away from the platinum and out of the C_2H_4 plane, the displacement being around 18 pm (see *Figure 11.52*). In view of this it is permissible to discuss the structure

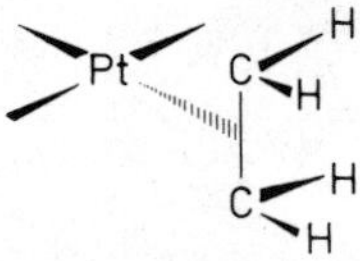

Figure 11.52 Distortion of olefin molecule in platinum complexes

on the basis of a cyclopropane-like three-membered PtC_2 ring, in which the platinum atom forms a (bent) σ bond to each carbon atom, rather than a σ and a π bond to the C_2 unit.

Readers may be helped if this approach is described in valence-bond terms. Thus we need two equivalent platinum orbitals, and these can be compounded from the dsp^2 (used for σ bonding) and the dp (used for π bonding); this is shown in *Figure 11.53* in simplified form. The ethene molecule is no longer planar, so we can regard the bonding as somewhere

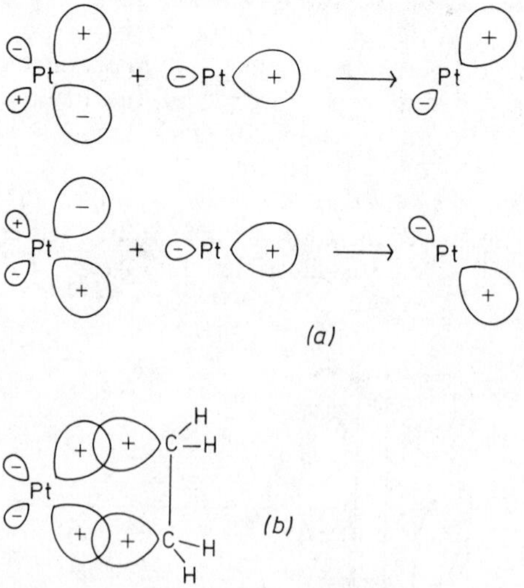

Figure 11.53 Bonding in olefin complexes: σ bonding description

between sp^2 (+ p) and sp^3. Hence there is an orbital available from each carbon that can overlap with each of the platinum orbitals.

Thus there are two extreme ways of looking pictorially at the Pt–C_2H_4

bonding, either a Pt $\rightleftarrows$ C_2H_4 σ–π system, or a pair of Pt–C σ bonds. In our simplified descriptions the same orbitals are used but grouped together differently.

It is interesting to note that hexa-1,5-diene forms a compound $C_6H_{10}PtCl_2$ (45) which is much more stable than the analogous *trans*-bisethene complex (44) because the diene is acting as a bidentate ligand, and each double bond can link with platinum.

(45)

Alkyne complexes can also be formed by platinum and neighbouring elements, and the infrared spectra again suggest a weakening of the carbon–carbon bond since the stretching frequency is lowered by around 200 cm^{-1}. In molecules such as Pt(PhCCPh),$2PPh_3$ (Glanville[108]) and Pd{$C_2(CO_2Me)_2$},$2PPh_3$ (McGinnety[109]) the structures are best considered as giving a square-planar environment to the metals, with the carbon atoms of the 'triple' bond occupying two of the four positions (*Figure 11.54*).

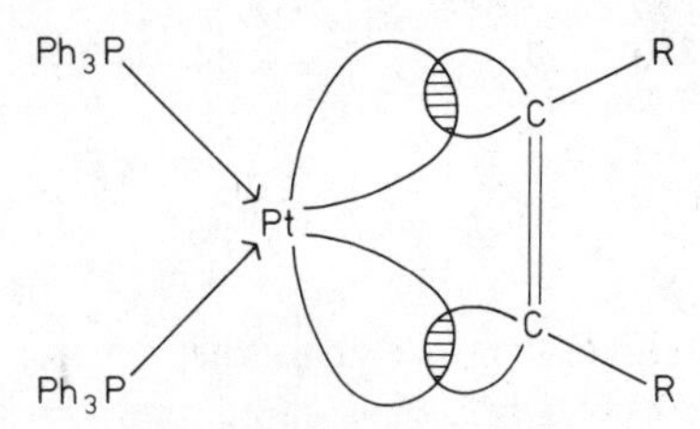

Figure 11.54 Bonding in ethyne complex of divalent platinum

Once again the groups linked to the carbon–carbon multiple bond are bent away from the metal atoms, and in *Figure 11.54* we have represented the bonding as two M–C σ bonds and a C=C double bond. This may be a somewhat exaggerated description in so far as the palladium compound is concerned, since the C–C bond distance of 128 pm is between the double and triple bond lengths. As with the alkene complexes, the bonding in alkyne complexes between the metal and the two carbon atoms can be described in terms of donor σ and acceptor π bonds. In molecular-orbital terms there are three (3-centre) bonding molecular orbitals embracing the metal and the two carbon atoms; they correspond roughly to donation from the alkyne π to a metal σ orbital, the C–C σ bond, and back-donation from a doubly filled metal *d* orbital to an antibonding (π*) orbital of the alkyne. In addition, there is an unmodified π bond between the two carbon atoms in a plane perpendicular to the MCC plane. The orbital scheme is illustrated in *Figure 11.55.*

11.19 Reactivity of Complex Compounds: Orbital Considerations[3]

When we come to consider the reactivity of complex compounds we find that the rate of reaction is often more important than stability, i.e. we consider

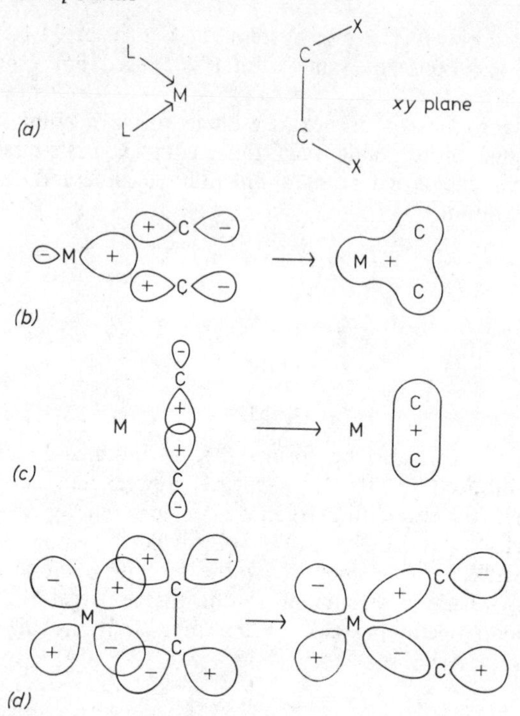

Figure 11.55 Simplified molecular-orbitals for ethyne complexes: (a) schematic formula; (b), (c), (d) molecular orbitals

the speed at which equilibrium is reached rather than the position of equilibrium. Although a thermodynamically stable compound is often slow to react and unstable complexes often react quickly, this is not always the case. Let us, for instance, consider reactions of some complex cyanide anions in which there is exchange of 'labelled' cyanide ions with unlabelled cyanide ions from the complex:

$$[M(CN)_n]^{x-} + CN^{-*} \rightleftharpoons [M(CN)_{n-1}(CN^*)]^{x-} + CN^-$$

Table 11.24 shows that the four-co-ordinate complexes exchange cyanide ions rapidly, whereas the six-co-ordinate complexes exchange only slowly. The exchange rates do not fall into the same order as the dissociation constants.

It should be noted that whereas the typical ionic reactions usually associated with inorganic compounds are 'instantaneous', complex compounds may react much more slowly. The reaction rates vary widely, however, and depend upon the nature of the reactant and the experimental conditions, but complexes can be grouped into two fairly distinct 'reactivity classes':

1. *Labile* complexes, which react quickly (within the time of mixing).
2. *Inert* complexes, which react only slowly.

Table 11.24 Reactivity of complex cyanide ions

Complex ion	*Dissociation constant*	*Exchange rate with* CN^-
$[Ni(CN)_4]^{2-}$	10^{-22}	Very fast
$[Fe(CN)_6]^{4-}$	10^{-37}	Very slow
$[Hg(CN)_4]^{2-}$	10^{-42}	Very fast
$[Fe(CN)_6]^{3-}$	10^{-44}	Very slow

The distinction is surprisingly clear-cut and there are relatively few borderline cases. Although the reactivity depends upon such factors as the nature of the substituting ion, its charge and its radius, we usually find that the electronic configuration of the complex and the nature of the bonding are more important considerations, and we will concentrate our discussion on these aspects. Readers interested in a more detailed discussion of the mechanism of such reactions are referred to the major book by Basolo and Pearson[3] or Tobe's shorter monograph[110].

11.19.1 VALENCE-BOND APPROACH

Initially, we will concern ourselves particularly with the qualitative aspects of reactivity, and here the valence-bond approach is quite helpful and easy to understand. Readers will recall our earlier discussion of bonding in octahedral complexes, when two distinct categories emerged, *inner* (strong-field), which used d^2sp^3 orbitals, and *outer* (weak-field) which used sp^3d^2 orbitals.

The inner complexes, which make use of the inner *d* orbitals, may be quite sharply divided into labile and inert compounds. If the metal ion has three or more electrons available for the *d* orbitals, it must, by Hund's rule, have at least one electron in each of the three *d* orbitals which are not used for bonding. Such ions are invariably inert. Those ions which have fewer than three electrons in the *d* orbitals must have at least one completely vacant *d* orbital. Such ions are always labile. *Table 11.25*

Table 11.25 Labile and inert inner complexes

Ion	*Nature of complexes*	*Electronic configuration**
Sc^{3+}, Ti^{4+}	Labile	d^0 d^0 d^0 [D^2 D^2 S^2 P^2 P^2 P^2]
Ti^{3+}	Labile	d^1 d^0 d^0 [D^2 D^2 S^2 P^2 P^2 P^2]
V^{3+}	Labile	d^1 d^1 d^0 [D^2 D^2 S^2 P^2 P^2 P^2]
Cr^{3+}	Inert	d^1 d^1 d^1 [D^2 D^2 S^2 P^2 P^2 P^2]
Mn^{3+}	Inert	d^2 d^1 d^1 [D^2 D^2 S^2 P^2 P^2 P^2]
Fe^{3+}	Inert	d^2 d^2 d^1 [D^2 D^2 S^2 P^2 P^2 P^2]
Co^{3+}	Inert	d^2 d^2 d^2 [D^2 D^2 S^2 P^2 P^2 P^2]

*The electronic arrangement is described using Taube's nomenclature, the small *d*'s representing orbitals not used in bonding and the capital *D, S* and *P*'s representing the orbitals that have accepted lone-pair electrons from the six ligands lists examples of labile and inert complexes, together with the electronic configurations of the metal ions.

A very sharp change occurs at Cr^{3+}, whose complexes undergo only very slow substitution reactions in contrast to the rapid reactions of all V^{3+} complexes. This distinction can be accounted for if it is assumed that a vacant *d* orbital can be used to accept an electron pair from the substituting ligand, thus providing an 'easy' reaction route through a seven-co-ordinate intermediate. This 'transition-state complex' then breaks down with the release of one of the originally co-ordinated groups, i.e.

$$MX_6 + Y \rightarrow [YMX_6] \rightarrow YMX_5 + X$$

Such a process (known in theoretical organic chemistry as an S_N2 process – substitution, nucleophilic, bimolecular) is likely to take place readily, since the activation energy needed should be low. Hence d^0, d^1 or d^2 transition metal complexes might be expected to be labile.

With a d^3 ion, such as Cr^{3+}, there is no vacant *d* orbital, and it can only be made available by electron pairing – a process requiring energy; alternatively the approaching group could make use of a higher-energy, outer orbital. Thus in such cases any S_N2 process which requires the formation of an intermediate of co-ordination number seven would require a high activation energy, and it would be slow. It is found experimentally that inert complexes react by an S_N2 mechanism only when the substituting group is a strong nucleophile, and that with other reagents the reaction goes by way of an S_N1 mechanism, in which the M–X bond breaks, leaving a five-co-ordinate transition-state complex that can take up the substituting group Y, i.e.

$$MX_6 \rightarrow [MX_5] + X \xrightarrow{+\ Y} YMX_5$$

Outer complexes of transition metals are generally labile, and the valence-bond approach predicts this. Thus the metal uses two of the outer *d* orbitals in sp^3d^2 bonding and the other three are available to accept electron pairs from approaching groups. Provided there is no steric hindrance, a seven-co-ordinate intermediate should form readily and the substitution reaction will be facilitated.

11.19.2 CRYSTAL-FIELD APPROACH

The valence-bond description is purely qualitative and cannot make predictions about relative rates of reaction of, say, d^1 and d^2 complexes. Moreover, all outer complexes are placed in one group, apart from considerations of ion size and charge, and there is no distinction related to electronic configuration.

Basolo and Pearson have put the problem on a more quantitative basis by the application of crystal-field theory. They point out that when octahedral complexes react, the transition state in an S_N1 reaction will have five co-ordination (being either trigonal-bipyramidal or square-pyramidal), and in an S_N2 reaction will have seven co-ordination (pentagonal-bipyramidal),

Table 11.26 Losses in CFSE for strong-field octahedral complexes forming pentagonal-bipyramidal or square-pyramidal intermediates

No of d *electrons*	*CFSE/Δ*			*Loss in CFSE/Δ*	
	Oct	*PBP*	*SP*	*Oct → PBP*	*Oct → SP*
0	0	0	0	0	0
1	0.4	0.53	0.46	−0.13	−0.06
2	0.8	1.06	0.91	−0.26	−0.11
3	1.2	0.77	1.00	0.43	0.2
4	1.6	1.30	1.46	0.30	0.14
5	2.0	1.83	1.91	0.17	0.09
6	2.4	1.55	2.00	0.85	0.4
7	1.8	1.27	1.91	0.53	−0.11
8	1.2	0.77	1.00	0.43	0.2
9	0.6	0.49	0.91	0.11	−0.31
10	0	0	0	0	0

and they assume that a change in CFSE in going from a co-ordination number of six to one of five or seven can be considered as a ligand-field contribution to the activation energy of the reaction. If the CFSE for the transition state is smaller than that for the octahedral complex, the activation energy contribution will be large, and the reaction will be slow. *Table 11.26* summarizes these changes in CFSE for strong-field complexes; the differences quoted (in Δ) are those between the CFSE for the octahedral configuration and that for either the square-pyramidal or pentagonal-bipyramidal arrangements. Values are not given for transition states involving the trigonal-bipyramidal configuration since they are much greater than those for the square-pyramidal arrangement; an S_N1 reaction would therefore always go by way of the square-pyramidal transition state.

From the values quoted in *Table 11.26* it can be seen that irrespective of whether the reaction proceeds by an S_N1 or an S_N2 mechanism, the d^0, d^1 and d^2 complexes will react quickly compared with the d^3, d^4, d^5 and d^6 complexes. Thus the ligand-field approach confirms the general predictions of the valence-bond method, but the CFSE changes quoted in *Table 11.26* further suggest a decreasing order of reactivity:

$$d^5 > d^4 > d^3 > d^6$$

Similar CFSE calculations can be made for spin-free complexes and the results are given in *Table 11.27* for d^4, d^5, d^6 and d^7 complexes; there is no distinction between spin-free and spin-paired for other complexes (see p.220). The data show that there is no loss in CFSE when the octahedral complex forms an intermediate, except for a very small amount for

Table 11.27 Losses in CFSE for weak-field octahedral complexes forming pentagonal-bipyramidal (PBP) or square-pyramidal (SP) intermediates

No. of d *electrons*	*CFSE/Δ*			*Loss in CFSE/Δ*	
	Oct	*PBP*	*SP*	*Oct → PBP*	*Oct → SP*
4	0.6	0.49	0.91	0.11	−0.31
5	0	0	0	0	0
6	0.4	0.53	0.43	−0.13	−0.06
7	0.8	1.06	0.91	−0.26	−0.11

a d^4 complex reacting by way of a seven-co-ordinate intermediate. Accordingly we expect all the d^4–d^7 spin-free complexes to be labile unless there are exceptional circumstances such as bulky polydentate ligands.

It can be seen that both valence-bond and crystal-field descriptions provide a qualitative explanation for the reactivity of octahedral transition metal complexes, although the crystal-field method is more satisfying since it provides at least a partial explanation of the relative rates of reaction of a series of complexes that might otherwise be labelled simply labile or inert. It must be emphasized, however, that the crystal-field approach is considerably oversimplified because it assumes perfect geometries for the various polyhedra, and it ignores such factors as ligand–ligand repulsions.

11.20 The Stability and Reactivity of Four-co-ordinate Complexes: the '*trans*' Influence and '*trans*' Effect[111]

Earlier in this chapter (Section 11.5) we discussed the part played by the electronic configuration of a metal in the stability of its octahedral complexes, and this was extended in the last section to include the reactivity of such complexes. Similar considerations apply to complexes of other co-ordination numbers, of course, and there has been considerable discussion about four-co-ordinate complexes, and particularly square-planar complexes. As far as reactivity goes, some square-planar complexes (e.g. $[Ni(CN)_4]^{2-}$) are labile and others (e.g. $[PtX_4]^{2-}$) are inert, although both examples quoted should be labile according to simple valence-bond ideas because in each case there is a vacant p orbital available to accept an electron pair from a would-be substituent group. However, the distinction is accounted for if the crystal-field approach is used because for a given type of ligand the value of Δ is very much greater for platinum than for nickel, so that the loss in CFSE in forming the transition state will be accordingly greater.

Much of the interest in square-planar complexes has been concerned with the influence one ligand has on the ligand *trans* to it in the complex. Formula (46) shows a general structure MLAXY, with L and A two ligands

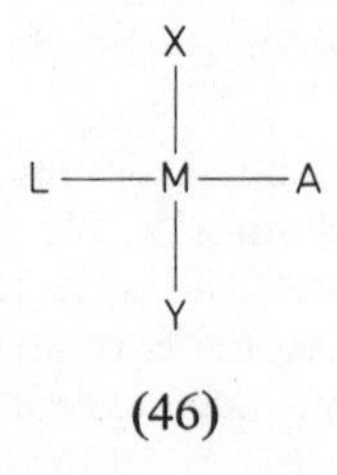

(46)

trans to one another. Unfortunately some confusion has arisen about the way in which L influences A, and the expression '*trans*-effect' has been used indiscriminantly in respect of the way in which L influences the stability of the *trans* M–A bond, and its effect on the reactivity of this bond. Venanzi and co-workers[112] have set out a very useful clarification. They have defined the '*trans*-influence' of ligand L as the extent to which it weakens the bond (M–A) *trans* to itself; this bond weakening refers to the normal equilibrium state of the molecule. The term '*trans*-effect', on the other hand, is defined as the effect of one ligand (L) on the rate at which the *trans* ligand (A) undergoes substitution reactions; hence the *trans*-effect describes a kinetic phenomenon and it is concerned with the transition state in substitution reactions.

It is worth outlining some of the available experimental evidence before we discuss the bonding aspects. As far as the equilibrium state of the molecule is concerned, there is quite a lot of X-ray structural evidence about the lengths of bonds *trans* to particular ligands. This evidence has to be treated with some caution because the variations in bond length are fairly small, and for earlier structural studies may be of the same order of magnitude as the experimental errors; moreover, crystal-packing factors may vary from compound to compound.

Readers may be helped if we quote just two complexes, $(PtCl_2,PPr^n_3)_2$ and $(PtCl_2,AsMe_3)_2$, which have analogous dimeric chlorine-bridged structures (47). *Table 11.28* lists the Pt–Cl bond lengths in the bridges, and

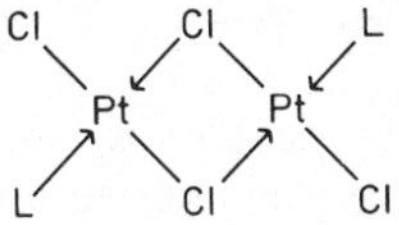

(47)

it can be seen that the longest bond is that *trans* to PPr^n_3; $AsMe_3$ has a slightly smaller effect and Cl^- appreciably less.

Table 11.28 Bond lengths of Pt–Cl bridging bonds in $(PtCl_2,L)_2$

Complex	*Pt–Cl/pm*	
	trans *to terminal Cl*	trans *to terminal L*
$(PtCl_2,PPr^n_3)_2$	232	243
$(PtCl_2,AsMe_3)_2$	231	239

A cautious consideration of the crystal structures of a considerable number of square-planar complexes suggests the following *trans* influence order:

$$R_3Si^- \approx CH_3^- \approx H > PR_3 > AsR_3 > CO > RNC > C{=}C \approx Cl^- \approx NH_3$$

Infrared spectroscopy has been used to devise a similar series. For this purpose a closely related series of square-planar complexes is chosen, the infrared spectra measured and bond-stretching frequencies assigned; thus for the MLAXY complexes of formula (46), assignments are made for the ν_{M-A} stretching frequency. It is assumed that a lowering of this stretching frequency reflects a weakening of the M–A bond as a result of the *trans*-influence of L. Typical data are given in *Table 11.29* about the ν_{Pt-Cl} frequency in two series of complexes, *trans*-$PtClL,2PMe_3$ and $[PtCl_3L]^-$, from which it can be seen that the lowering of this frequency follows the same general pattern as the lengthening of the *trans* Pt–Cl bond.

A *trans*-influence series may also be deduced from nuclear magnetic resonance spectra. One important approach considers variations in the metal–ligand coupling constants, these being a measure of the contribution

Table 11.29 ν_{Pt-Cl} (in cm^{-1}) for square-planar platinum(II) complexes

L	trans-$PtClL,2PMe_3$	$[PtCl_3L]^-$
Me_3Si^-	238	–
H^-	269	–
Et_3P	–	271
Ph_3P	298	279
Et_3As	–	280
Ph_3As	306	–
py	337	–
C_2H_4	–	309
Cl^-	340	–
CO	344	322

of the metal *s* orbital to the covalent metal–ligand bond. Consideration of Pt–P coupling constants give rise to the *trans*-influence series:

$$R_3Si^- > CH_3^- > PR_3 > AsR_3 > py > Cl^-$$

Readers interested in a full account of X-ray and spectroscopic approache to the study of *trans*-influence are referred to the excellent review by Appleton *et al.*[111].

When we come to consider the ease with which square-planar complexes undergo substitution reactions, it is apparent that some ligands have a particular ability to 'labilize' *trans* groups. This labilizing has an effect on the *trans*-effect, and from various qualitative observations the ligands can be placed in an order of decreasing *trans*-effect:

$$CH_3^- \approx H^- \approx C_2H_4 \approx CO \approx CN^- > PR_3 > I^- > Br^- > Cl^- > py > NH_3 > H_2O$$

Readers will notice a lot in common between the *trans*-influence and the *trans*-effect orders, with ligands such as CH_3^- and H^- at the top, PR_3 in the middle, and Cl^- and NH_3 at the bottom end. The main difference is that ligands such as CO and C_2H_4 have only a small *trans* influence but a large *trans* effect; it will be recalled (Sections 11.16 and 11.18) that these ligands were characterized by their ability to form d_π–p_π back bonding with the metal.

In developing simple theoretical explanations for the structures and behaviour of square-planar complexes we must recognize that while ligands having a big *trans*-influence also have a big *trans*-effect, there is a group of π-acceptor ligands such as CO and C_2H_4 that has only a small *trans*-influence but a big *trans*-effect. Initially we confine our discussion to *trans*-influence and omit the π-acceptor ligands to try and establish some simple explanation for the weakening of the *trans* bond.

Early theories were electrostatic ones and involved the mutual polarization of a ligand and the central metal ion. Thus the metal M was considered to induce a dipole in L, which in turn induced a dipole in M; the net effect was a negative charge build-up on the side of M facing A, such that negative charge in A was repelled and the M–A bond weakened; A had a similar effect on the M–L bond, but M–A was weakened most if L was more polarizable than A. The trouble with this approach is that it is concerned with point charges and dipoles and is inappropriate for complexes in which the bonding is predominantly covalent.

Accordingly, we look towards theories of covalent bonding that will explain why the M–A bond is weakened. If we refer for the moment to platinum(II) complexes, then the bonding is considered to involve the $5d_{x^2-y^2}$, the $6s$ and the $6p_x$ and $6p_y$ a.o.'s (taking the xy plane as the plane of the complex). Bonding in the Pt–L and Pt–A links can be interpreted in terms of the interaction of the platinum $5d_{x^2-y^2}$, $6s$ and $6p_x$ (x taken as the L–Pt–A axis) a.o.'s with x_σ a.o.'s of the L and A ligands. This σ bonding can be in terms of two localized two-electron bonds (Pt–L and Pt–A) or two three-centre, two-electron bonds, each covering the L–Pt–A system (see Chapter 8, p.89, for the related basic discussion of bonding in $BeCl_2$). L and A both compete for the same Pt orbitals, so if the L orbital overlaps well and forms a strong σ bond the platinum σ orbitals are less available for bonding to A. Ligands high in the *trans*-influence series, such as R_3Si^- and H^-, are those that overlap well, while ligands well down the series give poorer overlap. The π-acceptor ligands (CO and C_2H_4) form weak σ-bonds and make comparatively little demand on the Pt σ orbitals. Readers interested in a detailed discussion of the molecular-orbital explanation for the *trans*-influence are recommended to read the review by Shustorovich and co-workers[114].

If ligands have a big *trans*-influence, they will weaken the *trans* bond both in the ground and transition states, so that these ligands will also produce a big *trans*-effect, and facilitate the replacement of the *trans* ligand in substitution reactions.

The π acceptor ligands are believed to have a big *trans*-effect because they will become involved in d_π–d_π or d_π–p_π back bonding with a doubly filled d orbital of platinum (see p.205). The effect of such π bonding is to alter the d-electron charge density by dragging it into the region between the ligand (L) and the platinum atom. The gap which is created in the d-electron charge cloud permits a more ready approach of the substituting ligand (E) (see *Figure 11.56*). If the four ligands are in the xy plane, then L can π bond with the doubly filled d_{xz} orbital, and the approaching

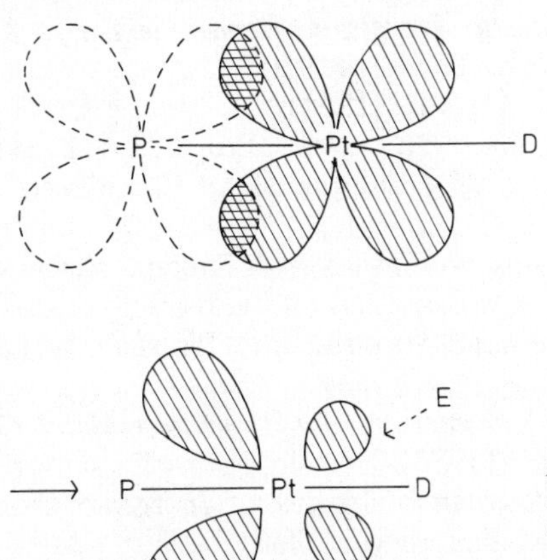

Figure 11.56 Effect of π bonding on d-*electron charge density*

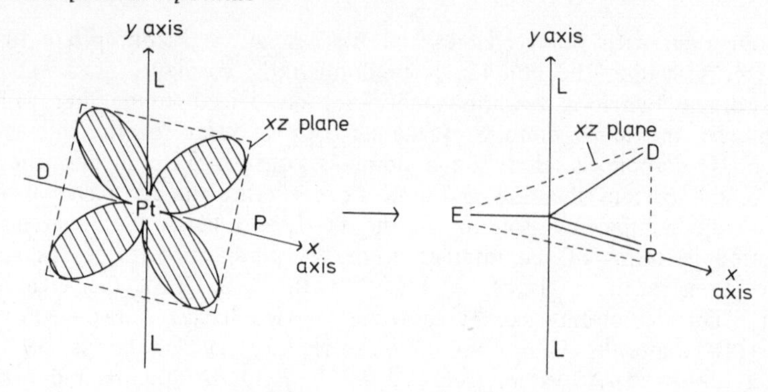

Figure 11.57 Formation of transition state in substitution reactions of square-planar complexes

ligand (E) approaches in the *xz* plane, giving a transition state with L, A and E all in this plane (see *Figure 11.57*). This π bonding also strengthens the transition-state complex, thus lowering the activation energy and speeding up the reaction.

11.21 References

1. *IUPAC Nomenclature of Inorganic Chemistry,* Butterworths, London (1971)
2. CAHN, R.S., *An Introduction to Chemical Nomenclature,* Butterworths London (1974)
3. BASOLO, F. and PEARSON, R.G., *Mechanisms of Inorganic Reactions,* 2nd Ed., Wiley, New York (1968)
4. GRAY, H.B., *J. Chem. Ed.,* **41**, 2 (1964)
5. BALLHAUSEN, C.J. and GRAY, H.B., *Molecular Orbital Theory,* p.92, W.A. Benjamin, New York (1964)
6. ORGEL, L.E., *An Introduction to Transition Metal Chemistry: Ligand-field Theory,* p.41, Methuen, London (1960)
7. COTTON, F.A., *J. Chem. Ed.,* **41**, 466 (1964)
8. FREEMAN, A.J. and WATSON, R.E., *Phys. Rev.,* **120**, 1254 (1964)
9. SUTTON, D., *Electronic Spectra of Transition Metal Complexes,* McGraw-Hill, London (1968)
10. BALLHAUSEN, C.J., *Introduction to Ligand-field Theory*, McGraw-Hill, New York (1962)
11. LEVER, A.B.P., *Inorganic Electronic Spectroscopy,* Elsevier, Amsterdam (1968)
12. MABB, F.E. and MACHIN, D.J., *Magnetism and Transition Metal Complexes,* Chapman and Hall, London (1973)
13. JOHNSON, D.A., *Some Thermodynamic Aspects of Inorganic Chemistry,* Cambridge University Press, Cambridge (1968)
14. BJERRUM, J., *Stability Constants,* The Chemical Society, London (1957)
15. GEORGE, P. and McCLURE, D.S., *Prog. Inorg. Chem.,* **1**, 381 (1959)

16. AVDEEF, A., COSTAMAGNA, J.A. and FACKLER, J.P., *Inorg. Chem.*, **13**, 1854 (1974)
17. ANDERSON, O.P., *J. Chem. Soc. Dalton Trans.*, 2597 (1972)
18. BERNSTEIN, P.K., RODLEY, G.A., MARSH, R. and GRAY, H.B., *Inorg. Chem.*, **11**, 3040 (1972)
19. NEWTON, G., CAUGHMAN, H.D. and TAYLOR, R.C., *J. Chem. Soc. Dalton Trans.*, 258 (1974)
20. EINSTEIN, F.W.B. and RESTIVO, R., *Acta Cryst.*, **B31**, 624 (1975)
21. DREW, M.G.B. and RIEDL, M.J., *J. Chem. Soc. Dalton Trans.*, 52 (1973)
22. BACH, R.D., WEIBEL, A.T., SCHMONSES, W. and GLICK, M.D., *J. Chem. Soc. Chem. Comm.*, 961 (1974)
23. ORGEL, L.E., *An Introduction to Transition Metal Chemistry: Ligand-field Theory*, p. 60, Methuen, London (1960)
24. WEINMYER, M.S., HUNT, G.W. and AMMA, E.L., *J. Chem. Soc. Chem. Comm.*, 1140 (1972)
25. TIETHOF, J.A., HETEY, A.T. and MEEK, D.W., *Inorg. Chem.*, **13**, 2505 (1974)
26. DAVIS, P.H., BELFORD, R.F. and PAUL, I.C., *Inorg. Chem.*, **12**, 213 (1973)
27. BAENZIGER, N.C., DITTEMORE, K.M. and DOYLE, J.R., *Inorg. Chem.*, **13**, 805 (1974)
28. TIETHOF, J.A., STALICK, J.K., CORFIELD, P.W.R. and MEEK, D.W., *J. Chem. Soc. Chem. Comm.*, 1141 (1972)
29. BRADLEY, D.C., *Adv. Inorg. Chem. Radiochem.*, **15**, 314 (1972)
30. HURSTHOUSE, M.B. and RODESILER, P.F., *J. Chem. Soc. Dalton Trans.*, 2100 (1972)
31. BRADLEY, D.C., HURSTHOUSE, M.B., SMALLWOOD, R.J. and WELCH, A.J., *J. Chem. Soc. Chem. Comm.*, 872 (1972)
32. LAUHER, J.W. and IBERS, J.A., *Inorg. Chem.*, **16**, 348 (1975)
33. CHURCHILL, M.R., COOKE, J., FENNESSEY, J.P. and WORMALD, J., *Inorg. Chem.*, **10**, 1031 (1971)
34. BECKETT, R. and HOSKINO, B.F., *J. Chem. Soc. Dalton Trans.*, 622 (1974)
35. BRADLEY, D.C., HURSTHOUSE, M.B., NEWING, C.N. and WELCH, A.J., *J. Chem. Soc. Chem. Comm.*, 567 (1972)
36. ALYEA, E.C., CASTIN, A., FERGUSON, G., FEY, G.T., GOEL, R.G. and RESTIVO, R.J., *J. Chem. Soc. Dalton Trans.*, 1294 (1975)
37. HARLOW, R.L., WELLS, W.J., WATT, G.W. and SIMONSEN, S.H., *Inorg. Chem.*, **13**, 2106 (1974)
38. TEXTOR, M., DUBLER, E. and OSWALD, H.R., *Inorg. Chem.*, **13**, 1361 (1974)
39. WILLETT, R.D. and CHOW, C., *Acta Cryst.*, **B30**, 207 (1974)
40. MUETTERTIES, E.L. and SCHUNN, R.A., *Quart. Rev.*, **20**, 245 (1966)
41. WOOD, J.S., *Prog. Inorg. Chem.*, **16**, 227 (1972)
42. GOODYEAR, J. and KENNEDY, D.J., *Acta Cryst.*, **B32**, 631 (1976)
43. SHELDRICK, G.M. and TAYLOR, R., *Acta Cryst.*, **B31**, 2740 (1975)
44. BUTTENSHAW, A.J., DUCHÊNE, M. and WEBSTER, M., *J. Chem. Soc., Dalton Trans.*, 2230 (1975)
45. BANCROFT, G.M., DAVIES, B.W., PAYNE, N.C. and SHAM, T.K., *J. Chem. Soc. Dalton Trans.*, 973 (1975)
46. ROSSI, A.R. and HOFFMANN, R., *Inorg. Chem.*, **14**, 365 (1975)
47. MORASSI, R., BERTINI, I. and SACCONI, L., *Co-ordination Chem. Rev.*, **11**, 343 (1973)

48. HOSKINS, B.F. and WHILLANS, F.D., *Co-ordination Chem. Rev.*, **9**, 365 (1972)
49. PAINE, R.T. and McDOWELL, R.S., *Inorg. Chem.*, **13**, 2346 (1974)
50. HENRICK, K., RASTON, C.L. and WHITE, A.H., *J. Chem. Soc. Dalton Trans.*, 26 (1976)
51. PHILLIPS, F.L. and SKAPSKI, A.C., *Acta Cryst.*, **B31**, 1814 (1975)
52. CIANI, G., MANASSERO, M. and SANSONI, M., *J. Inorg. Nuclear Chem.*, **34**, 1760 (1972)
53. STALICK, J.K., CORFIELD, P.W.R. and MEEK, D.W., *Inorg. Chem.*, **12**, 1668 (1973)
54. BRYAN, R.F. and SCHMIDT, W.C., *J. Chem. Soc. Dalton Trans.*, 2337 (1974)
55. GOLDFIELD, S.A. and RAYMOND, K.N., *Inorg. Chem.*, **13**, 770 (1974)
56. ANDERSON, D.P., *Inorg. Chem.*, **14**, 730 (1975)
57. BOORMAN, P.M., GARNER, C.D., MABBS, F.E. and KING, T.J., *J. Chem. Soc. Chem. Comm.*, 663 (1974)
58. PHILLIPS, F.L., SHREEVE, F.M. and SKAPSKI, A.C., *Acta Cryst.*, **B32**, 687 (1976)
59. DAWSON, J.W., McLENNAN, T.J., ROBINSON, W., MERLE, A., DARTINQUE-VAVE, M., DARTINQUEVAVE, Y. and GRAY, H.B., *J. Amer. Chem. Soc.*, **96**, 4428 (1974)
60. van de GRIEND, L.J., CLARDY, J.C. and VERKADE, J.G., *Inorg. Chem.*, **14**, 710 (1975)
61. ROBINSON, W.T. and SINN, E., *J. Chem. Soc. Dalton Trans.*, 726 (1975)
62. HUQ, F. and SKAPSKI, A.C., *J. Chem. Soc. Dalton Trans.*, 1927 (1971)
63. GHILARDI, C.A. and ORLANDINI, A.B., *J. Chem. Soc. Dalton Trans.*, 1698 (1972)
64. ORLANDINI, A.B., CALABRESI, C., GHILARDI, C.A., ORIOLI, P. and SACCONI, L., *J. Chem. Soc. Dalton Trans.*, 1383 (1973)
65. ZIOLO, R.F., ALLEN, M., TITUS, D.D., GRAY, H.B. and DOTI, Z., *Inorg. Chem.*, **11**, 3044 (1972)
66. CLEGG, W. and WHEATLEY, P.J., *J. Chem. Soc. Dalton Trans.*, 90 (1973)
67. BIANCHI, A., DAPPORTO, P., FALLONI, G., GHILARDI, C.A. and SACCONI, L., *J. Chem. Soc. Dalton Trans.*, 641 (1973)
68. MATTHEW, M., PALENIK, G.J., DYER, G. and MEEK, D.W., *J. Chem. Soc. Chem. Comm.*, 379 (1972)
69. HOSKINS, B.F. and WHILLANS, F.D., *J. Chem. Soc. Dalton Trans.*, 657 (1975)
70. DREW, M.G.B., *Prog. Inorg. Chem.*, in the press
71. EISENBERG, R., *Prog. Inorg. Chem.*, **12**, 313 (1970)
72. DREW, M.G.B., FOWLES, G.W.A. and LEWIS, D.F., *J. Chem. Soc. Chem. Comm.*, 876 (1969)
73. KEPERT, D.L., *The Early Transition Elements*, Academic Press, London (1972)
74. SINHA, S.P., *Structure and Bonding*, **25**, 69 (1976)
75. CORDEN, B.J., CUNNINGHAM, J.A. and EISENBERG, R., *Inorg. Chem.*, **9**, 356 (1970)
76. BOK, L.D.-C., LEIPOLDT, J.G. and BASSON, S.S., *Acta Cryst.*, **B26**, 684 (1970)
77. BOMBIERI, G., MOSELEY, P.T. and BROWN, D., *J. Chem. Soc. Dalton Trans.*, 1520 (1975)

78. BOMBIERI, G. and BAGNALL, K.W., *J. Chem. Soc. Chem. Comm.*, 188 (1975)
79. CAUO, F.H. and CRUICKSHANK, D.W.J., *Chem. Comm.*, 1617 (1971)
80. DREW, M.G.B., EGGINTON, G.M. and WILKINS, J.D., *Acta Cryst.*, **B30**, 1895 (1974)
81. BONAMICO, M., DESSY, G., FAREO, V. and SCARAMUZZA, L., *J. Chem. Soc. Dalton Trans.*, 1258 (1974)
82. PIOVESANA, O. and SESTILI, L., *Inorg. Chem.*, **13**, 2745 (1974)
83. GRAZIANI, R., BOMBIERI, G., FORSELLINI, E., DEGATO, S. and MARANGONI, G., *J. Chem. Soc. Dalton Trans.*, 451 (1973)
84. GRAZIANI, R., BOMBIERI, G. and FORSELLINI, E., *J. Chem. Soc. Dalton Trans.*, 2059 (1972)
85. COUNTRYMAN, R. and McDONALD, W.S., *J. Inorg. Nuclear Chem.*, **33**, 2213 (1971)
86. PARISH, R.V. and SIMMS, P.G., *J. Chem. Soc. Dalton Trans.*, 2389 (1972)
87. LIPPARD, S.J., *Prog. Inorg. Chem.*, **8**, 109 (1967)
88. BHANDARY, K.K., MANOHAR, H. and VENKATESAN, K., *J. Chem. Soc. Dalton Trans.*, 288 (1975)
89. AL-KARAGHOULI, A.R. and WOOD, J.S., *Inorg. Chem.*, **11**, 2293 (1972)
90. MUETTERTIES, E.L. and WRIGHT, C.M., *Quart. Rev.*, **21**, 109 (1967)
91. COTTON, F.A., *Chem. Soc. Rev.*, **4**, 27 (1975)
92. WALTON, R.A., *Prog. Inorg. Chem.*, **16**, 165 (1972)
93. KOKNAT, F.W. and McCARLEY, R.E., *Inorg. Chem.*, **13**, 295 (1974) (this gives related earlier references)
94. CHURCHILL, M.R. and BEZMAN, S.A., *Inorg. Chem.*, **13**, 1418 (1974)
95. HANLAN, L.A. and OZIN, G.A., *J. Amer. Chem. Soc.*, **96**, 6324 (1974)
96. COTTON, F.A. and TROUP, J.M., *J. Chem. Soc. Dalton Trans.*, 800 (1974)
97. COTTON, F.A., *Prog. Inorg. Chem.*, **21**, 1 (1976)
98. MASON, R., THOMAS, K.M. and MINGOS, D.M.P., *J. Amer. Chem. Soc.*, **95**, 3802 (1973)
99. CALABRESE, J.C., DAHL, L.F., CHINI, P., LONGONI, G. and MARTINENGO, S., *J. Amer. Chem. Soc.*, **96**, 2614 (1974)
100. PAUSON, P.L., Tilden Lecture, *Proc. Chem. Soc.*, 297 (1960)
101. KING, R.B., *Prog. Inorg. Chem.*, **15**, 287 (1972)
102. WILKINSON, G. and COTTON, F.A., *Prog. Inorg. Chem.*, **1**, 1 (1959)
103. FISCHER, E.O. and FRITZ, H.P., *Adv. Inorg. Chem. Radiochem.*, **1**, 56 (1959)
104. PROUT, K., CAMERON, T.S. and FORDER, R.A., *Acta Cryst.*, **B30**, 2290 (1974)
105. PARTENHEIMER, W. and DURHAM, B., *J. Amer. Chem. Soc.*, **96**, 3800 (1974)
106. RÖSCH, N., MESSMER, R.P. and JOHNSON, K.H., *J. Amer. Chem. Soc.*, **96**, 3855 (1974)
107. HAMILTON, W.C., KLANDERMAN, K.A. and SPRATLEY, R., *Acta Cryst.*, **A25**, S172 (1969)
108. GLANVILLE, J.O., *J. Organometallic Chem.*, **7**, 9 (1967)
109. McGINNETY, J.A., *J. Chem. Soc. Dalton Trans.*, 1038 (1974)
110. TOBE, M.L., *Inorganic Reaction Mechanisms*, Nelson, London (1972)
111. APPLETON, T.C., CLARK, H.C. and MANZER, L.E., *Co-ordination Chem. Rev.*, **10**, 335 (1973)

112. PIDOCK, A., RICHARDS, R.E. and VENANZI, L.M., *J. Chem. Soc. A,* 1707 (1966)

113. VENANZI, L.M., *Chem. Brit.,* 162 (1968)

114. SHUSTOROVICH, E.M., PORAI-KOSHITS, M.A. and BUSLAEV, Yu.A., *Co-ordination Chem. Rev.,* **17**, 1 (1975)

12
ELECTRONIC SPECTRA OF TRANSITION-METAL COMPLEXES

12.1 Introduction

In the last chapter we referred briefly to the electronic absorption of complexes of the first-row transition elements, and pointed out that such spectra had peaks that usually fall into two reasonably clear-cut categories. Relatively weak peaks found in the visible region are attributed to *d–d* transitions, since they result from the transition or promotion of an electron from one *d* orbital to another. The more intense bands that are mostly found at shorter wavelengths are also associated with the transition of electrons from one energy level to another, of course, but the orbitals representing the ground and excited states are normally associated with different parts of the complex, thus accounting for the bands being called *charge-transfer*. With the spectrum of $TiCl_3$,3py, for instance (see *Figure 11.9*, p.215), there is an intense peak at 24 300 cm^{-1} that has been attributed to the movement of the electron from a *d* orbital of Ti^{3+} to an anti-bonding π^* orbital of the pyridine ligand. In addition to *d–d* and charge-transfer bands, the spectrum may show bands (generally intense in the ultraviolet region) that can be attributed to transitions between orbitals involving only the ligands; the positions of such bands will be similar to those found in the spectra of the unco-ordinated ligands.

In this chapter we shall restrict further discussion to the *d–d* type of transition, and pay particular attention to the problem of accounting for the spectra of complexes of metal ions containing more than one *d* electron. Readers wishing to go further into this subject are referred to specialized books by Ballhausen[1], Lever[2] and Sutton[3].

In order to establish some basic ideas, particularly about nomenclature, we will briefly recapitulate and then extend our previous discussion (see Section 11.3) on d^1 complexes.

12.2 Electronic Spectra of d^1 Ions

If we consider complexes of tervalent titanium, such as $[TiCl_6]^{3-}$, in which the d^1 metal ion is co-ordinated by six identical ligands, we expect their spectra to show a single weak *d–d* band. This band is caused by the transition of the *d* electron between the t_{2g} and e_g orbitals, and is formally described as

$$e_g \leftarrow t_{2g}$$

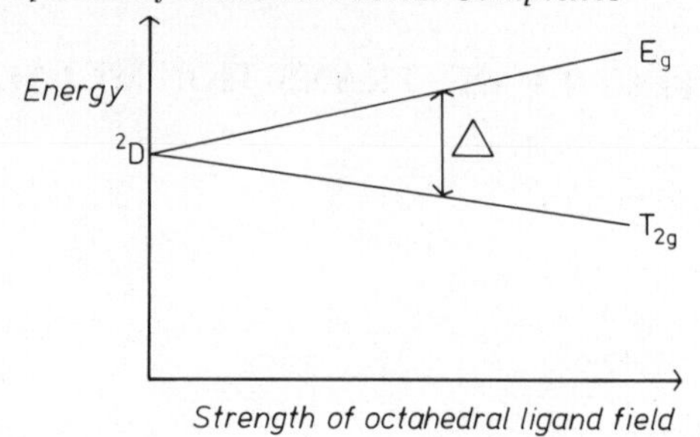

Figure 12.1 Splitting of ^{2}D term in an octahedral field

Now an ion with a d^1 configuration is said to be described by the *term* ^2D and in an octahedral environment the ^2D term splits into two levels T_{2g} and E_g, the extent of the splitting being related to the strength of the octahedral ligand field. This is illustrated in *Figure 12.1.* The word *term* is a new one for our discussion, and its full significance will become clear when we discuss d^n ions, but readers can see that ^2D refers to a situation in which there is a single d electron sitting in one of the five degenerate d orbitals of the free Ti^{3+} ion. The expression T_{2g} refers to the configuration $(t_{2g})^1(e_g)^0$ and E_g to the configuration $(t_{2g})^0(e_g)^1$. A level given the label T contains three orbitals of the same energy and is said to be triply degenerate; in the same way E levels are orbitally doubly degenerate. As we saw in the last chapter, Δ can be evaluated for a given set of six ligands, and the relative values of Δ provide a 'pecking order' or spectrochemical series for the ligands. This order is virtually independent of the metal ion, although the actual Δ value varies from ion to ion; thus we saw that Δ increases with the charge on the ion, and on going from an ion of a first-row element (e.g. Ti^{3+}) to the corresponding ion of the second-row element of the same group (i.e. Zr^{3+}).

We further saw that the splitting of the individual d orbitals becomes more complex as the environment of the metal ion becomes less regular. Thus with d^1 complexes such as $TiCl_3$,3L and VCl_4,2L, we get the splitting patterns given in *Figure 12.2* for the *trans* forms. It can be seen that the splittings are such that we would predict two and three peaks

Figure 12.2 d-Orbital splitting for non-octahedral d^1 *complexes*

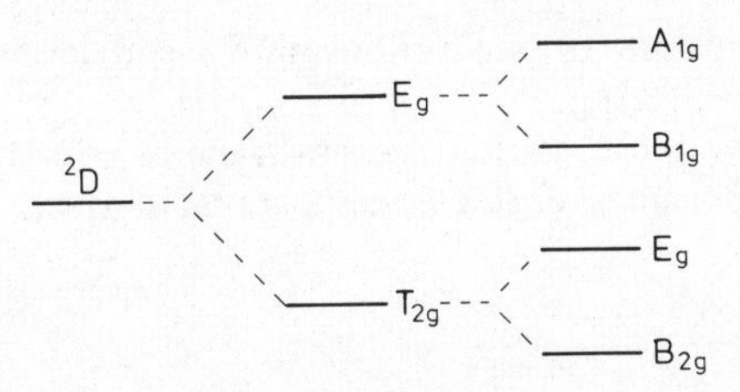

Figure 12.3 Splitting of the ^{2}D term for a VCl_4,2L complex

respectively for the Ti^{3+} and V^{4+} complexes. The V^{4+} ion is again described by a ^2D term, and this again splits into T_{2g} and E_g levels in an octahedral field; in the *trans*-VCl_4,2L complex these levels split further to give a set of E_g, A_{1g}, B_{1g} and B_{2g} levels (see *Figure 12.3*). Because we are dealing with a d^1 ion, these levels can be equated with the electron occupying particular orbitals. The particular labels define the orbital degeneracy and the symmetry properties. Thus A and B levels both relate to a single orbital, and differ only in the symmetry of the orbital with respect to a rotation about the z axis by $2\pi/4$ (i.e. 90°). The A level orbital, d_{z^2}, does not change sign on rotation about the z axis, whereas rotation of either of the B level orbitals (d_{xy} and $d_{x^2-y^2}$) by $2\pi/4$ results in a change of sign (see *Figure 12.4*). The subscripts 1 and 2 are

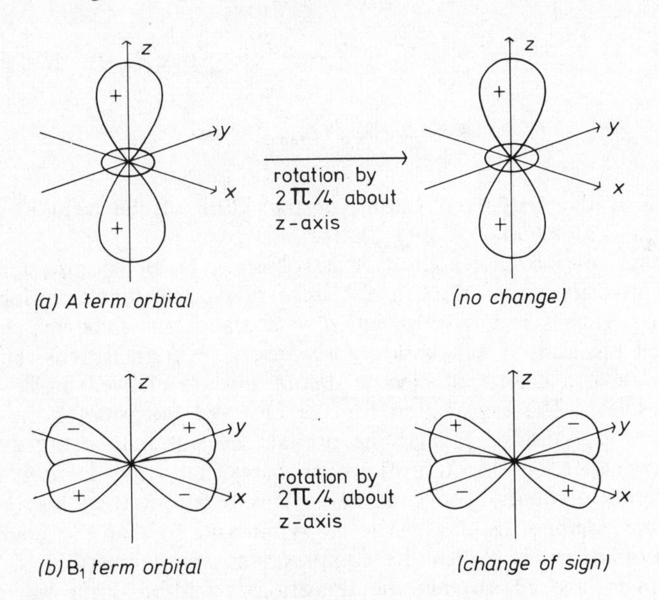

Figure 12.4 Symmetry properties of d *orbitals*

given to orbitals that are symmetrical and antisymmetrical, respectively, with respect to rotation by $\pi/2$ (180°) about either the x or y axes; thus d_{z^2} and $d_{x^2-y^2}$ have the subscript 1 because the sign is unchanged on rotation, but d_{xy} is given the subscript 2 because rotation changes + to − and vice versa. All the d orbitals are g (i.e. gerade) because they have a centre of symmetry. We shall see that these symmetry labels will occur in our descriptions of the various terms and levels of ions with more than a single electron.

Before we discuss these d^n ions, a few general comments should be made about the intensities of these comparatively weak *d–d* bands.

12.3 Intensity of *d–d* Bands

We saw in Chapter 11 (Section 11.3, p.212) that practical chemists usually describe intensities by extinction coefficients (ϵ), which are of the order of 1–50 for *d–d* transitions of first-row transition-metal ions. However, extinction coefficients only give a qualitative picture of peak intensity because solution spectra are frequently quite broad; thus each 'level' has many 'sub-levels' associated with vibrational modes. This broadening can be allowed for by plotting the extinction coefficient against wave number ($\bar{\nu}$) and measuring the area under the curve for each peak. The result is expressed in terms of the *oscillator strength* (*f*), where

$$f = 4.32 \times 10^{-9}\ \mathrm{d}\bar{\nu}$$

In practice it is much easier to replace $\mathrm{d}\bar{\nu}$ by $\epsilon_{max} \times \Delta\bar{\nu}$, giving

$$f = 4.32 \times 10^{-9}\ \epsilon_{max} \times \Delta\bar{\nu}$$

where $\Delta\bar{\nu}$ is the 'half band width', or the width of the band in cm^{-1} at half height (i.e. where $\epsilon = \frac{1}{2}\epsilon_{max}$).

Of course, *d–d* bands should not be observed at all because a peak should have no intensity unless there is a change of electric dipole during the transition. Thus $s \rightarrow s$, $p \rightarrow p$ and $d \rightarrow d$ transitions show no change of dipole and are said to be *Laporte forbidden*; $s \rightarrow p$ transitions, on the other hand, do show a change of electric dipole and are allowed under the Laporte selection rules. The explanation for the observed intensity of the Laporte-forbidden *d–d* transition is that the orbitals are not pure *d* but contain some *p* character, so that transitions are between *dp* hybrid orbitals that have different amounts of *p* character. This *p* contribution may arise because the complex lacks a centre of symmetry, so that the bonding orbitals are not identical. When the complex ions are symmetrical, a *vibronic* mechanism is invoked, so that the transition is said to occur when the vibrating complex is 'off-centre' and hence lacks a centre of symmetry.

In this brief discussion about intensity it is assumed that in the course of the transition there is no change of spin by the electron. We shall see later that in some cases, such as Mn^{2+} with a d^5 spin-free configuration, any transition must be between states with different numbers of unpaired electrons. Such changes of spin are also 'forbidden' and the resultant spin-forbidden peaks are very weak; hence the very pale colour of manganese(II) compounds.

12.4 Configurations for Free Ions (d^n)

12.4.1 A SINGLE ELECTRON

In earlier discussions of electrons in atoms we considered electrons to be in atomic orbitals, each atomic orbital being characterized by *three* quantum numbers:

n principal quantum number
l azimuthal quantum number
m magnetic quantum number

These quantum numbers take only a limited number of whole number values, which are inter-related. Thus for $n = 3$, l can take the values 0, 1 or 2, corresponding respectively to s, p or d orbitals; for $l = 0$, m has only one value (0), and there is only one 3s atomic orbital; for $l = 1$, m has three possible values (0, +1, −1) which give rise to the three 3p orbitals; for $l = 2$, m can take five values (0, ±1, ±2), so that there are five 3d orbitals. Strictly speaking we should use the symbol m_l rather than m, since it represents the component of the orbital angular momentum along the z axis.

An electron in such an orbital also requires a fourth quantum number to describe it. This is given the symbol m_s and termed the 'magnetic spin quantum number', and it can take only two values, +½ and −½. There is often some confusion over the symbolism of this fourth quantum number. Thus we also have the 'spin quantum number', with the symbol s, which has a single value (½) that is the length of the vector defining the spin angular momentum $\{(h/2\pi)[s(s+1)]^{1/2}\}$. The symbol m_s, on the other hand, is the component of the spin quantum number lying parallel or antiparallel to the magnetic field. Thus m_s is related to s in the same way as m_l is related to l. The distinction will become clearer as we consider multi-electron atoms.

12.4.2 MORE THAN ONE ELECTRON (RUSSELL–SAUNDERS COUPLING)

We now have to discuss the distribution of electrons amongst the various degenerate orbitals, and we have to consider the mutual influence these electrons have on one another. The position is much more complicated than with a single electron, and we shall see that the interactions (usually called coupling) of the magnetic fields produced by the electrons' orbital and spin angular momenta give rise to a series of energy levels, each energy level being defined by a set of quantum numbers.

Just as a single electron orbital is defined by l, m_l and m_s (s) quantum numbers, so an energy level arising from the interactions of two or more electrons is defined by new quantum numbers L, M_L and M_S (S); these quantum numbers for multi-electron atoms are obtained by summing vectorially the quantum numbers for the individual electrons. Vectorial addition is required since the quantum numbers define angular momenta which are vector quantities.

In the simplest coupling scheme it is assumed that simple summing of l, m_l, m_s and s values gives the overall L, M_L, M_S and S values:

$$\begin{aligned} \Sigma l &= L \\ \Sigma m_l &= M_L \\ \Sigma m_s &= M_S \\ \Sigma s &= S \end{aligned}$$

It is further assumed that these couplings are much more important than couplings such as those between individual orbital and spin momenta. For the lighter elements, including the first-row transition elements, this assumption is certainly valid, and we work on the basis of couplings being in the following order of importance:

spin–spin > orbital–orbital > spin–orbital

We shall return to the problem of spin–orbital coupling later, but for the moment we will concentrate on the first two types of coupling.

12.5 Coupling of Electron Spins

The spin quantum number S for a set of n electrons is obtained by the simple algebraic sum of the individual s values. The results are shown in *Figure 12.5*. When we have only two electrons the spins can be in the same direction (parallel), with $S = 1$, or they can cancel each other out to give $S = 0$. By similar arguments we see that three electrons give $S = 3/2$ or $1/2$. For four electrons $S = 2$, 1 or 0, and for five electrons $S = 5/2$, $3/2$ or $1/2$.

2 electrons	(a) ↑↑	$S = \frac{1}{2} + \frac{1}{2}$	$= 1$
	(b) ↑↓	$S = \frac{1}{2} - \frac{1}{2}$	$= 0$
3 electrons	(a) ↑↑↑	$S = \frac{1}{2} + \frac{1}{2} + \frac{1}{2}$	$= 3/2$
	(b) ↑↑↓	$S = \frac{1}{2} + \frac{1}{2} - \frac{1}{2}$	$= 1/2$
4 electrons	(a) ↑↑↑↑	$S = \frac{1}{2} + \frac{1}{2} + \frac{1}{2} + \frac{1}{2}$	$= 2$
	(b) ↑↑↑↓	$S = \frac{1}{2} + \frac{1}{2} + \frac{1}{2} - \frac{1}{2}$	$= 1$
	(c) ↑↑↓↓	$S = \frac{1}{2} + \frac{1}{2} - \frac{1}{2} - \frac{1}{2}$	$= 0$
5 electrons	(a) ↑↑↑↑↑	$S = \frac{1}{2} + \frac{1}{2} + \frac{1}{2} + \frac{1}{2} + \frac{1}{2}$	$= 5/2$
	(b) ↑↑↑↑↓	$S = \frac{1}{2} + \frac{1}{2} + \frac{1}{2} + \frac{1}{2} - \frac{1}{2}$	$= 3/2$
	(c) ↑↑↑↓↓	$S = \frac{1}{2} + \frac{1}{2} + \frac{1}{2} - \frac{1}{2} - \frac{1}{2}$	$= 1/2$

Figure 12.5 Coupling of spin angular momenta

The spin angular momentum for the system of electrons is given by the expression

$$\frac{h}{2\pi}\,[S(S+1)]^{1/2}$$

12.6 Coupling of Orbital Angular Momenta

We have seen that a single electron is described by its l quantum number, and with a set of n electrons the energy states are described by the related quantum number L. Thus while l is associated with the orbital angular momentum of a single electron, L is associated with the orbital angular momentum resulting from the coupling of the orbital angular momenta of the set of electrons. In practice we consider the electrons two at a time and vectorially couple their orbital momenta. Several resultants may emerge, depending upon the relative orientations of the l vectors. If our two electrons have l values l_1 and l_2, then the range of L values produced will be from $l_1 + l_2$ to $|l_1 - l_2|$, i.e.

$$l_1 + l_2,\ l_1 + l_2 - 1,\ l_1 + l_2 - 2,\ \ldots\ |l_1 - l_2|$$

It would clarify matters for readers if we give a specific example: coupling between two p electrons.

Both l_1 and l_2 have the value of unity, so that there are three resultant L values:

$$l_1 + l_2 = 1 + 1 = 2$$

$$l_1 + l_2 - 1 = 1 + 1 - 1 = 1$$

$$|l_1 - l_2| = 1 - 1 = 0$$

As with all quantum numbers, consecutive values must differ by unity. This coupling is best illustrated vectorially (see *Figure 12.6*).

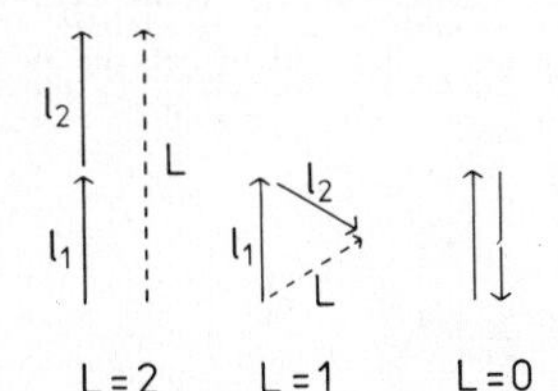

Figure 12.6 Vectorial coupling of the orbital momenta of two p electrons

Now with a single electron we used a shorthand way of describing the orbital, such that

$l = 0$	s orbital
$l = 1$	p orbital
$l = 2$	d orbital
$l = 3$	f orbital

In exactly the same way we can describe the *terms* that arise from coupling l's to give L values. Thus

$$L = 0 \quad S \text{ term}$$
$$L = 1 \quad P \text{ term}$$
$$L = 2 \quad D \text{ term}$$

If the coupling was between two d electrons, with $l_1 = l_2 = 2$, then five terms emerge, $S, P, D, F, G,$ corresponding to L = 0, 1, 2, 3 and 4. The coupling is shown vectorially in *Figure 12.7*.

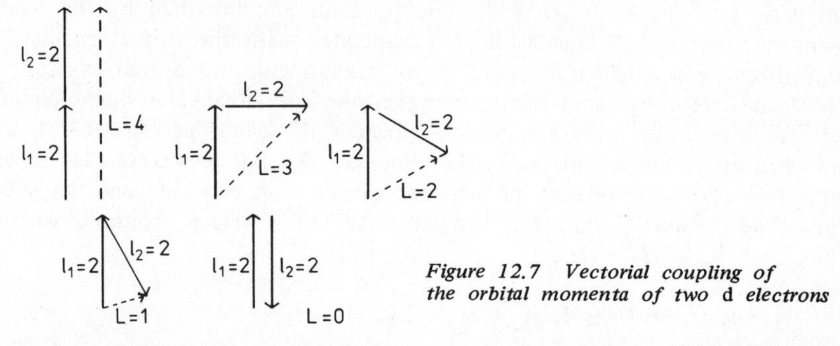

Figure 12.7 Vectorial coupling of the orbital momenta of two d *electrons*

Readers should note carefully that the symbol S is given to a term for which L = 0. This convention could cause confusion because S is also used for the resultant spin quantum number. However, in much the same way we used s for a single electron, both for the atomic orbital (l = 0) and for the spin quantum number.

12.7 Spin Multiplicity (2*S* + 1)

We have seen (Section 12.5) that when the spins of two electrons couple they do so with parallel spins (S = 1) or antiparallel spins (S = 0). We now introduce the expression *spin multiplicity*, which is the number of orientations of the spin vector (S) along the direction of the magnetic field, i.e. the number of M_S values (= $2S + 1$). When S = 1, there are three values of M_S, namely +1, 0 and −1. This is illustrated in *Figure 12.8*. Readers should note that M_S is the component of S in the z direction, i.e. the algebraic sum of individual m_s (the component of individual electron s in the z direction).

	m_s		M_s
↑ ↑	$\frac{1}{2}$, $\frac{1}{2}$	=	1
↑↓	$\frac{1}{2} - \frac{1}{2}$	=	0
↓ ↓	$-\frac{1}{2}$, $-\frac{1}{2}$	=	−1

Figure 12.8 M_S *values for* S = *1*

When we wish to describe the particular energy state, we have to take the spin multiplicity into account as well as the orbital coupling, so that each term (specified by the value of L) also has its spin multiplicity quoted. Hence the term is written ^{2S+1}L. This will be illustrated first for the simple cases of single electrons, and then for the p^2 configuration.

12.7.1 s^1 CONFIGURATION

$$S = \Sigma s = \tfrac{1}{2}$$

Therefore $2S + 1 = 2$, corresponding to ↑ and ↓

$$L = \Sigma l = 0, \text{ giving an } S \text{ term}$$

Hence we have a ^2S term, which is orbitally non-degenerate but doubly degenerate in respect of spin. The symbol ^2S is read as 'doublet S'.

12.7.2 p^1 CONFIGURATION

$$S = \Sigma s = \tfrac{1}{2}$$

Therefore $2S + 1 = 2$, corresponding to ↑ and ↓

$$L = \Sigma l = 1, \text{ giving a } P \text{ term}$$

Hence we have a doublet P (^2P) term, which has an orbital degeneracy of three and a spin degeneracy of two.

12.7.3 d^1 CONFIGURATION

$$S = \Sigma s = \tfrac{1}{2}$$

Therefore $2S + 1 = 2$, corresponding to ↑ and ↓

$$L = \Sigma l = 2, \text{ giving a } D \text{ term}$$

Hence we have a doublet D (^2D) term, which has an orbital degeneracy of five and a spin degeneracy of two. Thus there is a total degeneracy of $5 \times 2 = 10$, corresponding to the electron being in one of the five d orbitals, with one of two possible spin configurations; each one of the 10 possible combinations of orbital and spin is referred to as a 'microstate'.

12.7.4 p^2 CONFIGURATION

$S = 1$ and 0, so that $2S + 1 = 3$ and 1, referred to respectively as triplet and singlet. $L = 0$, 1 and 2, giving S, P and D terms. Hence a full description of the terms should provide the set ^1S, ^1P, ^1D, ^3S, ^3P, ^3D. Indeed, all six terms exist when the p electrons come from different shells, e.g. $2p$ and $3p$, but when the p electrons are equivalent (both $2p$ or both $3p$ for instance), then it turns out that only three of the six terms are allowed, namely ^1S, ^3P and ^1D. The reason for this limitation is that the Pauli Exclusion Principle must hold, so that both electrons cannot have an identical set of quantum numbers. This will be clear if we analyse the p^2 set of terms in more detail and relate them to the microstates.

12.8 Relationship Between Terms and Microstates for the p^2 Configuration

There are three available p orbitals to accommodate the two electrons, and there are 15 ways of arranging these electrons in the orbitals. This is

No.	$M_L = \Sigma m_l$	$M_S = \Sigma m_s$	m_l			Term
			1	0	−1	
1	2	0				^1D
2	1	1				^3P
3	1	0				^1D
4	1	0				^3P
5	1	−1				^3P
6	0	1				^3P
7	0	0				^1D
8	0	0				^1S
9	0	0				^3P
10	0	−1				^3P
11	−1	1				^3P
12	−1	0				^1D
13	−1	0				^3P
14	−1	−1				^3P
15	−2	0				^1D

Figure 12.9 Microstates for the p^2 *configuration (same* n *quantum number)*

illustrated in *Figure 12.9*, where ↗ represents $m_s = +½$ and ↘ represents $m_s = -½$. We can now assign microstates for each of the three terms ^1S, ^3P and ^1D.

12.8.1 ^1D TERM

A D term requires five M_L values, 2, 1, 0, −1 and −2. Microstates 1 and 15 have M_L values 2 and −2, respectively, and must be assigned to the ^1D term because there is no other microstate with $M_L = 2$ or -2. Since it is a singlet D term, we can take only microstates for which $M_S = 0$ and we take one from each of the required M_L values (+1, 0 and −1); we have taken microstates number 3, 7 and 12, but of course we could equally well have taken 4, 9 and 13.

Thus the ^1D term has a total degeneracy of 5 (5 microstates), there being an orbital degeneracy of 5, and it is non-degenerate in spin.

12.8.2 ^3P TERM

A P term needs three M_L values, +1, 0 and −1, and has an orbital degeneracy of 3; as it is a triplet term it has a spin degeneracy of 3 so that we have to pick out three microstates ($M_S = +1$, 0 and −1) for each M_L value. Hence we assign to the ^3P term the nine microstates 2, 4, 5, 6, 9, 10, 11, 13 and 14.

12.8.3 ^1S TERM

The remaining microstate (number 8), with $M_L = 0$ and $M_S = 0$, is assigned to the ^1S term, which is non-degenerate in both spin and orbital.

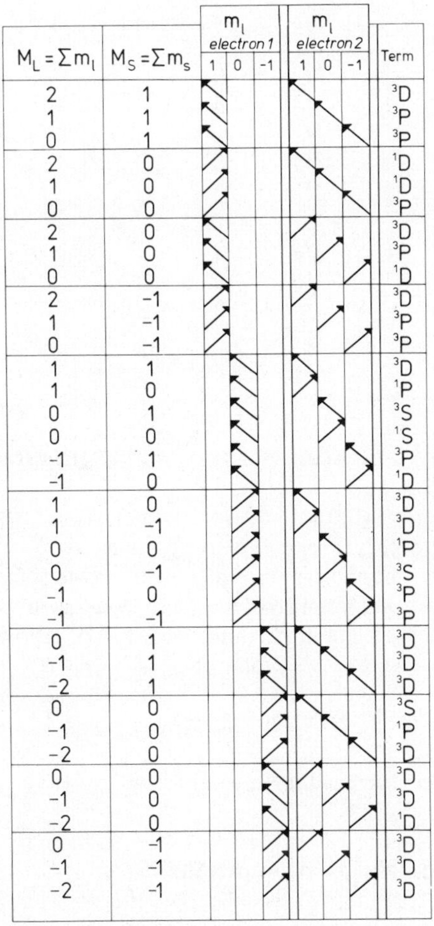

$M_L = \Sigma m_l$	$M_S = \Sigma m_s$	m_l electron 1			m_l electron 2			Term
		1	0	−1	1	0	−1	
2	1							^3D
1	1							^3P
0	1							^3P
2	0							^1D
1	0							^1D
0	0							^3P
2	0							^3D
1	0							^3P
0	0							^1D
2	−1							^3D
1	−1							^3P
0	−1							^3P
1	1							^3D
1	0							^1P
0	1							^3S
0	0							^1S
−1	1							^3P
−1	0							^1D
1	0							^3D
1	−1							^3D
0	0							^1P
0	−1							^3S
−1	0							^3P
−1	−1							^3P
0	1							^3D
−1	1							^3D
−2	1							^3D
0	0							^3S
−1	0							^1P
−2	0							^3D
0	0							^3D
−1	0							^3D
−2	0							^1D
0	−1							^3D
−1	−1							^3D
−2	−1							^3D

Figure 12.10 Microstates for the p^2 *configuration (different* n *quantum number)*

We pointed out that all six terms would exist if the two p electrons were non-equivalent, such as with a configuration $2p^1 3p^1$. A table can be drawn up similar to *Figure 12.9,* but treating each electron separately because each electron can take up any value of m_l and m_s, i.e. as the n quantum number is different for the two electrons the remaining quantum numbers can be the same if necessary. *Figure 12.10* shows the 36 possible microstates and assigns them to the six terms. Thus:

^3D
$$M_L = 2 \quad M_S = +1, 0, -1$$
$$M_L = 1 \quad M_S = +1, 0, -1$$
$$M_L = 0 \quad M_S = +1, 0, -1 \quad 5 \times 3 = 15 \text{ microstates}$$
$$M_L = -1 \quad M_S = +1, 0, -1$$
$$M_L = -2 \quad M_S = +1, 0, -1$$

^3P
$$M_L = +1 \quad M_S = +1, 0, -1$$
$$M_L = 0 \quad M_S = +1, 0, -1 \quad 3 \times 3 = 9 \text{ microstates}$$
$$M_L = -1 \quad M_S = +1, 0, -1$$

^3S	$M_L = 0$	$M_S = +1, 0, -1$	$1 \times 3 = 3$ microstates
^1D	$M_L = +2$	$M_S = 0$	
	$M_L = +1$	$M_S = 0$	
	$M_L = 0$	$M_S = 0$	$5 \times 1 = 5$ microstates
	$M_L = -1$	$M_S = 0$	
	$M_L = -2$	$M_S = 0$	
^1P	$M_L = +1$	$M_S = 0$	
	$M_L = 0$	$M_S = 0$	$3 \times 1 = 3$ microstates
	$M_L = -1$	$M_S = 0$	
^1S	$M_L = 0$	$M_S = 0$	$1 \times 1 = 1$ microstate

12.9 Microstates and Terms for a d^2 Configuration

We have seen (Section 12.6) that two d electrons ($l_1 = l_2 = 2$) couple to give L values at 0, 1, 2, 3 and 4 corresponding to S, P, D, F and G terms, respectively. Moreover, coupling of the two spins gives both singlets and triplets. There are 45 microstates, and a detailed consideration of them along the lines of the procedure used for p^2 shows that the allowed terms for two equivalent d electrons are:

$$^1S,\ ^3P,\ ^1D,\ ^3F,\ ^1G$$

These terms will be discussed in greater detail shortly.

12.10 Terms for d^n Configurations

In Section 12.6, which was concerned with the derivation of L, it was pointed out that the l values were coupled for two electrons at a time, and terms were derived for p^2 and d^2 configurations. By extension of these arguments, L values and the related terms may be derived for other spin-free d^n configurations, and the results are summarized in *Table 12.1*.

There are two general points to be made about this table. The first is that d^5 and d^{10} configurations involve half-full and completely full d orbitals, so that in each case every orbital has the same number of electrons

Table 12.1 Terms for spin-free d^n configurations*

Configuration	*Terms*†
d^1, d^9	^2D
d^2, d^8	^3F, ^3P, ^1S, ^1D, ^1G
d^3, d^7	^4F, ^4P, ^2P, ^2D, ^2F, ^2G, ^2H
d^4, d^6	^5D, ^3P, ^3D, ^3F, ^3G, ^3H, ^1S, ^1D, ^1G, ^1I
d^5	^6S, ^4P, ^4D, ^4F, ^4G, ^2S, ^2P, ^2D, ^2F, ^2G, ^2H, ^2I
d^{10}	^1S

*d electrons are equivalent, i.e. both $3d$ or $4d$ etc.
†Lowest energy (ground-state term) given first; remainder not given in order of energy

with the same spin. Accordingly, both configurations give rise to ground-state S terms, with $L = 0$. The second point is that the terms arising from a d^n configuration will also arise from the related d^{10-n} configuration, so that d^1 and d^9, for instance, both give rise to ^2D terms. This is often described by the 'hole' formalism, whereby d^9 is considered to have a single hole (or vacancy for an electron) in the e_g orbitals, so that the system is considered complete except for a positive hole or positron. When there is more than one 'positive hole', the mutual interactions should be the same as those between the same number of electrons, and the same terms should arise. Hence d^2 and d^8 both produce the set of terms ^1S, ^3P, ^1D, ^3F, ^1G.

In *Table 12.1* the lowest-energy term is listed first for each configuration, this being ^3F for d^2 or d^8. We can predict the lowest-energy term if we assume that S–S couplings are the most important. The lowest-energy term will then be the one corresponding to the highest multiplicity, i.e. the largest value of S and the maximum number of unpaired electrons. Hence, either ^3P or ^3F should be the term of lowest energy. The l–l couplings are the next most important, and for given multiplicity, the term with the biggest L value will have the lowest energy; hence ^3F is the ground state term. These 'rules' only help to establish the lowest-energy term and do not place the other terms in order. This coupling scheme, in which L–S coupling is assumed to be much less important than S–S or L–L coupling, is known as the Russell–Saunders scheme.

12.11 Coupling of Spin and Orbital Momenta

In the Russell–Saunders coupling scheme, L describes an energy state or term which is split into two or more levels or 'states' by L–S coupling. When S and L couple they give the total angular momentum of the system of electrons, and this is described by the total angular quantum number J.

J is obtained by summing vectorially the L and S values, such that the values J can take are:

$$L + S, L + S - 1, L + S - 2, \ldots L - S$$

We will now apply these coupling ideas to obtain J values for, first, a single p electron, and then for the terms arising from p^2 and d^2 configurations.

12.11.1 L–S COUPLING FOR A p^1 CONFIGURATION

In this instance $S = ½$ and $L = 1$, giving rise to a doublet P (^2P) term. When L and S couple they do so to give two levels of states corresponding to $J = 3/2$ and ½. The vectorial summation is shown in *Figure 12.11.* Thus the ^2P term splits into $^2P_{3/2}$ and $^2P_{1/2}$ levels under the influence of L–S coupling, the labels of the new levels having a subscript to reflect the J value. Now the ^2P term contains six microstates, corresponding to the

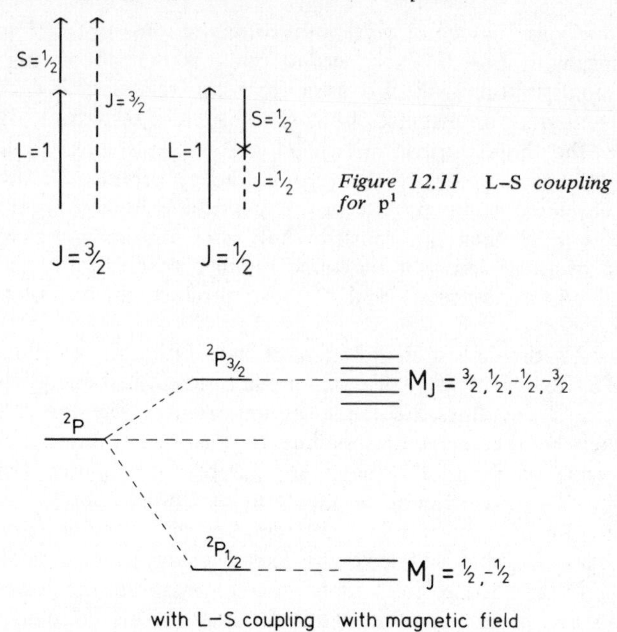

Figure 12.11 L–S *coupling for* p^1

Figure 12.12 *Splitting of* ^{2}p *term by* L–S *coupling and a magnetic field*

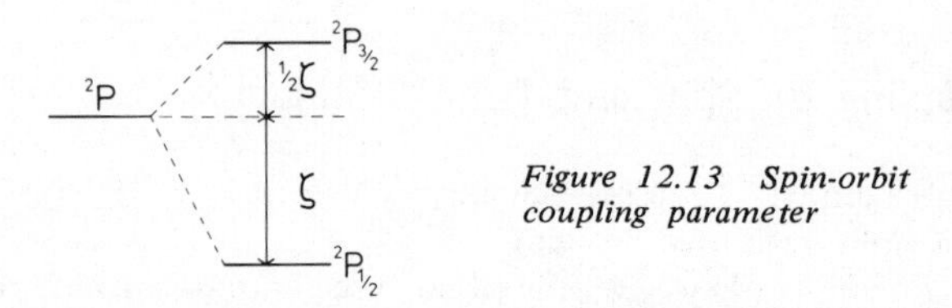

Figure 12.13 *Spin-orbit coupling parameter*

electron being in any one of the three *p* orbitals with either of the two possible spins. The $^2P_{3/2}$ and $^2P_{1/2}$ levels contain four and two micro-states, respectively, and if a strong magnetic field is applied these micro-states can be separated (see *Figure 12.12*). It can be seen that under the influence of the magnetic field *J* splits into $2J + 1$ levels, this being the multiplicity of the level or state. The extent to which the ^2P terms splits depends upon the strength of the *L–S* coupling, and it may be expressed in terms of a one-electron, spin–orbit coupling parameter (ζ) as shown in *Figure 12.13.* It should be noted that the 'centre of gravity' is maintained and that ζ is always positive, its value normally being expressed in cm^{-1}.

12.11.2 *L–S* COUPLING FOR A p^2 CONFIGURATION

L takes three values, 2, 1, and 0, corresponding to *D*, *P* and *S* terms, respectively, and *S* takes the values 1 and 0. Each value of *L* is coupled in turn with each value of *S*, this being illustrated in *Figure 12.14.* It can be seen that when $L > S$ there are $2S + 1$ *J* levels, so that for $S = 0$ and 1 there are 1 and 3 *J* levels, respectively. When $L < S$ there are $2L + 1$

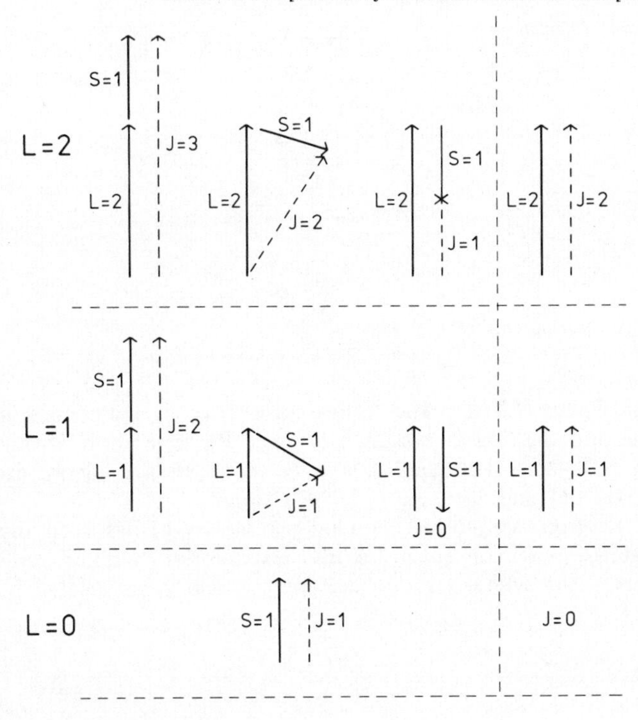

Figure 12.14 L–S *coupling for* p^2

levels. We have seen that when the p electrons are equivalent (i.e. both $2p$ or both $3p$, etc.) only the ^1D, ^3P and ^1S terms remain, and we can now illustrate how these terms split further as L–S coupling is allowed for and a magnetic field applied (see *Figure 12.15*). It can be seen how the 15 microstates are divided amongst the various terms.

Since we are now dealing with more than one electron, it will be worth looking a little more closely at the splitting of the triplet term ^3P, which

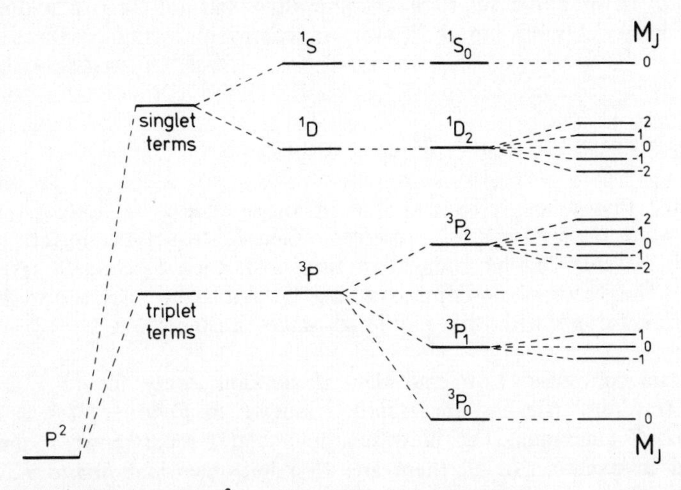

Figure 12.15 *Splittings for* p^2 *configuration (not to scale)*

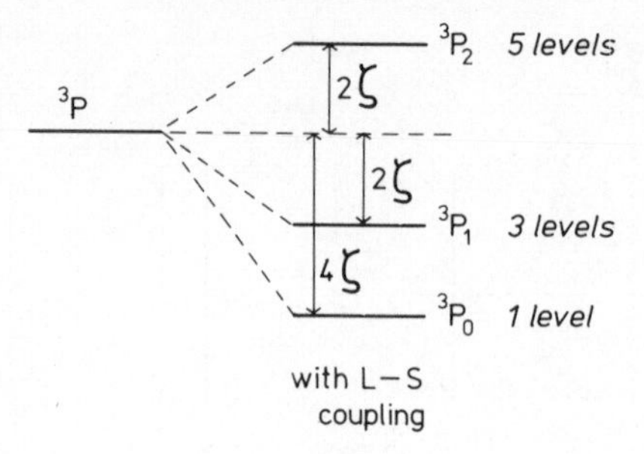

Figure 12.16 Splitting for the ^{3}P term

is shown in *Figure 12.16.* The 'centre of gravity' is maintained in the splitting because the five microstates of the $^3P_2$ state show an energy increase of $5 \times 2\zeta = 10\zeta$, and thus is balanced by the energy decrease of the $^3P_1$ ($3 \times 2\zeta$) and the $^3P_0$ ($1 \times 4\zeta$) states.

The one-electron, spin–orbit coupling parameter (ζ) has been used, but it is sometimes more convenient to use a spin–orbit coupling constant λ that represents the whole set of electrons. λ and ζ are related by the expression

$$\lambda = \pm \frac{\zeta}{2S}$$

This relationship takes the + sign when the term arises from a sub-shell that is less than half full; for sub-shells that are more than half full the − sign is needed. Thus ζ is always positive but λ can be positive or negative.

When *L–S* coupling produces a set of component levels such as that given in *Figure 12.16*, the energy sequence is that the levels with the highest J value are the highest in energy if the sub-shell is less than half full (e.g. p^2); when the sub-shell is more than half full (e.g. p^4) the levels with the higher J value are the lowest in energy.

12.11.3 *L–S* COUPLING FOR A d^2 CONFIGURATION

L may take the values 4, 3, 2, 1 and 0 corresponding to G, F, D, P and S terms, while S can be 1 or 0 corresponding to triplet or singlet. Once again each L value can be coupled in turn with each S value to give the J values. The vectorial description would be similar to that shown in *Figure 12.14* for p^2 with the additional values arising from $L = 3$ and $L = 4$.

As we are concerned with equivalent d electrons, only the ^1S, ^3P, ^1D, ^3F and ^1G terms arise and these split as shown in *Figure 12.17* as a result of *L–S* coupling. The individual microstates separate when the magnetic field is applied. In all there are 45 microstates and *Figure 12.17* shows how these are disposed between the various terms and states.

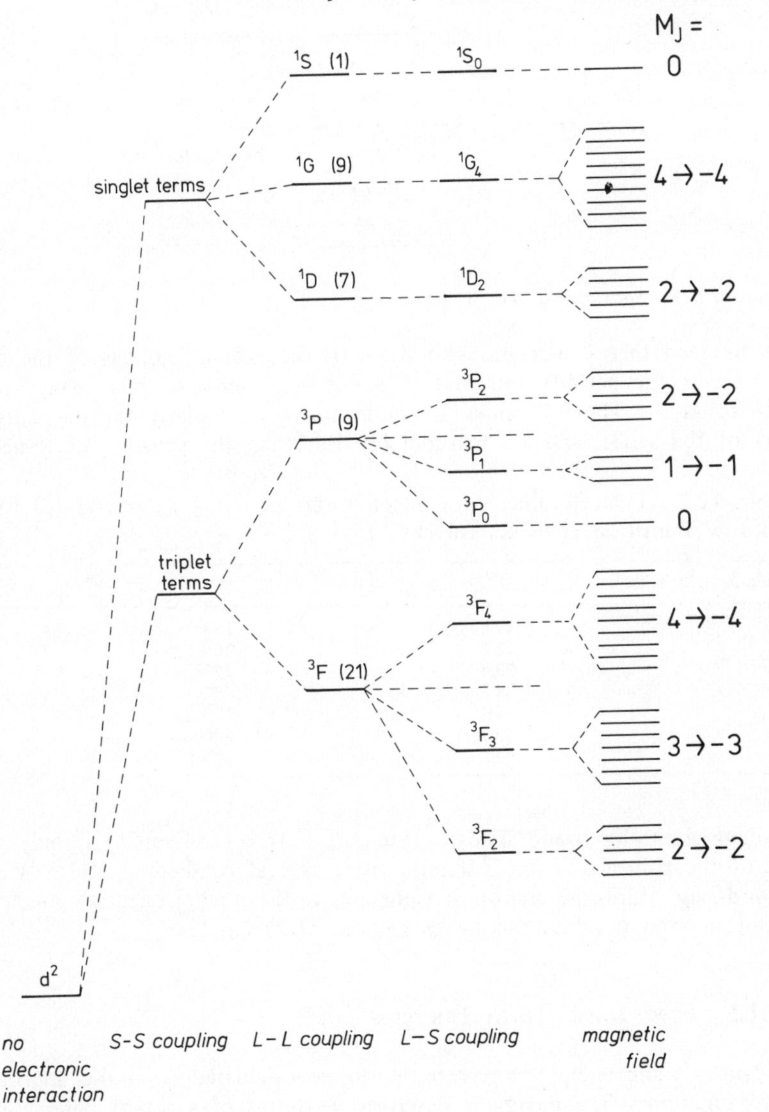

Figure 12.17 Splittings for d^2 *configuration (not to scale) (number of microstates given in parentheses)*

Figure 12.18 shows the splitting of the ^3F term under the influence of *L–S* coupling and once again it can be seen that the centre of gravity is maintained:

$$\begin{aligned} ^3F_4 \text{ (9 microstates)} &\rightarrow 9 \times (3/2)\zeta &= \; & 27\zeta/2 \\ ^3F_3 \text{ (7 microstates)} &\rightarrow 7 \times (-\tfrac{1}{2})\zeta &= \; & -7\zeta/2 \\ ^3F_2 \text{ (5 microstates)} &\rightarrow 5 \times (-2)\zeta &= \; & \underline{10\zeta} \\ & & & \underline{0\zeta} \end{aligned}$$

The extent of the splitting produced by *L–S* coupling is shown by the values of ζ listed in *Table 12.2* for the first-row transition elements. It

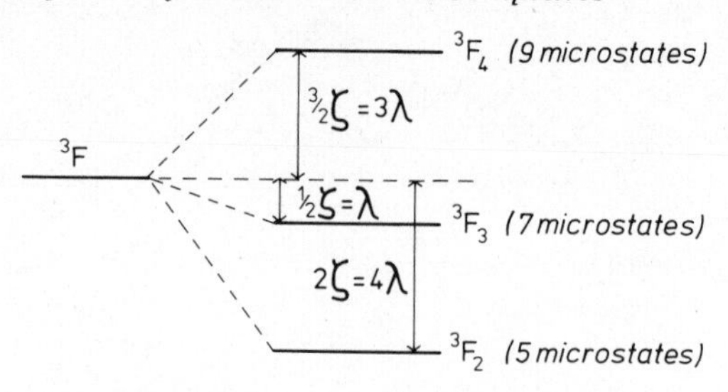

Figure 12.18 Splitting of the ^3F term

can be seen that ζ increases steadily with the atomic number of the element (i.e. across the period), but that it is not very sensitive to a change from M^{2+} to M^{3+}. These ζ values are quite small, particularly for the early members of the series, and are considerably less than the splitting of terms

Table 12.2 Typical values of the spin–orbit coupling parameter (ζ) for first-row transition elements (in cm^{-1})

Element	M^{2+}	M^{3+}
Ti	121	154
V	167	209
Cr	230	273
Mn	347	352
Fe	410	460
Co	533	580
Ni	647	705

brought about by ligand fields. Thus ζ for Ti^{3+} (154 cm^{-1}) is only around 1% of the splitting of the ^2D term by a typical octahedral field. With second- and third-row transition elements, L–S coupling becomes much more important and ζ values can be as high as 5000 cm^{-1}.

12.12 Free Ions: Term Energies

In principle the energy of any term can be calculated from the appropriate wave function. The energy is described in terms of a set of parameters (Slater–Condon–Shortley), which for d^n configurations are F_0, F_2 and F_4. F_0 involves only the radial wave functions $R(r)$, whereas F_2 and F_4 involve only angular wave functions $Y(\theta,\phi)$. *Table 12.3* lists the F parameters for the various terms arising from the d^2 configuration, from which it can be

Table 12.3 F parameters for d^2 configuration terms

Term	F *parameters*
^1S	$F_0 + 14F_2 + 126F_4$
^1G	$F_0 + 4F_2 + F_4$
^3P	$F_0 + 7F_2 - 84F_4$
^1D	$F_0 - 3F_2 + 36F_4$
^3F	$F_0 - 8F_2 - 9F_4$

seen that the expression for each term contains a single F_0, so that the differences between the energies of any two of these terms depend only on the F_2 and F_4 parameters. This is to be expected since it is the interaction of the angular components of the d orbital functions that produce the terms.

In practice the calculation procedure can be simplified by introducing two new parameters B and C, called Racah parameters, which are expressions in F_2 and F_4:

$$B = F_2 - 5F_4$$

$$C = 35F_4$$

This change of parameters has the very significant advantage that the energy differences between the lowest-energy (ground-state) term and other terms of the same multiplicity involve only the B parameter. Thus in the d^2 instance the difference in energy between the ground-state term ^3F and the other triplet term ^3P is $15B$ when we use Racah parameters, compared with $15F_2 + 75F_4$ in F parameters. When we wish to calculate energy differences between terms of different multiplicity then both B and C are needed, e.g. the $^1D - {}^3F$ difference is $5B + 2C$.

The values of B and C for a given metal ion are determined empirically for the experimentally obtained spectrum of the free ion. Assignments are made to account for the observed peaks and B and C chosen so as to give the best fit. The fit is rarely perfect, and one reason for this is that there may be more than one term (or level) of the same symmetry, and in such cases there will be interactions (*configurational interactions*) resulting in energy modifications. We will discuss this point more fully and provide some examples when we deal with the spectra of octahedral complexes of various M^{n+} ions.

12.13 Spectra of Complex Ions

We have seen that, for the free ion, inter-electronic repulsions give rise to terms, and that when the ion is complexed by ligands, some of the degeneracy of the terms may be removed and the terms split into levels.* This was illustrated in *Figure 12.1* for $[TiL_6]^{3+}$ when the ^2D term split into ^2E_g and $^2T_{2g}$ levels (the superscript 2 has now been included to describe the spin multiplicity, which was ignored in the simplified introduction).

Generally speaking there are two components to the energy of each level:

(a) crystal field energy (Δ)
(b) inter-electronic repulsion energy (B and C)

We discussed these factors in a rather more general way in the last chapter, where we saw that when Δ was small there tended to be a spin-free configuration, with the maximum separation of electrons into available orbitals

*Some authors describe the components of the split term still as terms, but we have used the word level since ^2E_g and $^2T_{2g}$ etc. can be determined experimentally from spectroscopic measurements. The word state may also be applied to the levels

as required by Hund's rule (see p.217). With large fields the electrons were forced to pair up to give a strong-field or spin-paired configuration. Now we wish to look more closely at the role played by the inter-electron repulsions. When Δ is small, by comparison with the energy gap between the terms, we again have a weak-field case, and we consider the energy gaps between the terms to dominate the energy diagram, so that the field produced by the surrounding ligands may be regarded as a relatively small perturbation. Furthermore, any effect of spin–orbital coupling will be still smaller.

When Δ is large by comparison with the gap between terms, it will dominate the energy diagram and we have a strong-field case. In these circumstances we start by considering the ways in which the t_{2g} and e_g atomic orbitals are occupied and then allow for inter-electronic repulsions. In practice, of course, Δ may be of the same order of magnitude as the energy gap between terms, and then we have the intermediate-field case; this can be discussed on the basis of modification to either the weak-field or strong-field approaches. Each method will now be reviewed in turn, with a fairly detailed consideration of the d^2 configuration, as illustrated by V^{3+}, together with a briefer survey of other configurations.

12.14 Weak-field Method

As this chapter is concerned primarily with the spectra of complexes of the first-row transition elements, for which spin–orbit coupling is relatively unimportant, the splitting associated with L–S coupling will be omitted,

Table 12.4 Splitting of terms by octahedral or tetrahedral fields

Term of free ion	*Levels produced by octahedral or tetrahedral fields**
S	A_1
P	T_1
D	$E + T_2$
F	$A_2 + T_1 + T_2$
G	$A_1 + E + T_1 + T_2$
H	$E + 2T_1 + T_2$

*These symbols will have g subscripts if the levels are produced by octahedral fields

and we will concentrate on the splitting of terms brought about by the imposition of octahedral or tetrahedral ligand fields. *Table 12.4* summarizes this splitting. The splitting of P, D and F terms will now be reviewed for the various d^n configurations.

12.14.1 D TERMS

We have seen already that in an octahedral field a D term splits into T_{2g} and E_g levels. *Figure 12.19* shows this splitting pattern for a D term of unspecified multiplicity, and it can be seen that for a given configuration we have a 'positron' in the e_g levels, and this positive centre

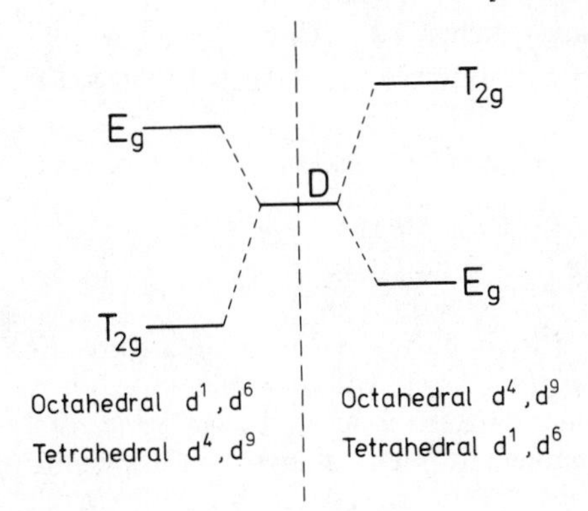

Figure 12.19 Splitting of D term in octahedral and tetrahedral fields; g subscripts dropped for levels arising in a tetrahedral field

will be the most stable in regions where the electron is least stable. Hence the ground state for d^9 is E_g. There is the same inverted splitting relationship between D terms arising from d^4 and d^6 configurations.

12.14.2 *P* AND *F* TERMS

The splitting of these terms in octahedral and tetrahedral fields is shown in *Figure 12.20*. These terms are included in the same diagram because they occur together for each of the d^n configurations quoted. It can be

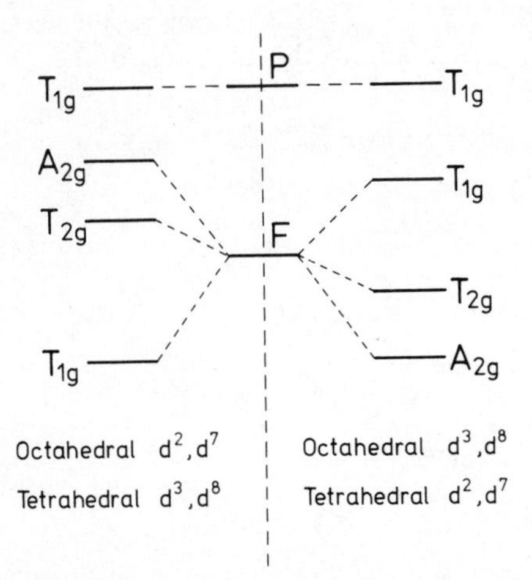

Figure 12.20 Splitting of P and F terms in octahedral and tetrahedral fields; g subscripts dropped for levels arising in a tetrahedral field

seen that the P term does not split in either an octahedral or a tetrahedral field, just as p orbitals would not split because p_x, p_y and p_z orbitals experience identical fields in either environment. For a given d^n configuration the F term splitting is inverted on going from an octahedral to a tetrahedral field, just as it was for the D term. The hole concept also holds, there being an inversion of term splitting on going from a d^n to a d^{10-n} configuration.

There is a final point of some importance that must be emphasized, namely that although the term splitting patterns may be inverted by a change of field (octahedral–tetrahedral) or configuration (d^n–d^{10-n}), the energy sequence of the terms remains unchanged. Thus d^2 and d^8 configurations both give rise to ^3P and ^3F terms, and in each instance the latter term is the lowest in energy.

12.15 Spectra of Octahedral d^2 Complexes (e.g. $[VL_6]^{3+}$): The Weak-field Approach

The energy diagram for octahedral d^2 complexes is shown in *Figure 12.21*, this being the appropriate part of *Figure 12.20* with the inclusion of Δ_o values. (Δ_o is the parameter for the splitting induced by an *octahedral* field – hence the subscript o.) Simple arithmetic shows that the splitting of the ^3F term is balanced about the 'centre of gravity':

$$1.2\Delta_o + 3 \times (0.2\Delta_o) + 3 \times (-0.6\Delta_o) = 0$$

Thus the $^3A_{2g}$ level is not degenerate and counts only once, whereas the two T levels are triply degenerate.

We can use this diagram to study the spectra of suitable complexes and evaluate Δ_o. A useful case is that of V^{3+} in Al_2O_3, where V^{3+} has an octahedral environment of oxygens and the spectrum shows three peaks, 17 400, 25 200 and 34 500 cm^{-1}.

The first step is the assignment of each peak to a suitable transition, from the ground state $^3T_{1g}(F)$ to one or other of the three excited levels $^3T_{2g}$, $^3A_{2g}$ and $^3T_{1g}(P)$. In *Figure 12.21*, $^3T_{1g}(P)$ is drawn as the level of highest energy, but if the field is fairly strong the splitting of the ^3F

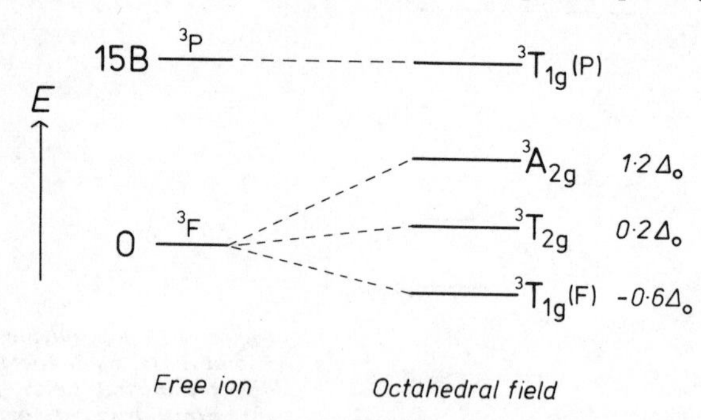

Figure 12.21 **Energy level diagram for octahedral d² complex**

term results in the $^3A_{2g}$ level being higher than the $^3T_{1g}(P)$ level. *Table 12.5* sets out the assignments. If the 17 400 cm^{-1} peak corresponds to $0.8\Delta_o$, then Δ_o must be 21 750 cm^{-1}. The transition involving the $^3A_{2g}$ level corresponds to $1.8\Delta_o$, and for consistency this cannot be the 25 200 cm^{-1} peak because it would yield a Δ_o value of 14 000 cm^{-1}. With the assignment given in the table, $1.8\Delta_o$ corresponds to the 34 500 cm^{-1} peak, giving Δ_o = 19 170 cm^{-1}.

The two values of Δ_o obtained from these assignments, 21 750 and 19 170 cm^{-1}, are not identical; our 'fit' is far from perfect. The reason is that our diagram is oversimplified and we have ignored any 'mixing' of the $^3T_{1g}(F)$ and $^3T_{1g}$ levels. This *configurational interaction*, which was

Table 12.5 Assignments for the spectrum of V^{3+} in Al_2O_3

Peak position/ cm^{-1}	*Assignment*	Δ_o *difference between levels*
17 400	${}^3T_{2g} \leftarrow {}^3T_{1g}(F)$	0.8
25 200	${}^3T_{1g}(P) \leftarrow {}^3T_{1g}(F)$	
34 500	${}^3A_{2g} \leftarrow {}^3T_{1g}(F)$	1.8

referred to earlier in Section 12.12, results in ${}^3T_{1g}(P)$ being raised slightly in energy, and ${}^3T_{1g}(F)$ lowered. The effect is relatively small, because ${}^3T_{1g}(F)$ and ${}^3T_{1g}(P)$ are fairly well separated, but it cannot be ignored completely if Δ_o is to be evaluated correctly.

12.16 Orgel Diagrams

By making the assumption that $15B$ has the same value in the complexed ion as in the free ion, Orgel was able to use the weak-field method to produce diagrams showing the variation of the energy of each level with increasing Δ. *Figure 12.22* shows such a diagram for a d^2 ion (e.g. V^{3+}) in an octahedral field. For completeness this diagram shows the splittings of both the singlet and triplet terms, but in practice we are concerned only with levels arising from the triplet terms (full lines) because any transitions between singlet and triplet levels are spin-forbidden and extremely weak.

There are two points worth noting in respect of this diagram. Firstly, the levels produced by the ligand field have the same spin multiplicity as the original terms of the free ion, and it is only the orbital degeneracy that changes as the field is applied. Secondly, levels with identical configurations never cross, e.g. ${}^3T_{1g}(F)$ and ${}^3T_{1g}(P)$.

The splittings shown in *Figure 12.22* merely put on to a quantitative basis the splitting scheme shown in *Figure 12.20* for the P and F terms. In the latter figure, we showed that terms arising from octahedral d^2 and d^7 ions had the same splitting and that the splitting was inverted for octahedral d^3 and d^8 ions (i.e. d^n–d^{10-n} relationship); for an ion of given d^n configuration the splitting was inverted when the ligand field changes from octahedral to tetrahedral. *Figure 12.23* shows how these splittings vary with Δ, and it can be seen that the same relationships hold. One point of interest brought out in *Figure 12.23* is that with octahedral d^3 and d^8 ions the inverted splitting of the F term brings the two T_{1g} levels [$T_{1g}(P)$ and $T_{1g}(F)$] much closer together, so that there is considerable configurational interaction and a consequent 'bending' of the energy–Δ plots. The first transition ($T_{2g} \leftarrow A_{2g}$) does not involve either of these levels so that it is a true measure of Δ_o.

It should be noted that the composite *Figure 12.23* does not specify the spin multiplicity; this is 3 for d^2 and d^8 and 4 for d^3 and d^7.

An Orgel diagram may also be drawn for the spin-free d^5 ion (see *Figure 12.24*), but the ground-state term is ${}^6S$ and there is no other sextet term, so that any transition from ${}^6S$ to one of the quartet terms

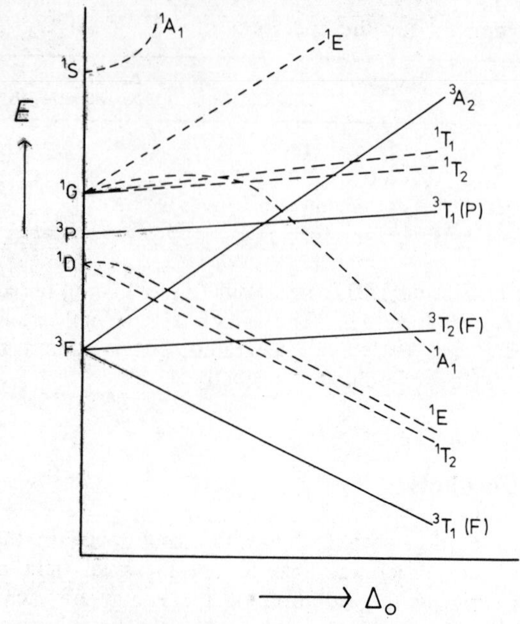

Figure 12.22 Orgel diagram for octahedral d^2 *complex. Full lines for each triplet level; dotted lines for each singlet level*

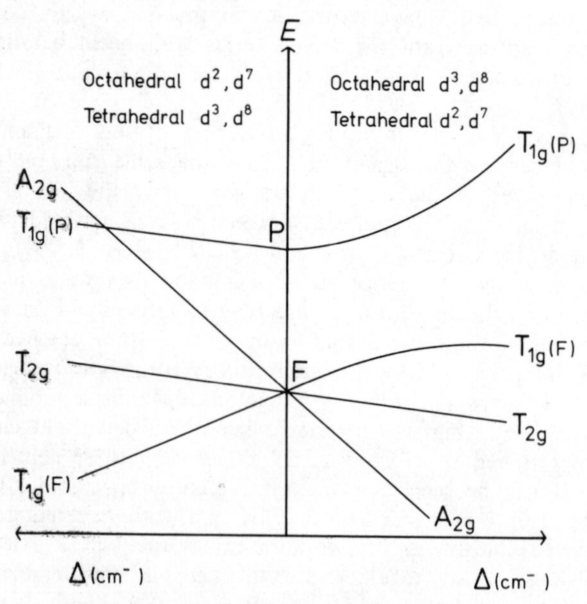

Figure 12.23 Orgel diagrams for octahedral and tetrahedral d^2, d^3, d^7 *and* d^8 *spin-free complex ions.* g *Subscripts dropped for levels arising in a tetrahedral field*

(^4G, ^4P, ^4D and ^4F) must involve a change in spin multiplicity. The absorption peaks will be very weak, and readers will remember that simple salts of divalent manganese, which contain the octahedral spin-free d^5, Mn^{2+} ion, are very pale pink in colour.

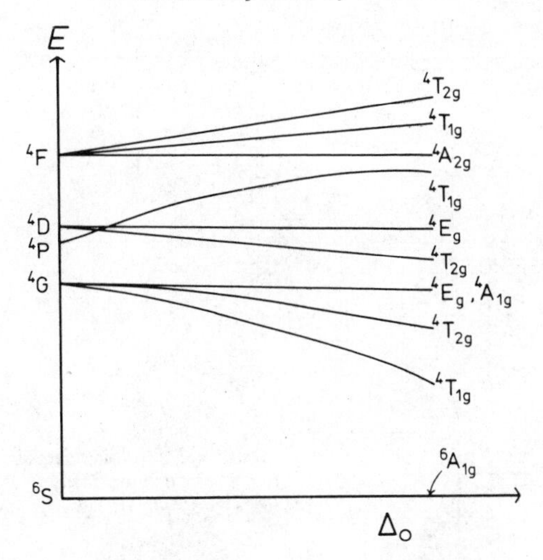

Figure 12.24 Orgel diagram for an octahedral d^5 *(spin-free) ion. The ground state* (^6S) *is shown as the* x *axis*

12.17 Strong-field Method (for Octahedral Fields)

Up until now we have concentrated on the weak-field approach in which the inter-electronic repulsions dominate. In the free ion these repulsions produce the terms, which split into levels when the ligand field is imposed. The extent to which this splitting takes place is measured by Δ_o (for octahedral fields).

With the strong-field approach, the ligand field is considered to dominate the energy, and the configuration depends primarily upon the filling of the t_{2g} and e_g orbitals, and the interaction between electrons is considered to be relatively unimportant. Hence the energy of an electron is said to be determined only by the energy of the orbital that it occupies, and in an octahedral environment this will be $-0.4\Delta_o$ if the electron is in an octahedral environment or $+0.6\Delta_o$ if it is in an e_g orbital. With a d^2 configuration there are three possible primary distributions for these two electrons:

$$(t_{2g})^2;\ \text{energy} = -0.8\Delta_o$$
$$(t_{2g})(e_g);\ \text{energy} = +0.2\Delta_o$$
$$(e_g)^2;\ \text{energy} = +1.2\Delta_o$$

In any real complex there will be some interaction between the electrons, of course, and this will result in the splitting of levels and the removal of some degeneracy.

At this point it is worth recalling the earlier discussion (Section 12.8) about the ways in which two electrons can arrange themselves in the available orbitals, and the relationship between microstates and terms. There are some 45 microstates, or ways of arranging the two electrons in the t_{2g} and e_g orbitals. Thus for the situation in which both electrons are in t_{2g} orbitals there are 15 microstates, just as for a p^2 configuration (see *Figure*

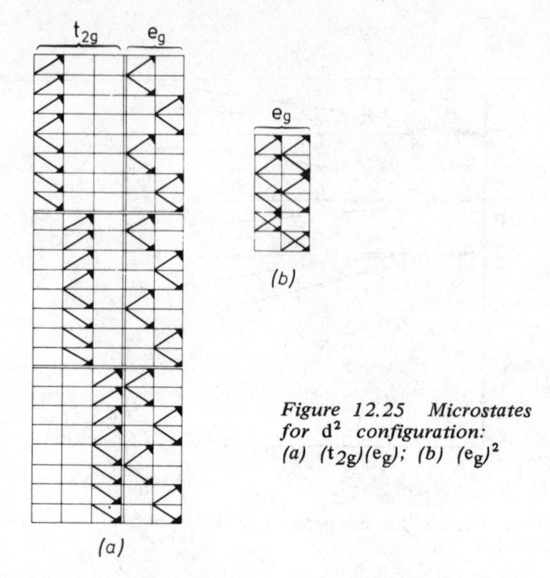

Figure 12.25 Microstates for d^2 configuration: (a) $(t_{2g})(e_g)$; (b) $(e_g)^2$

12.9), while the $(t_{2g})(e_g)$ and $(e_g)^2$ configurations provide 24 and 6 microstates, respectively. This is demonstrated in *Figure 12.25.* From this extreme strong-field viewpoint we can examine the effect of allowing for a measure of inter-electronic repulsion, when various levels are produced. As the field strength is reduced so the diagram gradually changes to the weak-field one already discussed. The relationship between the weak-field and strong-field approaches is shown by appropriate correlation diagrams.

12.18 Correlation of Weak-field and Strong-field Levels

To start with, the trivial case of a d^1 configuration is shown in *Figure 12.26*, and it can be seen that both approaches give the same result.

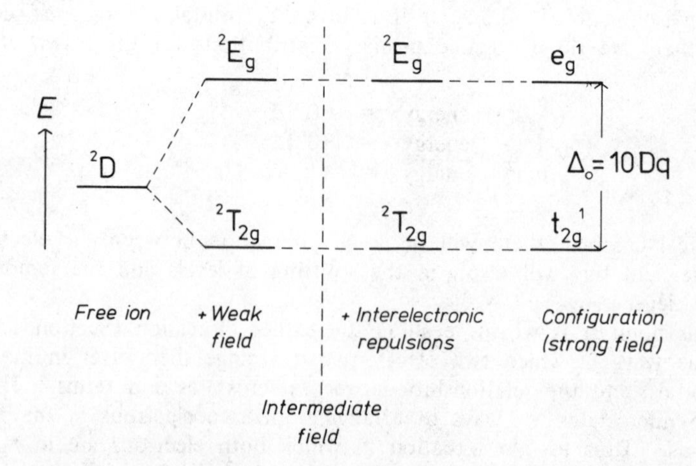

Figure 12.26 Weak-field–strong-field correlation diagram for d^1 configuration

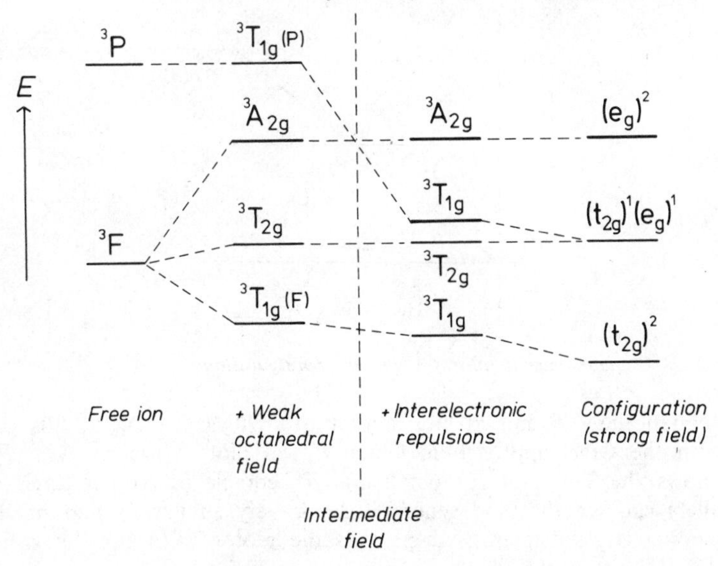

Figure 12.27 *Weak-field–strong-field correlation diagram for* d^2 *configuration*

Figure 12.27 shows the comparable diagram for the d^2 configuration, although the diagram is a simplified one which includes only the triplet terms. If the diagram is viewed from the strong-field side it can be seen that when inter-electronic repulsions are taken into account, so the $(t_{2g})^2$ configuration increases in energy and a $^3T_{1g}$ level emerges. This level corresponds to 9 microstates, there being a spin degeneracy of 3 and an orbital degeneracy of 3. The $(t_{2g})(e_g)$ configuration produces two levels, $^3T_{2g}$ and $^3T_{1g}$, each corresponding to 9 microstates, while the $(e_g)^2$ configuration gives rise to a $^3A_{2g}$ level of 3 microstates. Thus, in all, this partial correlation diagram corresponds to $9 + 18 + 3 = 30$ microstates, the remaining 15 being associated with the singlet terms that have been omitted from this simplified diagram.

It should be noted that lines do not cross if they connect terms (or levels) of the same spin and symmetry, but crossing is permitted if the levels have different symmetries (e.g. $^3T_{1g}$ and $^3A_{2g}$) or spin.

Analogous correlation diagrams can be drawn up for other d^n configurations and interested readers are referred to Figgis's book[4].

Tanabe and Sugano[5] have used the strong-field approach to produce semi-quantitative energy diagrams (T–S diagrams) that describe the levels resulting from intermediate octahedral fields, and similar diagrams have been developed by other workers for complexes of other symmetries. This discussion will be restricted to octahedral complexes and will concentrate on the particular case of d^2 configurations. The appropriate T–S diagram will be used to make self-consistent assignments for the peaks found in experimentally determined spectra, and Δ_o will be calculated.

As the discussion involves only triplet–triplet transitions, the only Racah parameter involved is B, and the energy of each level is described in terms of B and Δ_o. Now B is not normally the same for a complex as for the free ion, and Tanabe and Sugano allowed for this by dividing both sides

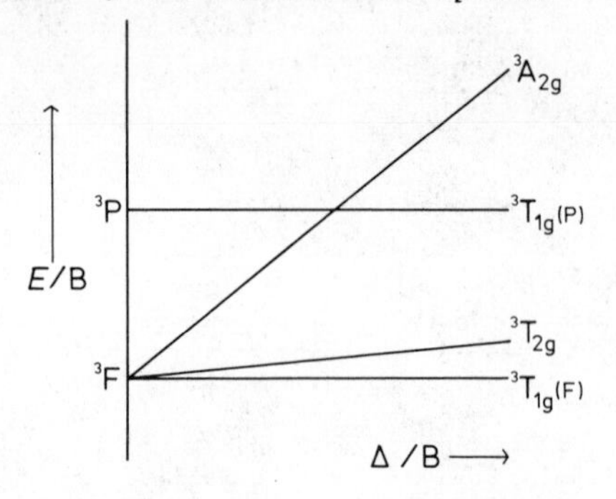

Figure 12.28 Tanabe–Sugano plot for the d^2 *configuration*

of the equation by B and making a plot of E/B against Δ_o/B; this contrasts with the Orgel approach in which E was plotted against Δ_o. *Figure 12.28* shows the T–S plot for octahedral d^2 complexes (triplet levels only), and readers can see that the general form is very similar to that of the corresponding Orgel diagram, except that the plot of E/B for the ground state [$^3T_{1g}(F)$] is taken as the x axis.

This diagram can be used to account for the spectrum of $[V(H_2O)_6]^{3+}$, which is illustrated in *Figure 12.29*; readers will note that the two peaks

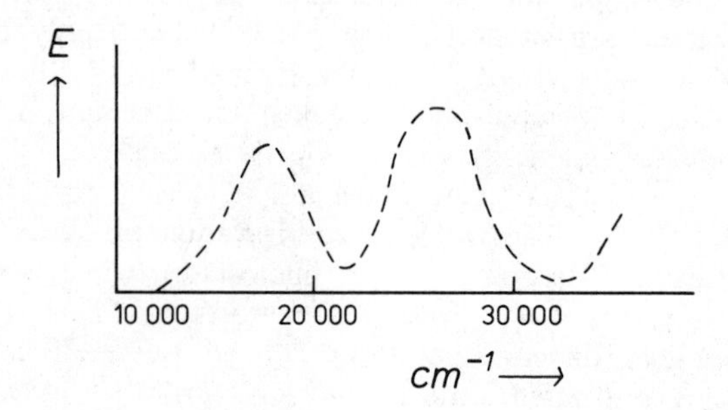

Figure 12.29 Spectrum of $[V(H_2O)_6]^{3+}$

are in very similar positions to those found in the spectrum of V^{3+} in Al_2O_3 (p.312), but the third peak found in the latter case can no longer be observed in $[V(H_2O)_6]^{3+}$, since it is obscured by charge-transfer bands. Assignments for the two peaks are listed in *Table 12.6*, and it should be noted that the 25 200 cm^{-1} peak involves promotion to $^3T_{1g}(P)$ rather than to $^3A_{2g}$ for the reasons discussed in Section 12.14.

The best value for Δ_o can now be evaluated by the following series of steps:

(i) The energy of each level is written down in terms of Δ_o and B.
(ii) The energy associated with each transition is calculated from the energies of the ground and excited levels.
(iii) A plot is made of the ratio of the energies of the two transitions (E_1/E_2) against Δ_o/B; the value of Δ_o/B corresponding to the experimental E_1/E_2 is determined.

Table 12.6 Assignments for the spectrum of $[V(H_2O)_6]^{3+}$

Peak position/ cm^{-1}	*Assignment*	*Strong-field assignment*	*Associated energy*
17 100	$^3T_{2g} \leftarrow {}^3T_{1g}(F)$	$(t_{2g})(e_g) \leftarrow (t_{2g})^2$	E_1
25 200	$^3T_{1g}(P) \leftarrow {}^3T_{1g}(F)$	$(t_{2g})(e_g) \leftarrow (t_{2g})^2$	E_2

(iv) The best value of Δ_o/B is substituted into the expression of E_1/B and E_2/B and B is determined.
(v) Δ_o is evaluated.

We can now look a little more closely at each of these steps.

12.18.1 ENERGIES OF LEVELS

The detailed procedure for the derivation of these energy expressions is well outside the scope of this book, and interested readers are referred to a brief summary article by Lever[6] and to an excellent full account (Reedjik *et al.*[7]) of the method adopted in respect of d^8 (Ni^{2+}) systems. The procedure for obtaining E involves the solution of a quadratic equation in Δ_o and B, which for d^2 systems lead to the expressions shown in *Table 12.7.*

Table 12.7 Energy of levels for d^2 systems

Level	*Energy (E)*
$^3T_{1g}(F)$	$0.5[15B - 0.6\Delta_o - (225B^2 + 18B\Delta_o + \Delta^2)^{1/2}]$
$^3T_{2g}$	$0.2\Delta_o$
$^3T_{1g}(P)$	$0.5[15B - 0.6\Delta_o + (225B^2 + 18B\Delta_o + \Delta_o^2)^{1/2}]$
$^3A_{2g}$	$1.2\Delta_o$

12.18.2 ENERGIES OF TRANSITIONS

The energy associated with each of the transitions listed in *Table 12.6* is obtained from the difference between the energies of the contributing ground and excited state levels, with the results shown in *Table 12.8.*

Table 12.8 Energies of E_1 and E_2 transitions

Transition	*Energy*
$^3T_{2g} \leftarrow {}^3T_{1g}(F)$	$E_1 = 0.5\{0.4\Delta_o - [15B - 0.6\Delta_o - (225B^2 + 18B\Delta_o + \Delta^2)^{1/2}]\}$
	$= 0.5[\Delta_o - 15B + (225B^2 + 18B\Delta_o + \Delta^2)^{1/2}]$
$^3T_{1g}(P) \leftarrow {}^3T_{1g}(F)$	$E_2 = 0.5[15B - 0.6\Delta_o + (225B^2 + 18B\Delta_o + \Delta^2)^{1/2}]$
	$- [15B - 0.6\Delta_o - (225B^2 + 18B\Delta_o + \Delta^2)^{1/2}]$
	$= (225B^2 + 18B\Delta_o + \Delta^2)^{1/2}$

12.18.3 DETERMINATION OF Δ_o/B

Expressions can now be written down for E_1/B and E_2/B:

$$E_1/B = 0.5[\Delta_o/B - 15 + (225 + 18\Delta_o/B + \Delta_o^2/B^2)^{1/2}] \tag{12.1}$$

$$E_2/B = (225 + 18\Delta_o/B + \Delta_o^2/B^2)^{1/2} \tag{12.2}$$

Equations 12.1 and 12.2 may be used to obtain the expression 12.3 for E_1/E_2:

$$\begin{aligned} E_1/E_2 &= \frac{\Delta_o/B - 15 + (225 + 18\Delta_o/B + \Delta_o^2/B^2)^{1/2}}{2(225 + 18\Delta_o/B + \Delta_o^2/B^2)^{1/2}} \\ &= f(\Delta_o,B) \end{aligned} \tag{12.3}$$

This function can now be plotted against Δ_o/B to give *Figure 12.30*. The experimental value for $E_1/E_2 = 17\,100/25\,200 = 0.678$, and from the graph

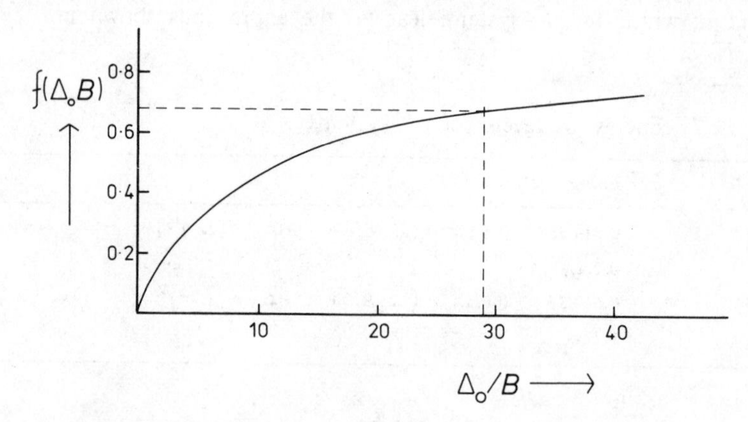

Figure 12.30 Plot of $f(\Delta_o,B)$ *against* Δ_o/B

this gives a corresponding value for Δ_o/B of 29.3. Readers can substitute 29.3 for Δ_o/B into expression 12.3 and check that it fits.

12.18.4 EVALUATION OF Δ_o/B

If this value of Δ_o/B is now put into equations 12.1 and 12.2, then B can be evaluated:

From equation 12.1, $E_1/B = 27.22$, giving $B = \dfrac{17\,100}{27.22} = 628\ \text{cm}^{-1}$

From equation 12.2, $E_2/B = 40.14$, giving $B = \dfrac{25\,200}{40.14} = 625\ \text{cm}^{-1}$

12.18.5 CALCULATION OF Δ_o

If B is taken as 628 cm^{-1}, then $\Delta_o = 29.3 \times B = 18\,400\ cm^{-1}$.

12.19 Other Tanabe–Sugano Diagrams

These diagrams have been produced for all d^n configurations (see Figgis[4]) and although strictly speaking any diagram is accurate for only a particular ion, it can be applied to isoelectronic ions provided only levels of the same multiplicity as the ground-state level are involved. Thus in our d^2 discussion for V^{3+}, we were interested only in triplet–triplet transitions, and the only Racah parameter involved was B. If the singlet levels are to be incorporated then the Racah parameter C must be included, and the C/B ratio specified. The T–S diagrams take the same general outline as the Orgel diagrams referred to earlier, the important and obvious distinction being that the x axis is always the E/B plot of the level of lowest energy. Accordingly, when there is a crossover of the corresponding E against Δ_o plots (Orgel diagrams), there will be a 'change of direction' in the E/B against Δ_o/B plots since the x axis of the crossover point corresponds to a different level. This is illustrated in *Figures 12.31* and *12.32*, which show T–S

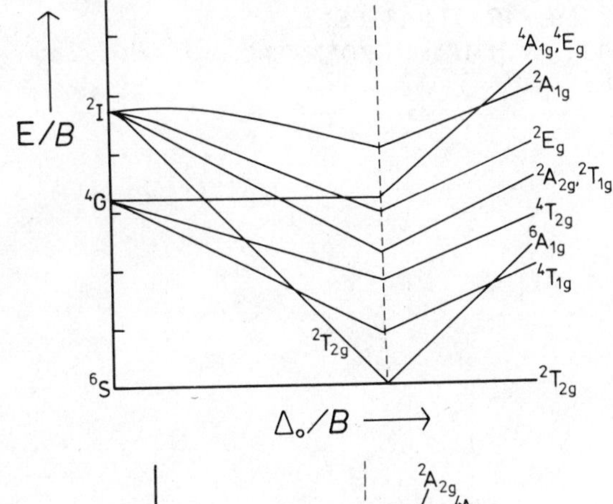

Figure 12.31 Partial Tanabe–Sugano plot for the d^5 *configuration*

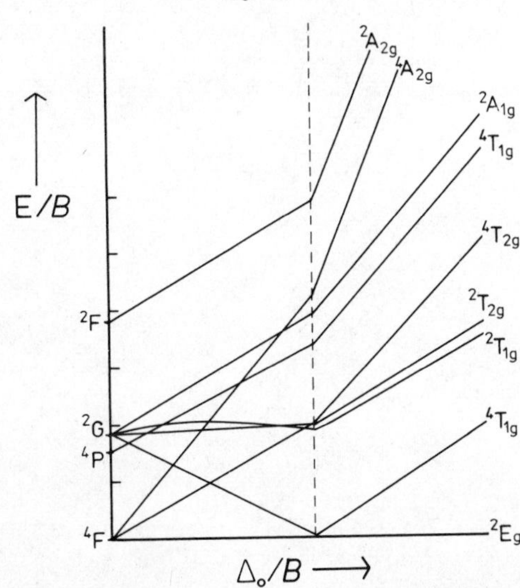

Figure 12.32 Partial Tanabe–Sugano plot for the d^7 *configuration*

plots for d^5 and d^7. Thus in the d^5 diagram, typified by Mn^{2+}, the ground state is a spin-free sextet level, but as Δ_o/B increases so the doublet level $^2T_{2g}$ (corresponding to just one unpaired electron) decreases in energy by comparison with $^6A_{1g}$, and at the critical point becomes the ground-state level. In the same way, with a d^7 configuration such as the Co^{2+}, the quartet level is the lowest for small values of Δ_o/B, but the doublet level becomes the ground state as Δ_o/B becomes large and spin-pairing occurs.

12.20 References

1. BALLHAUSEN, C.J., *Ligand-field Theory*, McGraw-Hill, London (1962)
2. LEVER, A.B.P., *Inorganic Electronic Spectroscopy*, Elsevier, Amsterdam (1968)
3. SUTTON, D., *Electronic Spectra of Transition Metal Complexes*, McGraw-Hill, London (1968)
4. FIGGIS, B.N., *Introduction to Ligand Fields*, p.153, Interscience, New York (1966)
5. TANABE, Y. and SUGANO, S., *J. Phys. Soc. Japan*, **9**, 753 (1954)
6. LEVER, A.B.P., *J. Chem. Ed.*, **45**, 711 (1968)
7. REEDJIK, J., Van LEEUWEN, P.W.N.M. and GROENWELD, W.L., *Rec. Trav. Chim.*, **87**, 129 (1968)

13

ELECTRON-DEFICIENT MOLECULES[1]

13.1 Introduction

We have seen that elements of main groups II and III, such as beryllium, boron and aluminium, have a considerable tendency to accept electrons, and so acquire a tetrahedral configuration whenever possible. Molecules such as BF_3,NH_3 are substances in which the acceptor atom (boron) receives electrons from a donor (nitrogen) with a suitable lone pair, thus producing a compound with normal two-electron bonds, one being represented formally as a co-ordinate bond.

This tendency for the acceptor atom to increase its co-ordination number from two or three to four or more is considerable, and in the absence of suitable donor molecules, simple compounds of these elements may dimerize. A typical example is $AlCl_3$, which in solution and low temperature gas phase is dimeric (Al_2Cl_6) with chlorine bridges; the solid is even more polymeric, each aluminium having an 'octahedral' environment with three pairs of chlorine bridges. In such halogen-bridged structures the halogen atoms donate lone pairs, so that normal two-electron bonds result. It is much more difficult to explain the bonding in dimers such as diborane, B_2H_6, and the aluminium alkyls, Al_2R_6, since there are not enough electrons to provide two-electron bonds between the atoms in these molecules. Such molecules are said, therefore, to be 'electron-deficient'.

We will discuss the classic case of diborane in some detail to develop bonding descriptions, and then extend the ideas to other hydrides and alkyls.

13.2 The Structure and Bonding of Diborane

We will discuss these problems in two stages. Firstly we will consider the evidence that has now established beyond doubt the relative positions of the atomic nuclei concerned, and, secondly, we will give a theoretical interpretation of the bonding.

13.2.1 THE NUCLEAR CONFIGURATION OF DIBORANE

This was a source of controversy for many years, and two main models were proposed, namely the so-called 'ethane' and 'bridge' models shown in

(1). The 'ethane' structure is analogous to that of C_2H_6 and the 'bridge' structure is comparable with that of Al_2Cl_6, but in neither case are there enough electrons available to provide normal two-electron bonds.

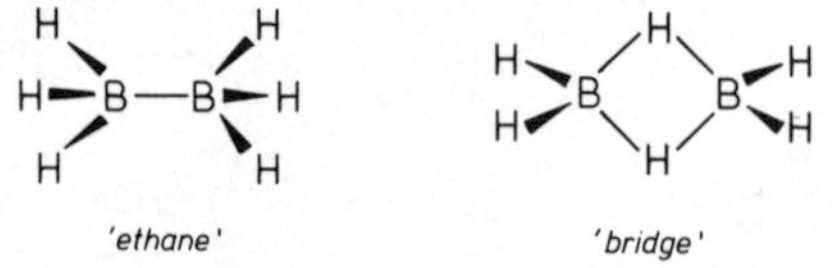

(1)

The essential difference between the two models is that whereas all six hydrogen atoms are equivalent in the ethane structure, this is not so in the bridge model, where two hydrogen atoms play a special role. Moreover, free rotation should be possible about the B–B link in the ethane structure but not in the bridge structure. This was indicated by the chemical behaviour of diborane, because only four of the six hydrogen atoms could be replaced by methyl groups. A wide range of physico-chemical studies, including considerations of infrared and nuclear magnetic resonance spectra, provided pretty conclusive proof of the bridge structure, and this is now established completely through electron-diffraction and X-ray diffraction studies. The two boron atoms and the four terminal hydrogen atoms (H_t) are coplanar, and the two bridge hydrogens (H_b) lie above and below the plane. The electron diffraction and infrared spectral studies, which are concerned with the gas-phase structure, yield very similar molecular parameters. Thus the 'outside' H_tBH_t bond angles are close to 120° while those in the bridge, H_bBH_b, are around 97°. A particularly significant point is the difference between the bridging and the terminal B–H bond lengths. Thus whereas the terminal B–H_t distances at 119 pm are virtually the same as those found for 'normal' single B–H bonds, as in BH_3,CO, the bridging B–H_b bonds are very much longer (133 pm) and hence weaker. X-ray studies confirm that the bridging bonds are very much longer.

13.2.2 THE BONDING IN DIBORANE

All the structural evidence establishes the bridge arrangement for B_2H_6 and we must now try to account for the nature of the bonding, bearing in mind the differences in the B–H_t and B–H_b distances. While the structure is no longer in dispute, the interpretation of the bonding is much more a matter of conjecture, and it seems unlikely that the last word has yet been said. We will briefly review some of the earlier, and rather simple, theories put forward to explain the bonding, and then discuss at rather greater length the more recent descriptions. For any theory we can start with the basic assumption that the terminal B–H_t bonds contain two electrons and that the 'electron deficiency' is in the bridging bonds.

13.2.2.1 *Electrostatic Interactions: Hydrogen Bonds*

It was suggested by Buraway that the two BH_3 fragments might be held together by hydrogen bonds, it being argued that B–H bonds would be somewhat polarized in the sense $\overset{\delta+}{B}$–$\overset{\delta-}{H}$, so that bridging would be of the type shown in (2). Even if we accept the hypothesis of a $-\overset{\delta+}{B}-\overset{\delta-}{H}\cdots\overset{\delta+}{B}-$ bonding system, there are serious objections to the assumption that these links are responsible for the bonding in diborane. It was pointed out in

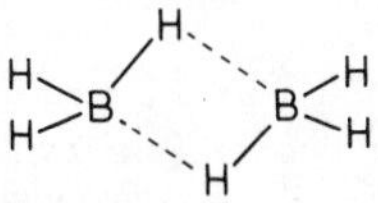

(2)

Chapter 9 that hydrogen bonds are fairly weak, and this is not the case with the bridge bonds in B_2H_6. Accordingly, this type of description can be eliminated from serious discussion.

13.2.2.2 *Valence-bond (Resonance) Description*

Diborane has been described as a resonance hybrid of the four canonical forms shown in (3), two forms (I and II) being so-called covalent structures and two (III and IV) being 'ionic' structures. This description is not

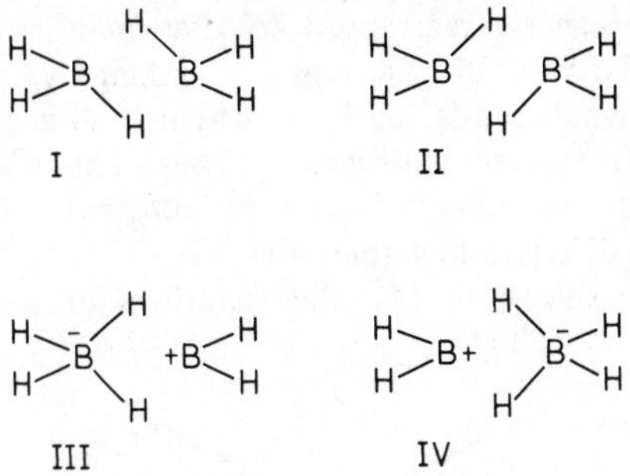

I II

III IV

(3)

a very helpful one since it disguises the nature of the bonding forces, and the bond angles in the canonical forms are considerably distorted from the theoretical, namely linear ($\overset{+}{B}H_2$), trigonal-planar (BH_3) and tetrahedral ($\overset{-}{B}H_4$).

13.2.2.3 *Protonated Double Bond Description*

Pitzer suggested an ethene-like structure for diborane, there being a $[B_2H_4]^{2-}$ planar arrangement with the protons embedded in the π-electron charge clouds of the double bond; the bridging system was described as a protonated double bond (see *Figure 13.1*). This description is an attractive one at first sight, since the ultraviolet absorption spectrum of diborane is ethenic in character, and the location of the bridging hydrogens is central as required; these protons would be screened by the π electrons and would not behave as acidic hydrogens. The B–B bond length of 177 pm is appreciably greater than the double bond in ethene (133 pm), however, and

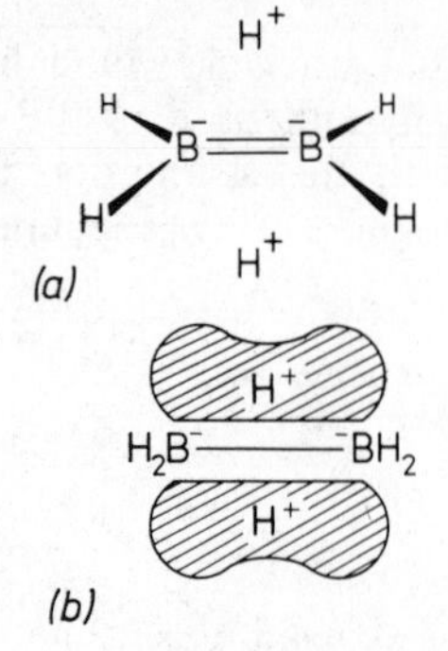

Figure 13.1 The 'protonated double bond' structure for diborane

although it can be accounted for through mutual repulsion of formally negative boron atoms, the length does imply comparatively poor π overlap of the $2p_z$ atomic orbitals of the boron atoms.

13.2.2.4 *Three-centre Orbital ('Banana' Bond) Description*

The Pitzer description had the particular merit of locating the hydrogens in the centre of the bridges, and allocated four electrons to hold together the B_2H_2 bridging system. Longuet-Higgins pointed out that the orbital functions used by Pitzer can be rearranged to give an alternative description of diborane in which the two BH_2 groups are linked by two equivalent 'three-centre' or 'banana' bonds, each of which embraces the two boron atoms and one of the bridge hydrogens. These equivalent orbitals resemble those discussed earlier for ethene (p.98), except that each orbital now covers three atoms (BHB) rather than two.

Using the labelling shown in (4), the functions of the Pitzer description may be summarized as follows:

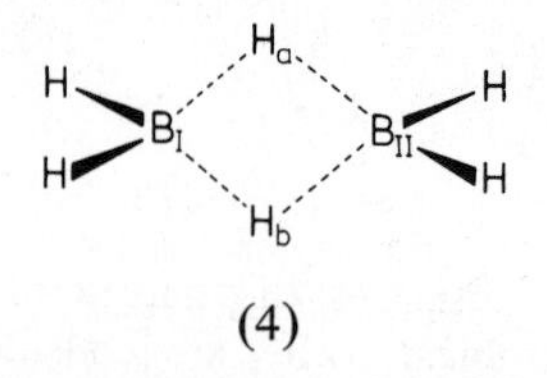

(4)

σ_1 = sp^2 orbital used by B_I for σ bonding
σ_2 = sp^2 orbital used by B_{II} for σ bonding
π_1 = $2p_z$ orbital used by B_I for π bonding
π_2 = $2p_z$ orbital used by B_{II} for π bonding
s_a = s orbital of bridging hydrogen H_a
s_b = s orbital of bridging hydrogen H_b

Two linear combinations of these atomic orbitals may be taken:

$$\psi_1 = N[\sigma_1 + \sigma_2 + \pi_1 + \pi_2 + 2s_a]$$

$$\psi_2 = N[\sigma_1 + \sigma_2 - \pi_1 - \pi_2 + 2s_b]$$

N is the normalization constant. An orbital outline for ψ_1 (see *Figure 13.2*) shows the large banana-shaped positive lobe embracing both boron

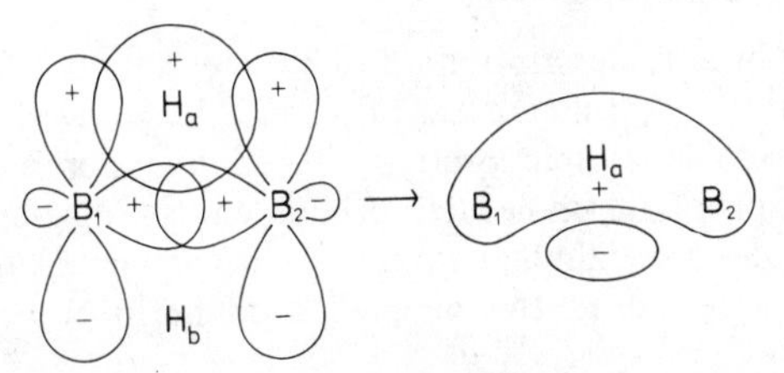

Figure 13.2 Orbital outline for a three-centre orbital for diborane

atoms and the bridging hydrogen, H_a. The shape of the orbital described by function ψ_2 is directly analogous to that of ψ_1 except that the positive lobe now covers the bridging hydrogen H_b. If the terminal B–H bonds are taken as normal two-electron bonds, then four electrons remain for the bridge bonding, one from each boron and one from each bridging hydrogen, so that each of the three-centre molecular orbitals will be doubly filled.

Just as with the alternative ($\sigma + \pi$ and equivalent orbital) description of ethene, the orbital outlines may be slightly misleading since they may suggest that in the Pitzer description there is a concentration of charge between the boron nuclei (as a result of the σ bond), but that the charge concentration in the banana orbitals is away from the boron–boron axis. This apparent difference arises through the oversimplification of the orbital-outline diagrams; the charge distributions arise from identical wave-mechanical functions and must therefore be the same. The three-centre orbital description is, perhaps, to be preferred to the protonated-double-bond one, especially since it may be considered as part of a series of polycentre delocalized molecular orbitals.

We can get a less accurate but simpler picture of the formation of the three-centre orbitals if we consider the boron atoms to be sp^3 hybridized – cf. the related discussion of ethene (p.102). The overlap of the 1*s* orbital of a bridging hydrogen by two sp^3 orbitals, one from each boron, gives a molecular orbital covering all three atoms (see *Figure 13.3*). This simplified derivation is less rigorous, and really amounts to superposing the two covalent contributing structures (I and II) to the resonance description of formula (3), but it can be readily visualized and it simplifies the discussion of bonding in the more complicated electron-deficient molecules such as the higher hydrides of boron and the metal alkyls.

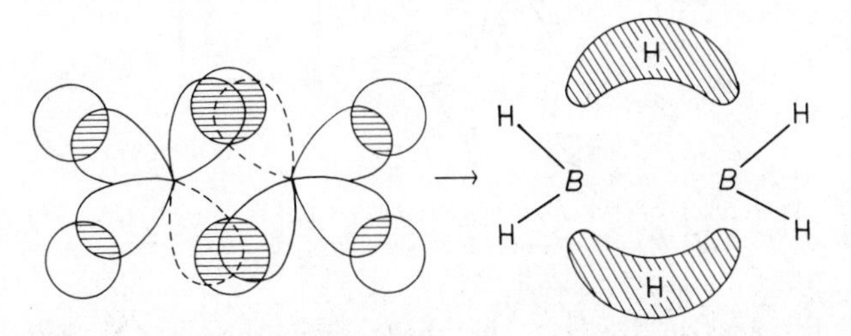

Figure 13.3 The three-centre orbital description of diborane. (The two sp^3 orbitals used by each boron atom in the bridge structure are equivalent, but since only one electron is available for the two orbitals then one is formally considered as vacant – a dotted outline – and the other singly occupied. A similar convention is used for subsequent diagrams for other bridge structures)

This molecular-orbital description of diborane is only partial, of course, and it is qualitative. A fuller description can be derived, which makes use of all the available atomic orbitals of the two boron and six hydrogen atoms, and an appropriate molecular orbital energy diagram drawn up. It is a fairly elaborate procedure and outside the scope of this book; interested readers are referred to the detailed account in Wade[1].

13.3 The Structures of the Higher Hydrides of Boron

The structures of B_4H_{10}, B_5H_9, B_5H_{11}, B_6H_{10} and $B_{10}H_{14}$, which have been established beyond doubt by X-ray, neutron diffraction, electron diffraction and nuclear magnetic resonance, are illustrated in *Figure 13.4.*

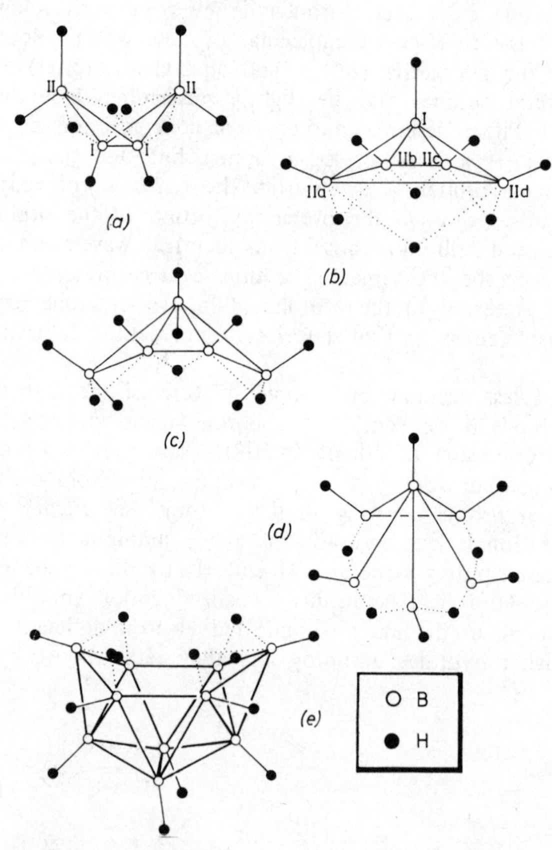

Figure 13.4 Structures of the higher hydrides of boron: (a) B_4H_{10}; (b) B_5H_9; (c) B_5H_{11}; (d) B_6H_{10}; (e) $B_{10}H_{14}$

The B–H distances in $B_{10}H_{14}$ have been established with a fair measure of precision through neutron diffraction studies, but in the other hydrides the B–H distances are not so well known. With some of the hydrides there are a number of similar B–B distances, and in such cases the average value is quoted in *Table 13.1.*

Table 13.1 Internuclear distance in the higher boron hydrides*

Hydride	*B–B distance*/pm	*Terminal B–H distance*/pm	*Bridging B–H distance*/pm
B_4H_{10}	173(1); 185(4)	119	133(4); 143(4)
B_5H_9	169(4); 180(4)	122	135(4)
B_5H_{11}	172(2); 176(3); 187(2)	109	124
B_6H_{10}	160(1); 177(9)	–	–
$B_{10}H_{14}$	177(19); 197(2)	118	130(4); 135(4)

*Numbers enclosed in parentheses give the number of such bonds for the particular hydride

There are three basic types of bond to be found in these hydrides:

(a) normal two-electron bonds;
(b) bridging bonds similar to those in diborane, covering one hydrogen and two boron atoms in a BHB system;
(c) skeleton delocalized bonds linking together several boron atoms.

The normal two-electron bonds are described in terms of the overlap of the hydrogen 1*s* orbitals with sp^3, sp^2 or *sp* orbitals of boron; the bridging B–H–B systems are two-electron 'banana' bonds as with diborane. In describing any particular higher hydride the simple approach involves assigning electrons and orbitals to B–B single bonds, and the terminal B–H and bridging B–H–B bonds, followed by the assignment of any remaining electrons to multi-centre orbitals that embrace boron atoms in the skeleton structure. When only three boron atoms are involved, the three-centre bonds may either be of a 'banana' or open type, much as in diborane, but with a boron atom taking the place of hydrogen, or it can be rather more like a 'steering wheel', or closed type, with each of the boron atoms contributing an orbital pointing towards the centre of the equilateral triangle formed by the three boron atoms. These two types of three-centre BBB orbitals are shown in *Figure 13.5*. When more than three boron atoms are involved the skeletal delocalized orbitals will be of the closed type, but correspondingly more complex.

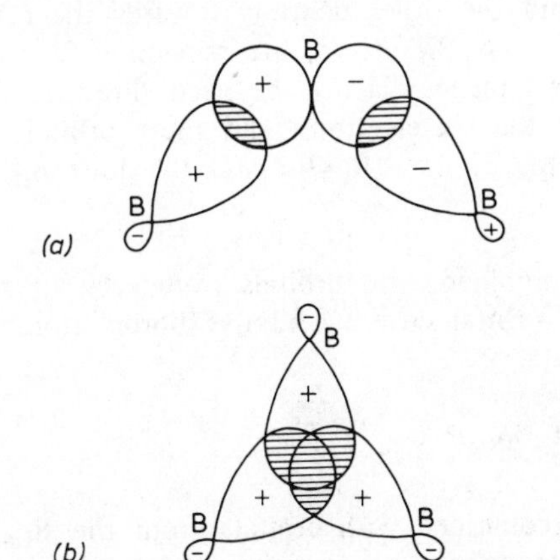

Figure 13.5 Three-centre orbitals for 'boron skeleton' in higher hydrides of boron; (a) 'banana' or open type; (b) 'steering wheel' or closed type

These principles have been applied to explain the bonding in each of the higher hydrides previously listed, but the discussion here will be restricted to two of the hydrides, namely B_4H_{10} and B_5H_9.

13.3.1 B_4H_{10}

There are 22 electrons to be placed in suitable orbitals, three from each boron and one from each hydrogen. For simplicity it is assumed that each boron has four sp^3 hybrid orbitals available for contribution to bonding and each hydrogen an s orbital. Each of the boron atoms labelled B_{II} [see *Figure 13.4(a)*] forms two terminal B–H bonds by overlap of the boron sp^3 and hydrogen s orbitals and two bridge B–H–B bonds. This uses up all four of the sp^3 orbitals of each B_{II} and accounts for 12 electrons. The B_I atoms, on the other hand, each form only one terminal B–H bond and two bridge B–H–B bonds, using up a further 8 electrons and leaving each B_I with an sp^3 orbital containing one electron. These singly filled B_I orbitals overlap to give a direct single B_I–B_I bond.

The bridge bonds are asymmetrical because of the non-equivalence of the B_I and B_{II} orbitals and the unfavourable angles ($HB_{II}H$) between the bridge bonds.

13.3.2 B_5H_9

In this hydride 24 electrons are available to hold the nuclei together. Once again the terminal B–H bonds will take two electrons each (making ten in all), as will the B–H–B bridges (making eight in all), which leaves six electrons to hold together the five boron nuclei. The four B_{II} atoms in the base [see *Figure 13.4(b)*] each form one terminal B–H and two bridge B–H–B bonds, and hence have left one sp^3 orbital and one electron to contribute to the framework. The single B_I atom, however, forms only one terminal B–H bond, leaving two electrons for the skeleton. This B_I atom is taken as sp hybridized (using the $2p_x$ orbital) with one orbital forming the terminal B–H bond and the other pointing towards the centre of the basal square formed by the four B_{II} atoms; the remaining $2p_y$ and $2p_z$ orbitals are parallel to the basal plane. Hence we need three bonding molecular orbitals to accommodate the six electrons, and these orbitals can be compounded from the B_I sp, B_I $2p_y$, B_I $2p_z$ and the four B_{II} sp^3 orbitals:

(a) The B_I sp orbital may be combined with orbitals from each of the four B_{II} atoms, giving a molecular orbital covering all five boron atoms, i.e. a five-centre closed type orbital:

$$\psi_1 = \frac{1}{\sqrt{2}}\phi_{I(sp)} + \frac{1}{2\sqrt{2}}[\phi_{IIa} + \phi_{IIb} + \phi_{IIc} + \phi_{IId}]$$

(b) The B_I $2p_y$ orbital may be combined with orbitals from the B_{IIa} and B_{IIc} atoms, giving a 'banana' or open type of three centre orbital:

$$\psi_2 = \frac{1}{\sqrt{2}} \phi_{I(2p_y)} + \tfrac{1}{2}[\phi_{IIa} - \phi_{IIc}]$$

(c) A similar construction to (b) uses the B_I $2p_z$ orbital and the B_{IIb} and B_{IId} sp^3 orbitals to give another 'banana' orbital:

$$\psi_3 = \frac{1}{\sqrt{2}} \phi_{I(2p_z)} + \tfrac{1}{2}[\phi_{IIb} - \phi_{IId}]$$

Lipscomb has used the basic concept of three-centre orbitals to develop a topological approach to predict which hydrides are capable of existence, and readers interested in this or in the detailed consideration of bonding in other higher hydrides and carboranes are referred to Lipscomb's review[2] or to the summary in Wade's book.

13.4 The Structures of the Borohydrides (Tetrahydroborates)

When diborane reacts with organometallic compounds, borohydrides (tetrahydroborates) of the general formula $M(BH_4)_n$ are formed, which often have the electron-deficient characteristics of diborane itself.

There is an interesting gradation in the physical and chemical properties of the borohydrides of lithium, beryllium and aluminium; *Table 13.2* gives

Table 13.2 Borohydrides of Li, Be and Al

	M.p./°C	*B.p./°C*
$LiBH_4$	275 (decomposition)	–
$Be(BH_4)_2$	123 (under pressure)	91.3 (sublimes)
$Al(BH_4)_3$	–64.5	4.5

the melting and boiling points of these substances, from which it can be seen that the volatility increases Li < Be < Al. A significant difference is also found in the reactivity of these borohydrides towards typical electron-donor molecules such as trimethylamine; lithium borohydride does not react, whereas both beryllium and aluminium borohydrides react readily even at low temperatures. It seems, therefore, that lithium borohydride is not electron-deficient, but is ionic and contains Li^+ cations and tetrahedral $[BH_4]^-$ anions. Similar ionic structures can be postulated for the analogous lithium aluminium hydride, $Li^+[AlH_4]^-$, and lithium gallium hydride, $Li^+[GaH_4]^-$.

The borohydrides of beryllium and aluminium, on the other hand, are predominantly covalent, judging from their volatility, and electron-acceptors, as indicated by their reactions with trimethylamine. It is known that both compounds contain M–H–B bridges. The beryllium compound was originally believed to have a linear structure (5a) with unsymmetrical Be–H–B bridges, but this is not the case since although it is monomeric in the gas

(a)

(b)

(c)

(5)

phase it has a dipole moment of around 2.50 D, indicating a non-linear arrangement of the three heavy atoms. The gas-phase structure is still uncertain, but it may be of the form shown in (5b). A crystal-structure study of the solid (Mayrick and Lipscomb[3]) has shown it to have a polymeric structure in which H_2BH_2Be units are linked by BH_4 units into a helical arrangement (5c). The BH_2Be bridges are unsymmetrical with B–H ≈ 110 pm, and Be–H ≈ 150 pm. It can be seen that each beryllium atom has an environment of six (bridging) hydrogens, the arrangement being approximately a trigonal prism. The trigonal prism arrangement is also found for aluminium in the borohydride (6), the B–H bridging distance (128 pm) again being much the shorter (Al–H, 180 pm). With the zirconium borohydride, $Zr(BH_4)_4$, the bridging involves three of the four hydrogens, so that the zirconium atom is surrounded by twelve hydrogens; this structure (7) is best regarded as a tetrahedral disposition of borons about zirconium with three bridging hydrogens symmetrically disposed about this tetrahedral axis.

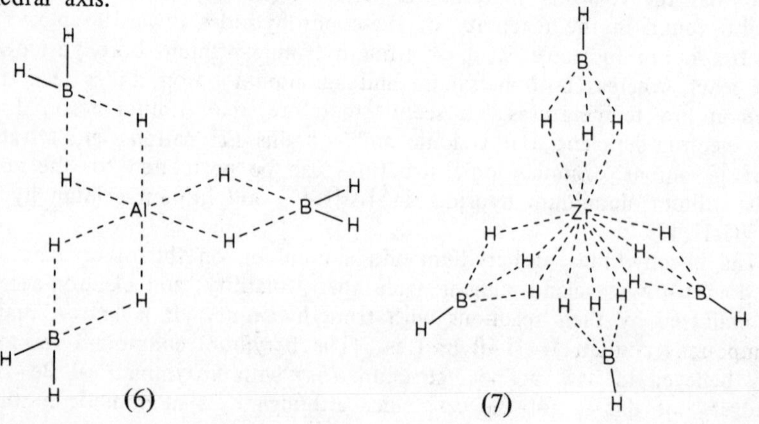

(6) (7)

In all of these borohydrides, the bridge M–H–B bonding can be described most simply in terms of 'banana' type three-centre orbitals, with some measure of asymmetry in the orbital to account for the unequal bridge bonds.

13.5 The Metal Alkyls

Alkyls formed by elements of the first three periodic groups (alkali metals, alkaline earth metals and the boron group) are particularly interesting from a valency point of view because many of them are polymeric. We shall see that the forces holding them together are similar to those in diborane. The detailed structures of a number of alkyls have been evaluated by X-ray and electron-diffraction methods and significant physical data are available on many others.

First of all the structures of the alkyls will be reviewed and then the bonding will be discussed.

13.5.1 ALKALI METALS

Lithium alkyls are polymeric and generally contain tetrameric units. With lithium methyl for instance, there is a tetrahedral set of four lithium atoms with a methyl group sitting symmetrically above each Li_3 face. This is illustrated in (8). The structure clearly involves a methyl group bridging

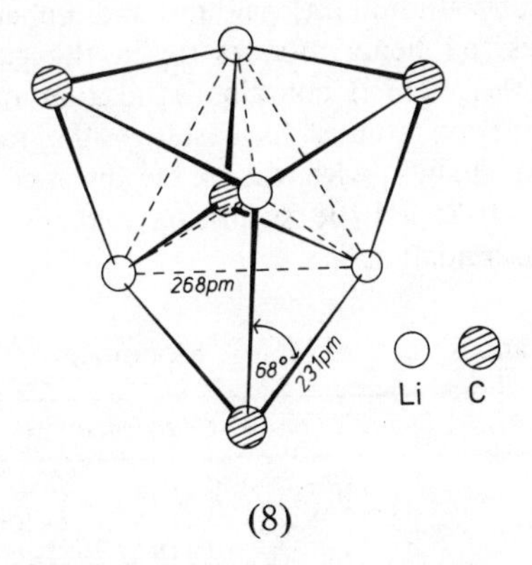

(8)

three lithium atoms. Alkyls of the heavier alkali metals are much more ionic, and with potassium methyl for instance, the structure is best regarded as arising from the packing of K^+ and CH_3^- ions.

13.5.2 ALKALINE EARTHS

In the solid state the dimethyls of both beryllium and magnesium have polymeric structures (9) containing chains of beryllium (or magnesium) atoms bridged by methyl groups. The bridging MCM angles are quite acute, being 66° for BeCBe and 75° for MgCMg.

The vapour of $BeMe_2$ appears to be principally the dimer Be_2Me_4, together with smaller amounts of monomer and trimer. While the structures of these species have not been established, it seems likely that the monomer has the expected linear Me–Be–Me structure, and the dimer and trimer species the bridged structures shown in (9b) and (9c).

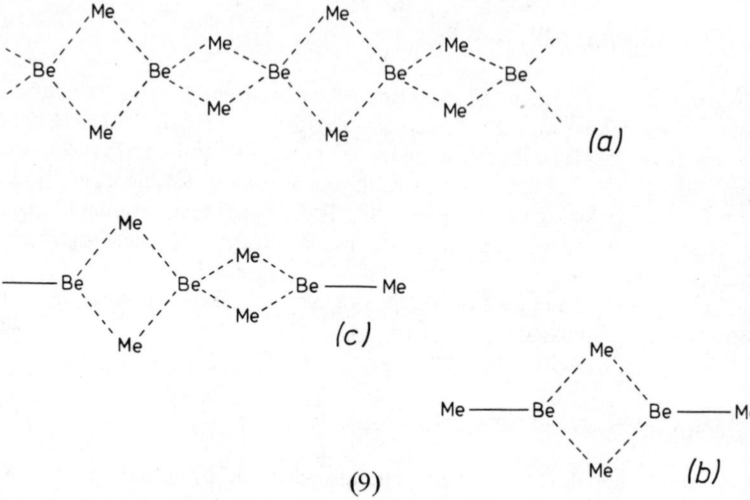

(9)

13.5.3 THE BORON GROUP

The trimethyls of boron, aluminium, gallium, indium and thallium form a very interesting series, all being monomeric in the gas phase (see Beagley *et al.*[4]) except for $AlMe_3$, which contains a mixture of monomers and dimers. Electron diffraction studies have shown the monomers to have the expected trigonal-planar shapes, with the M–C distance increasing along the series (see *Table 13.3*); it is a little surprising perhaps that the Al–C and Ga–C distances are so similar.

Table 13.3 M–C distances in the MMe_3 monomers

Trimethyl compound	*M–C distance*/pm
B	158
Al	196
Ga	197
In	216
Tl	222

In the solid phase BMe_3 is monomeric, $AlMe_3$ is dimeric and both $InMe_3$ and $TlMe_3$ are polymeric, albeit through rather long bonds. The dimeric structure of aluminium trimethyl is essentially the same in the gas phase (Almenningen *et al.*[5]) and the solid phase (Vranka and Amma[6]), the structure (10) being similar to that of Al_2Cl_6, but with bridging methyl groups; the terminal Al–C distances are 196 pm and the bridging distances 214 pm; the bridging AlCAl angle is 75°.

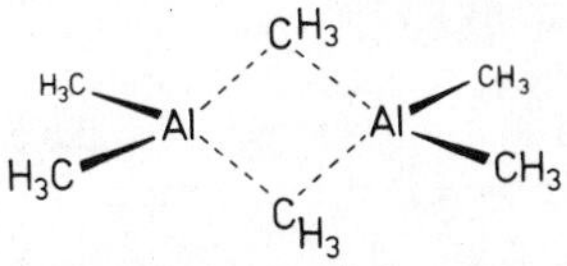

(10)

Other AlR_3 molecules are also bridged dimers in the solid state except when R is very bulky, such as with tertiary butyl. The crystal structure of $AlPh_3$ has shown (Malone and McDonald[7]) that the phenyl group will bridge as well as a methyl one, the terminal and bridging Al–C distances being 196 pm and 218 pm, respectively; in this dimeric structure the bridging phenyl groups are roughly at right angles with the Al_2C_2 bridging plane. Even more interesting is the dimeric structure of $AlPhMe_2$, in which it is the phenyl group and not the methyl group that bridges (Malone and McDonald[8]), the terminal and bridging distances being similar to those in Al_2Ph_6.

The indium and thallium trimethyls are not simple dimers but contain MMe_3 units linked through linear, but unsymmetrical, $M–CH_3 \cdots M$ bridges into polymeric structures. With $TlMe_3$, for instance (Sheldrick and Sheldrick[9]), the roughly trigonal-planar unit uses two of its methyl groups to link to neighbouring units, while two methyl groups from other units form long bonds along the *z* axis to give each thallium atom a trigonal-bipyramidal configuration (11). The bridging is approximately linear but very unsymmetrical, as the bond lengths in (11) indicate.

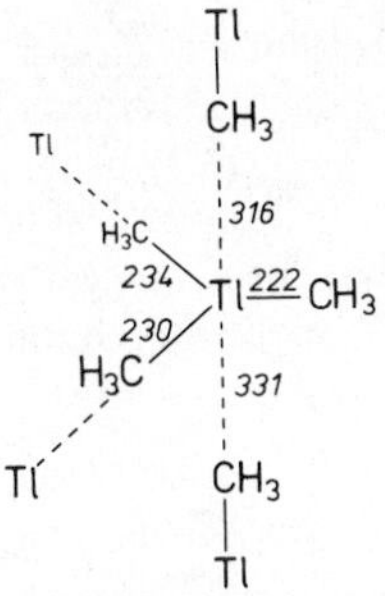

(11)

There is an interesting mixed compound, $MgAl_2Me_8$ (12), in which magnesium is linked to the two aluminium atoms through methyl bridges.

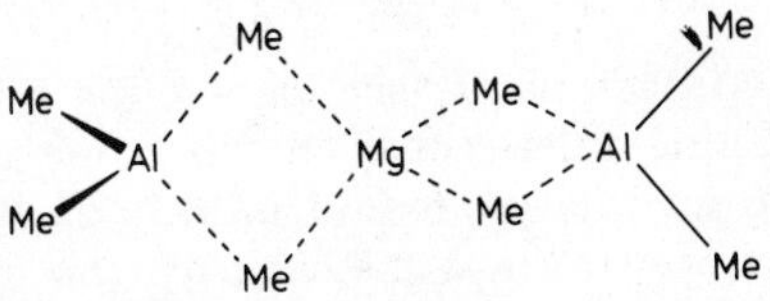

(12)

Since the two metal atoms are not the same it is not surprising that the bridges are slightly unsymmetrical, with Mg–C = 220 pm and Al–C = 210 pm.

13.5.4 BONDING IN METAL ALKYLS

It will be noticed that all the metal methyls have a structural feature in common, namely the bridging of two or more metal atoms by a methyl group, and in this respect there is at least a formal resemblance to the boron hydrides. If aluminium trimethyl is considered, it can be seen that the molecule is electron deficient, just like diborane, and if the terminal bonds are taken to be normal two-electron bonds, then the electron deficiency is associated with the bridging system where there are only four electrons available to provide the four Al–C bonds. Another feature requiring explanation is the formation of five bonds by the bridging carbon.

The bonding interpretation follows the pattern of those previously discussed for diborane, and hydrogen bonding and resonance descriptions [between forms I and IV, formula (13)] have been invoked. By analogy with the protonated double bond description of diborane, a 'methylated

I II

III IV

(13)

double bond' structure (14) has been suggested for Al_2Me_6, in which CH_3^+ groups are embedded in the π electron charge clouds of an Al–Al double bond.

$$\overset{+}{C}H_3$$
$$Me_2\bar{Al} = \bar{Al}Me_2$$
$$\overset{+}{C}H_3$$

(14)

Probably the most satisfactory explanation for the bridge bond in these alkyl structures uses central three-centre orbitals. The molecular orbitals may be compounded from the sp^3 hybrid orbitals of aluminium and carbon; each molecular orbital covers two aluminium and one carbon atom, and is completely filled [see *Figure 13.6(a)*]. If there is to be maximum overlap of the carbon sp^3 hybrid orbital by the two aluminium sp^3 orbitals, then the Al–C–Al angle must be small, as emphasized in *Figure 13.6(b)*.

This bonding description also accounts for the lower degree of polymerization found with the higher alkyls, since the heavier and bulkier alkyl groups would be much less effective in bridging the metal atoms. For effective bridging it has been suggested that there should be (i) a large electronegativity difference between M and C, (ii) a low promotion energy ($s \rightarrow p$) for M, (iii) a high M–C bond energy, and (iv) minimum inner-shell repulsions between the two metal atoms. The first two factors are unfavourable for boron and partly account for the monomeric nature of BMe_3. The last two factors apply to indium and thallium and explain why the MC_2M bridge system does not form.

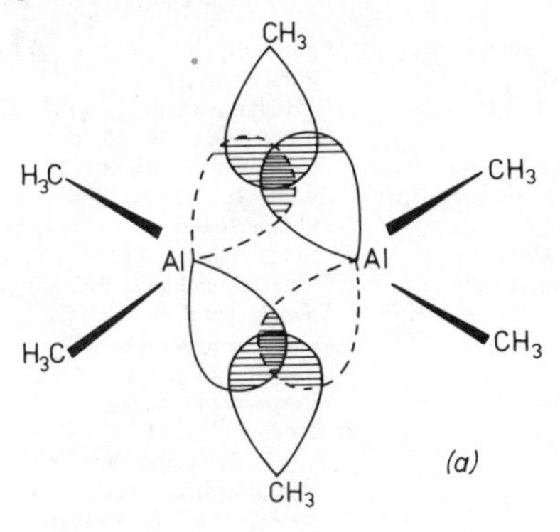

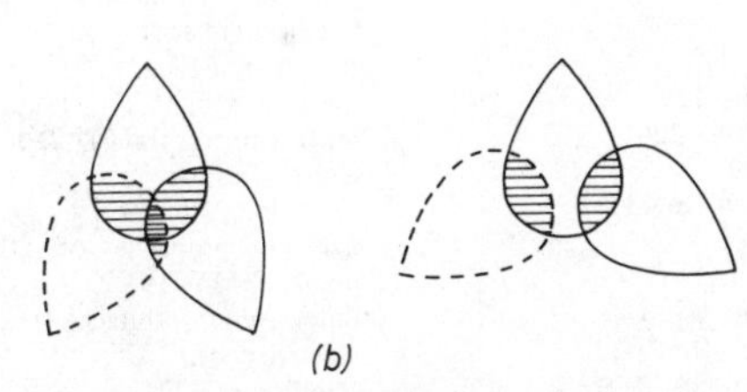

Figure 13.6 (a) The formation of 'three-centre orbitals' in $Al_2(CH_3)_6$. (b) The overlap of a carbon sp^3 orbital by two aluminium sp^3 orbitals: the acute-angled structure gives greater overlap, and hence a stronger bond

A comparison of the hydrides of boron, aluminium and gallium with the corresponding methyls shows that the bridged hydrides are more polymerized than the alkyls. We might expect this since the bridging hydrogen atoms use *s* orbitals that can give good overlap with the metal orbitals and be less dependent on the bond angle in the bridge.

13.6 References

1. WADE, K., *Electron Deficient Compounds*, Nelson, London (1971)
2. LIPSCOMB, W.N., *Adv. Inorg. Chem. Radiochem.*, **1**, 118 (1959)
3. MAYRICK, D.S. and LIPSCOMB, W.N., *Inorg. Chem.*, **11**, 820 (1972)
4. BEAGLEY, B., SCHMIDLING, D.G. and STEER, I.A., *J. Mol. Structure*, **21**, 437 (1974)
5. ALMENNINGEN, A., HALVORSEN, S. and HAALAND, A., *Acta Chem. Scand.*, **25**, 1937 (1971)
6. VRANKA, R.G. and AMMA, E.L., *J. Amer. Chem. Soc.*, **89**, 3121 (1967)
7. MALONE, J.F. and McDONALD, W.S., *J. Chem. Soc. Dalton Trans.*, 2646 (1972)
8. MALONE, J.F. and McDONALD, W.S., *J. Chem. Soc. Dalton Trans.*, 2648 (1972)
9. SHELDRICK, G.M. and SHELDRICK, W.S., *J. Chem. Soc. (A)*, 28 (1970)

INDEX